Engineering Materials

This series provides topical information on innovative, structural and functional materials and composites with applications in optical, electrical, mechanical, civil, aeronautical, medical, bio- and nano-engineering. The individual volumes are complete, comprehensive monographs covering the structure, properties, manufacturing process and applications of these materials. This multidisciplinary series is devoted to professionals, students and all those interested in the latest developments in the Materials Science field, that look for a carefully selected collection of high quality review articles on their respective field of expertise.

Indexed at Compendex (2021) and Scopus (2022)

K. Anantha Padmanabhan
S Balasivanandha Prabu • Akash Bhattacharjee
B P Gautham

Thermodynamics and Kinetics of Materials

Springer

K. Anantha Padmanabhan
Department of Mechanical Engineering
Anna University
Chennai, Tamil Nadu, India

S Balasivanandha Prabu
Department of Mechanical Engineering
Anna University
Chennai, Tamil Nadu, India

Akash Bhattacharjee
Tata Consultancy Services
Pune, Maharashtra, India

B P Gautham
Tata Consultancy Services
Pune, Maharashtra, India

ISSN 1612-1317 ISSN 1868-1212 (electronic)
Engineering Materials
ISBN 978-3-032-14526-0 ISBN 978-3-032-14527-7 (eBook)
https://doi.org/10.1007/978-3-032-14527-7

This Springer imprint is published by the registered company Springer Nature Switzerland AG
The registered company address is: Gewerbestrasse 11, 6330 Cham, Switzerland

Preface

This subject is often considered too vast for a one-semester course (necessitated by the introduction of the concept that 15% of the subjects taught in UG programs in engineering in the technologically more advanced countries should consist of electives in humanities and social sciences). An attempt has been made here to succinctly present an undergraduate-level narrative of thermodynamics and kinetics of materials based on lectures delivered by one of the authors (K. A. Padmanabhan) at the Indian Institutes of Technology in Varanasi, Madras, and Kanpur and Anna University, Chennai, and student feedback. In the teaching activities at Anna University, another author (S. Balasivanandha Prabu) also participated. The subjects of thermodynamics and kinetics of materials are presented here as a level one course.

First, the essential concepts of thermodynamics are discussed to explain a large part of human macro-level experience using a few simple and intuitive postulates presented in everyday terms. This is essential to develop the concepts of statistical mechanics, which provide a rigorous basis. The classical kinetic theory, perhaps the only case in which thermodynamics can nearly be derived from first principles, is discussed next, which leads one to molecular dynamics. Such a description makes one understand why statistical mechanics works.

Further treatment initially focuses on structure, the most important factor in determining the properties and industrial uses of materials. Application areas considered are (1) motion of atoms and molecules by diffusion, (2) motion of dislocations and interfaces, (3) morphological evolution due to capillary and mechanical forces, and (4) phase transformations. These topics are arranged in increasing order of complexity. Analytical foundations as well as approximations are highlighted. Topics concerning fluid flow are omitted. It is hoped that with the knowledge gained from this book, the students should be equipped to analyze topics that are not examined here specifically. The development of the topics in kinetics parallels a similar approach. Many interesting topics, like diffusion in simple and multicomponent materials, diffusion in materials containing crystal imperfections, grain boundaries and their influence on diffusion, diffusion in noncrystalline materials, motion of dislocations and interfaces, motion of crystal interfaces, role of interface reaction control in diffusion, evolution of surface morphology, role of capillary and mechanical forces, evolution and coarsening of microstructures (3D) due to capillary and mechanical forces, and kinetics of phase transformations, are considered in detail.

The real world proceeds at much faster rates than equilibrium reactions would allow. Therefore, irreversible thermodynamics and kinetics are examined. One of the chapters also examines how the use of digital computers has accelerated research and development in thermodynamics and kinetics of materials.

The background assumed on the part of the reader is knowledge of the structure of materials (including crystal imperfections), as taught in undergraduate courses, vector calculus and differential equations, elementary matrix theory, and classical elasticity/mechanics. Some ideas on the elementary notions of quantum mechanics are also used.

At the beginning of the book, a list of notations and symbols used is collated. Problems based on the topics discussed and their solutions are also available. The authors (in particular, K. A. Padmanabhan, B. P. Gautham, and A. Bhattacharjee) thank Mr. Ananth Krishnan, former CTO, Tata Consultancy Services (TCS), for encouraging them to write a book of this kind including work done by these authors at TCS, India's largest and one among the world's top computer service companies. The two chapters that present these findings emphasize that the future belongs to an interdisciplinary approach driven to a large extent by computational science and engineering.

Los Angeles, USA K. Anantha Padmanabhan
Chennai, India S. Balasivanandha Prabu
Pune, India A. Bhattacharjee
Pune, India B. P. Gautham

Summary of Concepts Needed to Study This Book

Systems are away from equilibrium if kinetic processes take place at a fast rate. Then, the systems should be away from equilibrium, slightly or significantly. For this situation, equilibrium should be understood first. Approximations help reduce complexity: for example, assumption of equilibrium in localized regions, kinetics that obey linear relations, and approximations that homogenize a given field. Then, one can predict the rates of processes/reactions, e.g., diffusion of a chemical species, heat conduction, growth of various kinds of grains, powder sintering, and solidification.

Structure is very important in materials science: for instance, crystal structure, microstructure, vacancies, interstitials, grain boundaries, dislocations, etc. and their correlations to properties. Mechanisms determine kinetics. And so, understanding the mechanisms is important. But the approach to a steady state at which a system is stable can be understood only using thermodynamics. In other words, the two subjects are closely connected.

1. In thermodynamics, a quantity is either "intrinsic" or "extrinsic." Intrinsic quantities are independent of the mass/volume. Extrinsic quantities are dependent on the volume/mass. For an extensive quantity, equilibrium is reached if the "conjugate potential," an intensive quantity, has equal magnitude in the two bodies.
2. Spatial variations are not permitted, while the "system" is in equilibrium. For example, chemical potential gradient is not allowed to change. Therefore, only a few homogeneous potentials are present at equilibrium in a nonuniform system. Then the variables needed to define an equilibrium system are rather small. For a nonequilibrium condition, numerous ways of characterizing the potential, different from the equilibrium condition, are available. An analysis of this pertains to the kinetics of materials, and this is more difficult than describing an equilibrium condition.
3. If a free energy function, G, of a closed system is minimized, it is in equilibrium with reservoirs of constant potentials. Then, δG should be nonnegative for that system to be in equilibrium (discussed later).

For agreement between theories and experiments, kinetic theories are empirically developed. These theories are frequently extended to predict new events in new phenomena. If the atomic processes are clearly understood, the predictions can be more confidently depended upon, particularly if they apply to other phenomena also.

Statistical mechanics, MD simulations, and *ab initio* calculations of interatomic potentials may be regarded as advances over traditional thermodynamics. However, according to present forecasts, the chances of developing such techniques to levels relevant to engineering applications are not in the foreseeable future. Many equations and results in thermodynamics involve continuous functions, e.g., spatially variable concentration, c $(\vec{r})$.

But materials systems do not have continuous structure. A rigorous mathematical definition of a suitable composition function is not easy. Using a continuous position vector $\vec{r}$ along with atomic locations that are not continuous confuses the beginner. The concept of a continuum, the mathematics needed, and its variations are described later.

4. Onsager's unified theory of irreversible processes (1931) applies to systems *near* equilibrium only. If the system is in nonequilibrium, there will be changes in its variables that are extensive, for example, total entropy and volume. They will change other fixed values following the second law of thermodynamics. If a system is not in equilibrium, the chemical potential (at equilibrium, the reversible work necessary for an infinitesimal amount of the same component to be added to a given system) spatially varies. In kinetic models, the concept of the chemical potential is used when a system is not in equilibrium, but close to it.

   Temperature is also an equilibrium concept. Some doubt its relevance for nonequilibrium situations. Let its two ends experience dissimilar temperatures in a bar. At equilibrium, the heat absorbed by the hotter end will equal the heat emitted by the colder end. Then, temperature can be treated as a continuous function, $T(x)$. At any point, an infinitesimal region around that point in the bar can be taken to exist in equilibrium. To have undisturbed measurement, the heat capacity of the measuring device should be of negligible heat conductivity and capacity. Such a thermometer does not exist. As a consequence, the intuition about such a function, $T(x)$, need not be wrong (a continuum limit). The temperature at a point is defined by such an idealized measurement.

Necessary Mathematical Background

Essential ideas needed are presented with examples. The necessary mathematical background is what is taught in usual undergraduate programs.

Fields

A physical quantity could be a scalar, e.g., mass, M; density, ρ; and temperature, T. A vector, which has both magnitude and direction, e.g., velocity, v, and acceleration due to gravity, a. Both Cartesian x, y, and z, i.e., $\vec{r}$ (x, y, z), and polar coordinates are defined in terms of the position vector, $\vec{r}$ (r, θ, φ). θ is the angle between $\vec{r}$ and direction x. The plane is x–y; its angle with the origin is φ (the position vector).

A vector field, e.g., a force or a flux, stress, strain, and strain rate, has not only a magnitude and direction but also a plane of action (i.e., the plane normal in a fixed frame of reference). They are second-order/rank tensors and are represented as, e.g., σ_{xy}. In the notation we shall follow, the suffix x indicates the direction of action, while y reveals the normal to the plane (plane normal) on which the stress, strain, etc. act. Higher-order tensors, e.g., elastic or shear modulus, also exist. The last two are fourth-order/rank tensors presented as, e.g., μ_{wxyz}. It is possible to shift from one frame of reference to another following some transformation rules. Such tensors follow some transformation rules. Tensor of the first order (rank) is known as a vector. A tensor of order zero is a scalar. These quantities could also be time dependent, e.g., T ($\vec{r}$,t) and ρ ($\vec{r}$, t). Then, even scalars (T and ρ) are no longer treated as scalars.

Variations

When the concentration changes with a velocity, $\vec{v}\,(x, y)$ (scalar field), the value of i in "c" *at time t* is obtained as

$$= \mathbf{c}\left(\vec{\boldsymbol{r}}\right) + \nabla \mathbf{c}.\vec{\boldsymbol{v}}\ |_{t=0}\mathbf{t} + \ldots \tag{1}$$

where ∇c is the gradient of c. This indicates that ∇c is the sum of the first derivatives of c along the three orthogonal axes, whose unit vectors are i, j, and k.

The maximum rate increase direction pointed by ∇c parallels the direction of $c\left(\vec{r}\right)$. This rate increase matches that of the gradient vector. The instantaneous rate change of c with t is

$$\frac{dc}{dt} = \nabla \mathbf{c}.\vec{\boldsymbol{v}}\,. \tag{2}$$

Equation (2) can be generalized into a time-dependent field c ($\vec{r}$, t). The velocity $\vec{v}\left(\vec{r}\right)$ and the instantaneous change in the rate of c are related as

$$\frac{dc}{dt} = \nabla \mathbf{c}.\vec{\boldsymbol{v}} + \frac{\partial c}{\partial t}. \tag{3}$$

The divergence of a vector (a different type of derivative) is also present.

Coarse Graining and Limits on a Continuum

Suppose in a limited volume (e.g., a cubic box), where ΔV is the material drawn around a vector terminus $\vec{r}$, a concentration, say, c, of a species i, which is in local equilibrium with a diffusion potential, is present. In the real case, atoms and molecules, discretely populated, make up the material. Then, defining concentration locally becomes difficult. To make sense, ΔV *must* $\rightarrow 0$; that is, this volume should be made infinitesimally small.

If the number of atoms of species, i, defines the concentration of i per volume, $c\left(\vec{r}\right)$, as $\Delta V \rightarrow 0$, the above limiting value is used to define the concentration $c\left(\vec{r}\right)$. This is an ill-defined limit of the function c. This is so because here a discrete quantity, e.g., concentration of atoms/molecules, is converted into a description in terms of a continuum. It is assumed that such continuum limits can be obtained.

Concept of a Flux

Let $\overrightarrow{J_i\left(\vec{r}\right)}$ describe the flow rate of i past a fixed unit area. $\Delta\vec{A} = \vec{n}\,\Delta A = (A_x, A_y, A_x)$ is an area oriented appropriately in the same Cartesian system. If $\dot{M}_i$ is a smooth function,

$$\dot{M}_i\left(\Delta\vec{A}\right)\alpha\,|\Delta\vec{A}\,|. \tag{4}$$

The proportionality factor is a vector field $\vec{J_i}$ or

$$\dot{M}_i\left(\Delta\vec{A}\right) = \vec{J}_i.\Delta\vec{A}. \tag{5}$$

The localized flux, $\vec{J}_i\left(\vec{r}\right)$, is the limit of the continuum. That is,

$$\frac{\dot{M}_i\left(\overrightarrow{\Delta A}\right)}{\Delta A} = \vec{J}_i\left(\vec{r}\right).\hat{n}, \tag{6}$$

where $\hat{n}$ is the plane normal.

Accumulation

The amount of i that accumulates in a volume $\Delta V = dx\,dy\,dz$; i accumulates in time δt as

$$\begin{aligned}\Delta M_i = {} & (i \text{ that flowed in time, } \delta t) - (i \text{ that flowed out in time } \delta t)\\ & + (i \text{ created in time, } \delta t)\end{aligned} \tag{7}$$

The flux $\vec{J}$, generalized for all three flux components. Then,

$$\begin{aligned}\delta M_i = {} & -\vec{J_\iota}(x+(dx/2),0)\cdot i dy dz \delta t + \vec{J}(x-(dx/2),0,0)\cdot i dy dz \delta t\\ & -\vec{J_\iota}(0,y+(y/2),0)\cdot\hat{J}(dz dx)\delta t - \hat{J_\iota}(0,(y-(y/2),0)\cdot i dz dx \delta t\\ & -\vec{J_\iota}\left(0,0,(dz+(dz/2))\cdot i dx dy \delta t + \vec{J_\iota}\left(0,0,(dz-(dz/2))\cdot i dy dx \delta t + \dot{\rho}_\iota\left(\vec{r}\right)\Delta V \delta t.\right.\right.\end{aligned} \tag{8}$$

Here, the density rate in volume, ΔV, of i is defined as $\dot{\rho}_i$. Using a first-order expansion and the continuum limit, one gets

$$\frac{\partial c_i}{\partial t} = -\nabla \cdot \overrightarrow{J_\iota} + \dot{\rho}_\iota. \tag{9}$$

The divergence of the quantity $\overrightarrow{J_i}$ is $\nabla \overrightarrow{J_i}$. Therefore, generalizing, one can write

$$\nabla \cdot \vec{J} = \frac{\partial\left(\overrightarrow{J_\iota} \cdot i\right)}{\partial x} + \frac{\partial\left(\overrightarrow{J_j} \cdot j\right)}{\partial y} + \frac{\partial\left(\overrightarrow{J_k} k\right)}{\partial t} = \frac{\partial J_x}{dx} + \frac{\partial J_y}{dy} + \frac{\partial J_z}{\partial z}. \tag{10}$$

Equation (11) predicts the rate of decrease in the density of the flux of that quantity that comprises the flux decreases. This is an extensive quantity/property (depends on the amount). The rate of change of its density accumulation is, therefore, equal to its rate of production minus the divergence of the flux. Or, Eq. (10) can be derived directly from

$$\begin{aligned} \dot{M}_\iota &= -\int_{B(\Delta V)} \cdot \overrightarrow{J_{\vec{r}}} \cdot \overrightarrow{dA} + \int_{\Delta V} \dot{\rho}_\iota dV = \int_{B(\Delta V)} \overrightarrow{J_\iota} \cdot \hat{n} dA + \int_{\Delta V} \dot{\rho}_\iota \\ &= \int_{\Delta V} \left(-\nabla \overrightarrow{J_\iota} + \rho_\iota \right) \mathbf{dV}. \end{aligned} \tag{11}$$

Here, $B(\Delta V)$ is the oriented surface around (ΔV).

The last equality follows from Gauss' theorem as

$$\int_{8(\Delta V)} \vec{J} \cdot \hat{n} dA = \int_{\Delta V} \nabla \vec{J} \, dV. \tag{12}$$

A geometric interpretation of the divergence theorem: If the volume comprises several adjacent cells, the accumulation is obtained as the sum of accumulations in all the cells (Eq. (11)). The surface flux results in each cell accumulation. When an interface is shared by two cells, opposing normal vectors result. Then, $\vec{J} \cdot \vec{n}$, which constitute the flux terms cancel. When cells abut, interior interface fluxes cancel out. Only the exterior surfaces contribute.

Conservative and Nonconservative Quantities

A conserved quantity cannot be created or destroyed (popularly known as the law of conservation of mass). Therefore, there are no sources or sinks for conserved quantities; for example, atom species i and internal energy U are given by

$$\frac{\partial c_i}{\partial t} = -\nabla \overrightarrow{J_\imath}. \tag{13}$$

$$\frac{\partial U}{\partial t} = -\nabla \cdot \overrightarrow{J_u}. \tag{14}$$

For non-conserved quantities, e.g., entropy, S,

$$\frac{\partial S}{\partial t} = -\nabla \cdot \overrightarrow{J_u} + \dot{\sigma}. \tag{15}$$

Here, the rate of entropy generation per unit volume is represented as $\dot{\sigma}$.

Matrices, Tensors, and Eigensystem

Consider the following linear equations:

$$y_n = M_{1n}x_1 + \ldots \cdots + \ldots \cdots + M_{nm}x_n \tag{16}$$

Or

$$y_i = \sum_{j=1}^{j=m} M_{ij}x_j \text{ for i} = 1, 2, \ldots.., \text{n}. \tag{17}$$

A matrix, $\underline{\text{M}}$, has elements M_{ij}, which multiplies a vector, $\vec{x}$, and results in

$$\begin{bmatrix} y_1 \\ y_2 \\ .. \\ .. \\ .. \\ y_n \end{bmatrix} = \begin{bmatrix} M_{11} & M_{12} & .. & .. & .. & M_{1m} \\ M_{21} & M_{22} & .. & .. & .. & M_{2m} \\ .. & .. & .. & .. & .. & .. \\ .. & .. & .. & .. & .. & .. \\ .. & .. & .. & .. & .. & .. \\ .. & .. & .. & .. & .. & .. \\ M_{n1} & M_{n2} & .. & .. & .. & M_{mn} \end{bmatrix} \begin{bmatrix} x_1 \\ x_2 \\ .. \\ .. \\ .. \\ x_n \end{bmatrix}. \tag{18}$$

Vectors like $\vec{x}$ and $\vec{y}$ are known as "column vectors." If needed, the transpose, e.g., $\vec{b}^r$, is used to indicate the "row vectors." If $\vec{x}$ and $\vec{y}$ define the orthogonal coordinate components in *(X, Y, Z)*, with "right," "forward," and "up" already defined, the square matrix $\underline{\text{M}}$ is a tensor of the second order. In this book, tensors of rank two and above are represented by underlined symbols. If $\vec{x}$ is the force applied and $\vec{y}$ is the response to the force applied, $\underline{\text{M}}$ is a material property second-order tensor. In anisotropic Ohm's law, for example, the charge flux $\overrightarrow{J_q}$ in an applied electric field,

$\vec{E}$, is given by

$$\vec{J_q} = \chi\vec{E}$$
$$\begin{bmatrix} J_{q1} \\ J_{q2} \\ J_{q3} \end{bmatrix} = = \begin{bmatrix} \chi_{11} & \chi_{12} & \chi_{13} \\ \chi_{21} & \chi_{22} & \chi_{23} \\ \chi_{31} & \chi_{32} & \chi_{33} \end{bmatrix} \begin{bmatrix} E_1 \\ E_2 \\ E_3 \end{bmatrix} \tag{19}$$

The conductivity tensor χ relates the "effect" $\vec{J_q}$ to the "cause" $\vec{E}$ ($\vec{J_q}$ need not be parallel to $\vec{E}$).

The physical law is usually expressed in an inverse relationship of Eq. (20), i.e.,

$$\vec{E} = \rho\vec{J}$$
$$\begin{bmatrix} E_x \\ E_y \\ E_z \end{bmatrix} = \begin{bmatrix} \rho_{xx} & \rho_{xy} & \rho_{xz} \\ \rho_{yx} & \rho_{yy} & \rho_{yz} \\ \rho_{zx} & \rho_{zy} & \rho_{zz} \end{bmatrix} \begin{bmatrix} J_{qx} \\ J_{qy} \\ J_{qz} \end{bmatrix} \tag{20}$$

ρ (the resistivity tensor of order two) is the inverse of χ (the conductivity tensor).

Materials often have properties that are not isotropic; that is, the properties are direction dependent. If the material and the coordinate axes are fixed, the properties in every direction have fixed/constant values or the material properties remain unchanged, regardless of the coordinate system. But the component values in any given direction may be different. The tensor components transform according to certain rules when the coordinates change. Such laws distinguish tensors from matrices.

Coordinate systems are chosen based on convenience. To achieve rotations, reflections, and affine distortions (parallel lines remain parallel, but perpendicular corners are distorted), Cartesian orthogonal coordinate system is preferred, e.g., force–response relations. Often, the coordinate systems are so chosen as to have only the diagonal elements nonzero (eigenvalues). Transformation of Eq. (20)—Ohm's law—illustrates how the material property χ^{old} transforms to χ^{new} and vice versa. This is the rule for rank two tensor transformation:

$$\begin{aligned} \vec{J_q}^{old} &= \chi^{old}\vec{E}^{old} \quad \textbf{(old coordinates.)} \\ \vec{J_q}^{new} &= \chi^{new}\vec{E}^{new} \quad \textbf{(new coordinates).} \end{aligned} \tag{21}$$

[This follows from the transformation rule

$$\boldsymbol{A}^{old \rightarrow new} = \boldsymbol{A}^{new \rightarrow old^{-1}}. \tag{22}$$

That is, $A^{old \rightarrow new}$ and $A^{new \rightarrow old}$ are the "inverse" of each other.

In other words,

$$\chi^{new} = \boldsymbol{A}^{old \rightarrow new} \chi^{old} \boldsymbol{A}^{new \rightarrow old} \text{ and } \chi^{old} = \boldsymbol{A}^{new \rightarrow old} \chi^{new} \boldsymbol{A}^{old \rightarrow new}. \tag{23}$$

That is, a vector is subjected to multiplication by a single matrix. A second-order tensor is subjected to matrix multiplication twice. A similar rule applies to tensors of any rank. If the inverse of a matrix is its transpose, it constitutes an orthogonal transformation. To present an example, for a rotation, $R_{ij}R_{ji} = \delta_{ij}$, where δ_{ij} is known as "Kronecker delta":

$$\delta_{ij} = \begin{Bmatrix} \mathbf{1}\ \boldsymbol{when}\ \boldsymbol{i} = \boldsymbol{j} \\ \mathbf{0}\ \boldsymbol{when}\ \boldsymbol{j} \neq \boldsymbol{i} \end{Bmatrix}$$

Acknowledgments

This book is an outcome of several years of teaching by K. A. Padmanabhan (KAP) and S. Balasivanandha Prabu (BSP) and the personal research of KAP, Dr. B. P. Gautham, and Mr. A. Bhattacharjee at Tata Consultancy Services (TCS). They profusely thank all their former students and the TCS management in this regard.

The contact point and senior editor at the publisher, Springer Verlag, Germany, Dr. Zachary Evenson, has been a pillar of strength, who supported us at the various stages and promptly resolved all issues. The book could not have been completed without his help and support. We are grateful. We also thank Mr. Niyati Naudiyal, editorial project manager; Antje Endemann, production contact; and Padmavathi Jagadeishkumar, production editor, for their help and guidance during the book production process.

All the figures in Chaps. 1–14 were drawn by D. Sugan. K. T. Kowshic prepared the questions and answers. KAP and BSP are extremely grateful to them for their help.

B. P. Gautham and Mr. A. Bhattacharjee would like to express their deepest gratitude to Mr. K. Ananth Krishnan, former Chief Technology Officer (CTO), and Dr. Harrick Vin, the present CTO of TCS, for their invaluable support. Additionally, B. P. Gautham and A. Bhattacharjee extend their heartfelt appreciation to their colleagues, Dr. Gerald Tennyson, Mr. Surya Ardham, Mr. Srimannarayana Pusuluri, Mr. K. R. Harisankar, Mr. Sandeep Pusuluri, Mr. Pravin Kumar, Mr. Danish Khan, Mr. Ayush Suhane, and Dr. Pramod Zagade, for generously providing their research findings and results that have enriched the examples presented in Chaps. 15–16 of this book.

Competing Interests The authors have no competing interests to declare that are relevant to the content of this manuscript.

Acknowledgments

List of Symbols

V	Volume
dV	Increase in volume
P	Pressure
T	Temperature
B	Magnetic field
dW	Infinitesimal transformation in work
ΔQ	Heat produced/added
C	Heat capacity
C_V	Heat capacity when heating is done at constant volume
C_P	Heat capacity when heating is done at constant pressure
k	Boltzmann's constant = 1.38×10^{16} erg/deg = 1.380649×10^{-23} joules.K^{-1}
N	Number of molecules
n	Number of moles of the gas
R	Gas constant; $R = 8.315$ J/deg. = 1.986 cal/deg. = 0.0821 L-atm/degree
A	Avogadro's number = 6.205×10^{23} atoms/mole
ΔQ	Net amount of heat absorbed
ΔW	Net amount of work done by the system
U	Internal energy
H	Enthalpy
η	Efficiency
Θ_1 and Θ_2	Absolute temperatures
S	Entropy
ν_D	Debye frequency
λ_{min}	Minimum wavelength
n_i	Number of atoms in the ith energy level
ϵ_i	Energy of the ith energy level of a harmonic oscillator
N_0	Avogadro number
E*	Activation energy
$r_{e(A)}$	Intrinsic evaporation rate

P_f	Partition function
X_A	Mole fraction of A
f_i	Fugacity of the component i in the solution at temperature T
$\overline{G}_A$ *and* $\overline{G}_B$	Partial molar Gibbs free energies of A and B, respectively
ΔH^M	Molar heat of formation of a solution
G^{XS}	Excess molar free energy of the solution
G	Molar Gibbs free energy of the solution
G^M	Free energy of mixing
p_A, p_B, p_C	Partial pressures of A, B, and C, respectively
K_X	Equilibrium constant expressed for the relevant mole fractions
F	Faraday's constant
ξ	EMF of the cell
μ_k	Chemical potential
n_k	Number of moles of component k
N_V	Vacancy mole fractions
c_i	Concentration of (ionic) component i
u_i	Electrochemical mobility of i
$c_i^\alpha . b_i^\alpha$	Transport coefficient of *i* in α
j_k; $\beta_{i,k}$	Friction (cross) coefficients
∇p_i^α	Potential of component *i* in α
ΔU	Charge in the interfacial voltage drop relative to its equilibrium value
J	Flux
$\overrightarrow{J_S}$	Entropy flux
ψ_j	Tensor
$\dot{\sigma}$	Rate of entropy production per unit volume
F_Q	Conjugate force
D	Diffusivity
$\tilde{D}$	Inter-diffusivity
$\overrightarrow{F_{app}}$	Applied tensile stress
L_{ik}	Onsager coefficients
Υ_{kj}	Thermodynamic factors that couple chemical potentials to concentrations
λ_i	Eigenvalues
f	Mechanical force
ρ	Medium density
μ	Shear modulus
Φ	Diffusion potential
I	Diffusion current
R or R	Particle growth rate with radius
Ω	Interaction parameter
x_A, x_B	Mole fraction of components *A* and *B*
v	Velocity of the grain boundary
M	Grain boundary mobility
γ	Grain boundary energy

κ	Local curvature of the grain boundary
F_0	Free energy of a disordered phase
Π	Product of a group
F	Faraday's constant
η	Overpotential
α	Charge transfer coefficient

Contents

Chapter 1
Introduction

Contents

1.1 On Analyzing a Problem

Broadly speaking, the questions generally raised in engineering/engineering sciences are expected to be answered in quantitative terms based on the following simple questions: (1) How? (2) How far? (3) How fast? (4) Why? and (5) Are there any losses? If so, what are the losses?

In addition, one may also write down a general rule applicable to almost all situations:

$$\text{Input–Output–Losses} = \text{Accumulation} \tag{1.1}$$

But the four questions listed are not independent/unrelated. For example, the speed with which one intends to execute a task on hand will be determined by how far one can carry the action forward or vice versa. The question "How", so far as physical processes are concerned, is answered by Unit Operations, e.g., comminution, crushing, grinding, cutting, vibrating, etc. In the case of chemical processes, one has recourse to synthesis, dissociation, chemical reaction, catalysis, and so on. Catalysis is not considered in this book. The question "How Far" is answered by Thermodynamics, "How Fast" by Kinetics, and the fundamental question "Why" is the bedrock of Engineering Sciences, cutting across all the disciplines. In this book, the authors intend to address the question "How Far and How Fast" at a basic level, in all its ramifications, to the extent possible.

K. A. Padmanabhan et al., *Thermodynamics and Kinetics of Materials*, Engineering Materials, https://doi.org/10.1007/978-3-032-14527-7_1

1.2 Working Definitions

Thermodynamics is a phenomenological theory of matter. Here, the concepts are derived from experiments.

A thermodynamic system is a macroscopic system (any part of the universe) under investigation.

Thermodynamic parameters are macroscopic quantities that are defined experimentally. For example, V, the volume, P, the pressure, the temperature T, B, the magnetic field.

A thermodynamic state is a set of values of all the thermodynamic parameters needed to describe a system.

Equation of state is a functional relationship among the thermodynamic parameters that are in equilibrium. For example, f (P, V, T) = 0. Here, there are 3 variables. Such an equation of state helps reduce the number of independent variables from 3 to 2. Following the measurements made, 'f' is given as a part of the specification of the system.

Such a state in the P-V-T space is represented by a point.

Therefore, a state equation defines a surface (Fig. 1.1) and every point on this surface defines an equilibrium state. ≡ a state in equilibrium, unless specified otherwise. A change of state means a thermodynamic transformation.

In an equilibrium state, transformations are brought about by changing the external conditions.

In a Quasi-Static system, if the change in the external condition is slow, at any moment, the system is nearly in equilibrium.

If the transformation retraces its history in time when the external condition retraces its history in time, it is a reversible transformation.

A quasi-static transformation is reversible. The converse is not necessarily true, e.g., expansion of a gas into successive infinitesimal volume elements.

A P-V diagram is a representation of an equilibrium state.

A continuous path on the P-V diagram represents a reversible transformation.

There are certain types of reversible transformations. Some processes are completed at a constant temperature—the isothermal transformations. Certain others take

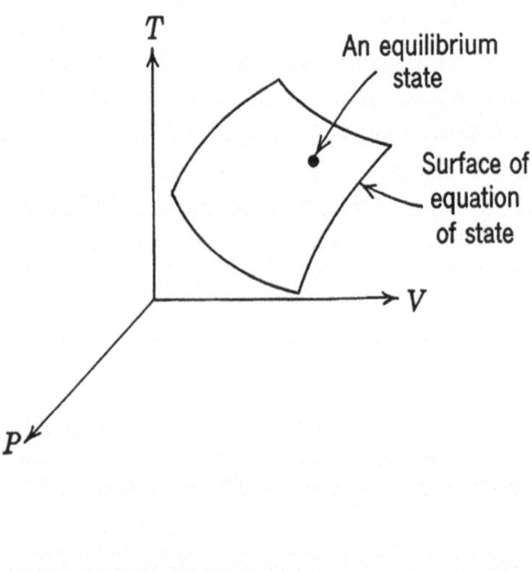

Fig. 1.1 Graphical illustration of an equation of state

place under hermetically sealed conditions, which do not allow heat to either enter or leave the system. These are adiabatic transformations.

Irreversible transformations cannot be represented in a diagram, as a reversible transformation can be.

An example of an infinitesimal transformation is work, dW = PdV. dV is the increase in volume at constant pressure. Generalization to different cases is obvious.

When the temperature of a homogeneous system increases, while no work is done, it is said to have absorbed heat. Heat produced/ added is defined as $\Delta Q = C\Delta T$, where C is the heat capacity. A comprehensive description of the system will define C.

ΔQ is path-dependent. It could be different for the same ΔT. Therefore, the way the heating is done defines the C, heat capacity. C_V is when heating is done at constant volume. C_P is obtained when heating is at constant pressure.

Specific heat is defined as the heat capacities (mentioned in the plural because both C_V or C_P are possible) per unit mass or per unit mole.

(Heat) Reservoir is defined as a large system whose temperature is not altered by any addition or extraction of a finite amount of heat.

In a thermally isolated system, there is no exchange of heat between the system and the outside. Thermal isolation is by an adiabatic wall. This will lead to an adiabatic transformation.

An extensive property of a thermodynamic quantity is proportional to the amount of substance present. An intensive property is independent of the amount of substance present.

To a good approximation, thermodynamic quantities are either extensive or intensive.

An Ideal Gas is a concept of a thermodynamic system. Here, the behavior of all gases is similar as they are dilute. It assumes that all gases behave in a universal way when they are sufficiently dilute. It represents a limiting behavior.

Ideal gas behavior is understood in terms of pressure (P), volume (V), temperature (T), and the number of molecules (N). The temperature is expressed in degrees Kelvin (K).

Boyle's law is an Equation of State of an ideal gas. T, in turn, is defined by the equation PV = NkT. Here, k is Boltzmann's constant = 1.38×10^{16} erg/deg. = 1.380649×10^{-23} joules.K^{-1}.

Value of T is determined by a conventional choice of temperature intervals, e.g., degrees K or degrees C. (Note: degrees Kelvin = degrees Centigrade so far as *temperature intervals* are concerned.) It is a universal concept because ideal gas behavior assumes that all gases behave identically.

T = 0 is chosen arbitrarily. An absolute meaning is attributed based on the second law of thermodynamics (please see later).

1.3 An Application

Measure $\left(\frac{PV}{Nk}\right)$ of an *ideal gas* at the temperatures at which water boils and freezes at sea level. A straight line is drawn through these points—see Fig. 1.2.

The line connecting the above-mentioned experimental points is extrapolated to intersect with the abscissa, which is defined as zero (the origin). Between the freezing point and the boiling point of water, 100 equal divisions are made. Each division corresponds to a degree—Fig. 1.2. The resulting scale is known as the Kelvin scale.

An example of an ideal gas is a gas of low density. $\left(\frac{PV}{Nk}\right)$ of the ideal gas is measured. T is measured from Fig. 1.2. An equivalent form of the equation of state of an ideal gas results, viz., $PV = (nRT)$ is an equivalent form of the equationof the state of an ideal gas. Here, $n =$ number of moles of the gas; $R =$ gas constant. $R = 8.315$ *J*/ deg. = 1.986 cal/deg. = 0.0821 L-atm/deg. $n = \left(\frac{N}{A}\right)$; where A is Avogadro's number. $A = 6.205 \times 10^{23}$ atoms/mole.

Thermodynamic Laws are mathematical axioms that define a mathematical model. The rigorous consequences of these axioms can be deduced. This model may not be a rigorous representation of the physical world.

Molecular Laws are fundamental laws of the physical world. In contrast, *Thermodynamic laws* are phenomenological, which conveniently summarize the macroscopic experience. Therefore, they are at least approximately true for the physical world. The relation between thermodynamic laws and molecular laws leads to kinetic theory. Thermodynamics may be defined as a mathematical model that can be axiomated in the best traditions of mathematics but contributes little to the understanding of the physics of the problem at hand.

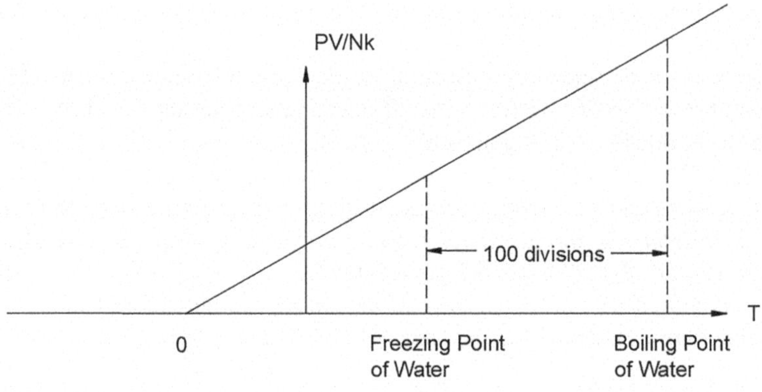

Fig. 1.2 Temperature scale for an ideal gas

Chapter 2
The Laws of Thermodynamics

Contents

2.1 Zeroth Law of Thermodynamics

This is mentioned in some textbooks only.

When two bodies have thermal equilibrium (equality of temperature) with a third body, they, in turn, have thermal equilibrium (equality of temperature) with each other. This fact, based on day-to-day experience, appears to be obvious. It is not derivable from the other laws but logically precedes the First and the Second laws. Therefore, it is referred to as the "Zeroth law of thermodynamics".

K. A. Padmanabhan et al., *Thermodynamics and Kinetics of Materials*, Engineering Materials, https://doi.org/10.1007/978-3-032-14527-7_2

2.2 The First Law of Thermodynamics

Consider an arbitrary thermodynamic transformation. Let heat ΔQ be the net amount absorbed and ΔW the net amount of work done by the system. *The first law of thermodynamics* states that the quantity, ΔU, is defined as

$$\Delta U = \Delta Q – \Delta W. \tag{2.1}$$

It is the same for all transformations that start from a given initial state to a predefined final state.

Equation (2.1) defines the state function U, called internal energy. Its value for any state is estimated as follows. Choose an arbitrary fixed state to be used as a reference state. Then the internal energy of any state, $= (\Delta Q - \Delta W)$, is the energy of any transformation leading from the reference state to the state under examination. It is known only up to an arbitrary additive constant.

U is an extensive property. This is known from the saturation properties of molecular forces, viz., the energy of a quantity is doubled when its mass is twice the original value. Joule's demonstration of the equivalence between heat and mechanical energy, i.e., mechanical energy can be completely converted into heat, provided the experimental foundation of the first law.

Acceptance of heat as a form of energy allows the inclusion of heat in the law of conservation of energy. The first law is stated as follows. The infinitesimal transformation is represented as,

$$dU = dQ – dW. \tag{2.2}$$

This is an exact differential, i.e., the differential of the function U is dU or $\int dU$ is independent of the integration path and depends only on the limits of integration. dQ or dW do not share this property.

If in $df = g\ (A, B)\ dA + h\ (A, B)\ dB$, df is exact, then $(\partial g/\partial B) = (\partial h/\partial A)$.

Some of the consequences of the exactness of dU are as follows.

Let P, V, T (pressure, volume, temperature) be the system parameters. Out of these 3 parameters, any two may be chosen as the independent variables to completely specify the state of the system. Then the other parameter is obtained using the equation of state.

Consider $U = U\ (P,\ V)$. Then,

$$dU = \left(\frac{\partial U}{\partial P}\right)_V dP + \left(\frac{\partial U}{\partial V}\right)_P dV \tag{2.3}$$

If dU is exact,

$$\frac{\partial}{\partial V}\left[\left(\frac{\partial U}{\partial P}\right)_V\right]_P = \frac{\partial}{\partial P}\left[\left(\frac{\partial U}{\partial V}\right)_P\right]_V. \tag{2.4}$$

If heat is absorbed by a system during an infinitesimal reversible transformation in which,

$$\mathrm{dW} = \mathrm{PdV},$$

$$dQ = \left(\frac{\partial U}{\partial P}\right)_V dP + \left[\left(\frac{\partial U}{\partial V}\right)_P + P\right] dV \tag{2.5}$$

$$dQ = \left[\left(\frac{\partial U}{\partial T}\right)_P + P\left(\frac{\partial U}{\partial T}\right)_P\right] dT + \left[\left(\frac{\partial U}{\partial P}\right)_T + P\left(\frac{\partial U}{\partial T}\right)_T\right] dP. \tag{2.6}$$

$$dQ = \left(\frac{\partial U}{\partial T}\right)_V dT + \left[\left(\frac{\partial U}{\partial V}\right)_T + P\right] dV \tag{2.7}$$

These are called the *dQ* equations. They are of little practical use. The partial derivatives that appear in the equations are usually unknown and cannot be estimated by direct measurement.

They can be transformed to more useful forms using the second law of thermodynamics (see later).

From the *dQ* equations,

$$C_V \equiv \left(\frac{\Delta Q}{\Delta T}\right)_V = \left(\frac{\partial U}{\partial T}\right)_V \tag{2.8}$$

$$C_P \equiv \left(\frac{\Delta Q}{\Delta T}\right)_P = \left(\frac{\partial H}{\partial T}\right)_P. \tag{2.8a}$$

Here $H = U + PV$ and is referred to as the enthalpy of the system.

Examples of the application of the first law of thermodynamics are:

(a) Joule's free-expansion experiment, which when analyzed leads to free expansion of an ideal gas into a vacuum (Fig. 2.1).

It is found experimentally that $T_1 = T_2$.

The following deductions emerge. Since the gas performs no work on the surroundings, $\Delta W = 0$.

As $\Delta \mathrm{T} = 0$, $\Delta Q = 0$. So, from the first law of thermodynamics $\Delta \mathrm{U} = 0$. Hence, a conclusion emerges that two states with the same *T*, but different volumes, have the same internal energy. As U is a state function and *T* and *V* may be taken as independent parameters, *U* is a function only of *T*. (This conclusion emerges, without experiment, from the second law of thermodynamics—see later.)

(b) Internal energy of ideal gas: Since *U* is only a function of *T*,

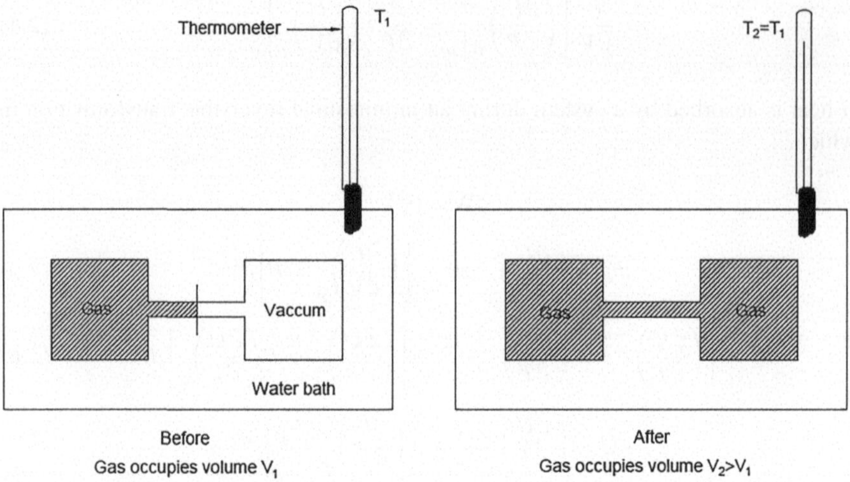

Fig. 2.1 Free-expansion experiment of Joule

$$C_V = \left(\frac{\partial U}{\partial T}\right)_V = \frac{dU}{dT} \tag{2.9}$$

Assuming C_V to be independent of T, one gets, $U = C_V T + Constant.$
Set, constant = 0;

(Arbitrary action to match the results of the analysis with the experimental results.)

(c) The quantity $(C_P - C_V)$ for an ideal gas: The enthalpy of an ideal gas depends only on T.

$$H = U + PV = (C_V + Nk)\,T. \tag{2.10}$$

From Eq. (2.8),

$$C_P = \left(\frac{\partial H}{\partial T}\right)_P = \frac{dH}{dT} = C_V + Nk. \tag{2.11}$$

Or

$$C_P - C_V = Nk. \tag{2.12}$$

Conclusion: Heating an ideal gas, keeping the volume constant, is more efficient than heating it keeping the pressure constant. It is intuitively obvious that no work is done at at constant volume. All the heat energy is used to increase the internal energy.

2.3 Second Law of Thermodynamics

The second law of thermodynamics is based on experience. There are processes in Nature that are expected to obey the law of conservation energy, yet they never occur. For example, A hot stove on the floor never jumps up to the ceiling and becomes cooler spontaneously (i.e., heat energy gets transformed to potential energy). The second law incorporates such experiential/ experimental facts into thermodynamics. All such experimental facts are consistent with common sense, and they form the foundation.

Kelvin statement, K: According to (Lord) Kelvin (Kelvin scale of temperature fame), in no thermodynamic transformation heat from a given heat reservoir can be converted into work entirely.

Clausius statement, C: Clausius put it differently that in no thermodynamic transformation the entire heat can be extracted from a colder reservoir and delivered entirely to a hotter reservoir.

For instance, when an ideal gas is expanded reversibly and isothermally work is done by the gas. Since for this process $\Delta U = 0$, work done = heat absorbed by the gas during the expansion. In other words, the entire heat is converted into work. But this is not the only effect of the transformation as gas occupies a larger volume in the final state. The second law of thermodynamics allows this process.

It will be proved by *Reductio ad Absurdum* that the above statement of Kelvin and Clausius' statement are equivalent, i.e., if Kelvin's statement is false, Clausius' statement is false, and the other way round.

The proof that K is false implies that C is false.

If K is false, one can extract heat from a reservoir at T_1 and convert it entirely to work. There is no other effect. Then one can convert this work into heat and deliver it to a reservoir at temperature $T_2 > T_1$, with no other effect. (e.g., Joule's experiment on the equivalence of heat and energy.[1] In the present 2-step process an amount of heat from a colder reservoir is transferred to a hotter one without any other effect.) Hence C is false.

The proof that C is false implies that K false.

Definition: Engine is a thermodynamic system that undergoes a cyclic transformation, i.e., a transformation whose final state is identical to the initial state. The system does only the following things.

[1]The potential energy of two falling masses was converted to heat energy. The experimental setup consisted of two masses attached to a paddle wheel through a rope.

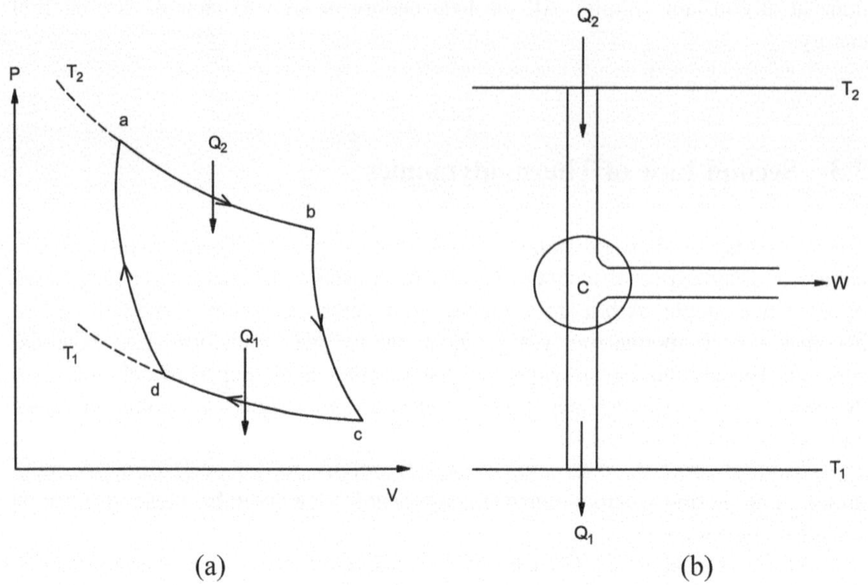

Fig. 2.2 The Carnot engine: (**a**) and (**b**) are alternate ways of representing what is stated in the text

(a) Absorbs heat $Q_2 > 0$ from reservoir T_2.
(b) Rejects heat $Q_1 > 0$ to reservoir T_2; $T_1 < T_2$.
(c) An amount of work $W > 0$ is performed.

If the Clausius statement is false: Extract Q_2 from reservoir T_1 and deliver it to reservoir T_2; $T_2 > T_1$. The engine operates between T_2 and T_1 for one cycle. Arrange the engine so that the amount of heat extracted by the engine from T_2 is exactly equal to Q_2. The net result: heat extracted from T_1 is entirely converted into work, with no other effect. Hence K is false. This completes the proof of equivalence.

The Carnot engine functions in a reversible way. In a reversible cyclic transformation a substance goes through the process illustrated in the P–V diagram below.

The path 'ab' is followed isothermally at temperature T_2, during which the system absorbs heat Q_2. 'bc' is a path that is followed adiabatically. 'cd' is a path followed isothermally at temperature T_1; $T_1 < T_2$, during which time the system rejects heat Q_1. The path 'da' is followed adiabatically (Fig. 2.2a). The above description may also be represented as on Fig. 2.2b.

Work done by the system, according to the first law, is given by $W = Q_2 - Q_1$, since $\Delta U = 0$ in a cyclic transformation. The efficiency of the engine is given as,

$$\eta = \frac{W}{Q_2} = 1 - \frac{Q_1}{Q_2}. \tag{2.13}$$

2.3.1 Demonstrate that if W > 0, then $Q_1 > 0$ and $Q_2 > 0$

Proof: $Q_1 \neq 0$; otherwise, the Kelvin statement is violated.

Suppose $Q_1 < 0$, this implies that engine absorbs Q_2 from T_2 and $-Q_1$ from T_1 and converts the heat $Q_2 - Q_1$ into work.

Now, we may convert this amount of work, which by assumption is positive, into heat and deliver it to the reservoir at T_2, with no other effect.

Since by assumption $T_2 > T_1$, this is impossible according to Clausius. Therefore, $Q_1 > 0$.

Similarly, one can show that if $W > 0$ and $Q_1 < 0$, then $Q_2 < 0$.

In this case, the engine operates in the reverse and becomes a "refrigerator".

Carnot engine is important because of the following theorem.

Carnot's theorem: No engine operating between two given temperatures is more efficient than a Carnot engine.

Proof: A Carnot engine C and an arbitrary engine X are operated between reservoirs T_2 and T_1 ($T_2 > T_1$)—Fig. 2.3.

From I law it follows that,

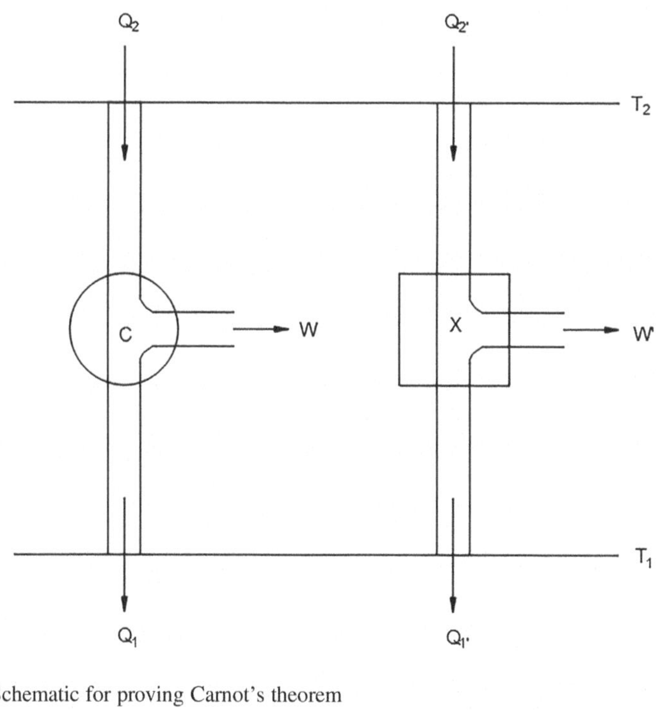

Fig. 2.3 Schematic for proving Carnot's theorem

$$W = Q_2 - Q_1.$$
$$W' = (Q_2' - Q_1').$$

Let

$$\left(\frac{Q_2}{Q_2'}\right) = \frac{N'}{N}. \tag{2.14}$$

Here N' and N are two integers.

If N′ and N are sufficiently large, the equality is satisfied to any desired accuracy.

C engine is operated *N* cycles in the reverse and *X* engine N' cycles. After this operation,

$$W_{Total} = N'W' - NW \tag{2.15}$$

Since, $\frac{Q_2}{Q_1} = \frac{N'}{N}$

$$(Q_2)_{Total} = N'Q'_2 - NQ_2 = 0.$$
$$(Q_1)_{Total} = N'Q'_1 - NQ_1.$$

We can also write $W_{Total} = (Q_2)_{Ttotal} - (Q_1)_{Total} = -(Q_1)_{Total}$, because $(Q_2)_{Total} = 0$.

The net result is a violation of the Kelvin statement, unless $W_{Total} < 0$. This implies $(Q_1)_{Total} > 0$.

Or

$$N'Q'_1 - NQ_1 \geq 0; Q_2Q_1' - Q_2'Q_1 > 0; \frac{Q_1}{Q_2} \leq \frac{Q_1'}{Q_2'}.$$

Therefore,

$$\left(1 - \frac{Q_1}{Q_2}\right) \geq \left(1 - \frac{Q_1'}{Q_2'}\right). \tag{2.16}$$

X is an arbitrary engine. So, it can also be a Carnot engine. A trivial corollary results: All Carnot engines operating between two given temperatures have the same efficiency.

2.3.2 *Absolute Scale of Temperature*

The corollary that results from Carnot's theorem results in the definition of the absolute scale of temperature.

Definition: If efficiency of a Carnot engine operating between two reservoirs of absolute temperatures Θ_1 and $\Theta_2 (\Theta_2 > \Theta_1)$ is η, then

$$\frac{\Theta_1}{\Theta_2} = 1 - \eta. \tag{2.17}$$

Since $0 < \eta \leq 1$, the absolute temperature of the reservoir is always greater than zero.

Let a series of Carnot engines so arranged are performing the same amount of work W. The heat rejected by one Carnot engine is absorbed by the next (Fig. 2.4).

For all n we have, $Q_{n+1} - Q_n = W$.

Therefore,

$\frac{Q_n}{Q_{n+1}} = \frac{\Theta_n}{\Theta_{n+1}}$. The above equation can be rewritten as, $\frac{\Theta_n}{Q_n} = \frac{\Theta_{n+1}}{Q_{n+1}} = x$.

Hence, $x \equiv \frac{\Theta_n}{Q_n}$, which is independent of n.

Therefore, $\Theta_{n+1} - \Theta_n = xW \left(\Theta_{n+1} = \frac{Q_{n+1}}{Q_n} . \Theta_n \right)$

Therefore,

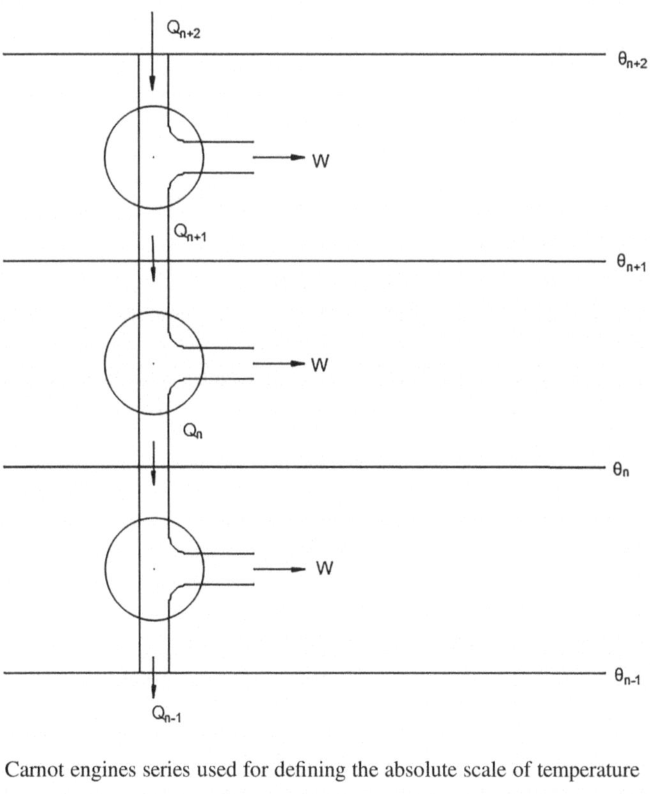

Fig. 2.4 A Carnot engines series used for defining the absolute scale of temperature

$$\Theta_{n+1} - \Theta_n = \Theta_n \left(\frac{Q_{n+1}}{Q_n} - 1 \right) = x.(Q_{n+1} - Q_n) = xW. \tag{2.18}$$

which is independent of n. The net result is a temperature scale of equal intervals. When $xW = 1°K$, the absolute Kelvin scale of temperature results.

Therefore, the following points emerge.

(a) Definition of the absolute scale of temperature is independent of the specific properties of any substance. It depends only on the second law of thermodynamics, which applies to all substances.
(b) The hint $\Theta = 0$ is the greatest lower bound of the temperature scale. This is referred to as absolute zero. As the second law would be violated, in Nature no Carnot engine with absolute zero as the temperature of the lower reservoir would exist. Absolute zero is a limiting case.
(c) The absolute Kelvin scale θ is identical with the ideal gas temperature scale T if $T > 0$. Using an ideal gas to form a Carnot engine this can be proved. No distinction is made between the two scales. The absolute temperature is denoted by T.

The Second Law allows the definition of a state function S, which is referred to as **Entropy**.

Clausius' Theorem: If the temperature is defined in a cyclic transformation by the following inequality:

$$\oint \frac{dQ}{T} \leq 0, \tag{2.19}$$

the integral extends over one cycle of transformation. If the cyclic transformation is reversible, equality holds.

Proof Let a cyclic transformation is denoted by O.

The cycle is divided into n infinitesimal steps. The temperature may be constant in each step.

The system is brought successively into contact with a heat reservoir at temperatures $T_1, T_2, \ldots, T_n$. From heat reservoir at temperature T_i, Q_i amount of heat be absorbed by the system.

To prove:

$$\sum_{i=1}^{n} \left(\frac{Q_i}{T_i} \right) \leq 0. \tag{2.20}$$

The theorem is obtained as $n \to \infty$.

Construct a set of Carnot engines $\{C_1, C_2, \ldots, C_n\}$ such that C_i

(a) Is operated between T_i *and* T_O ($T_O \geq T_i$ for all i),
(b) absorbs amount of heat Q_i from T_O,
(c) rejects amount of heat Q_i to T_i.

By the definition of the temperature scale,

$$\frac{Q_i^{(0)}}{Q_i} = \frac{T_0}{T_i}. \tag{2.21}$$

Consider one cycle of the combined operation,

$$0 + (C_1 + \ldots + C_n)$$

Therefore, an amount of heat

$$Q_0 = \sum_{i=1}^{n} Q_i^{(0)} = T_0 \sum_{i=1}^{n} \left(\frac{Q_i}{T_i}\right) \tag{2.22}$$

is absorbed from the reservoir T_O and converted entirely into work. There are no other effect.

According to the second law, this is impossible unless $Q_0 \leq 0$.

Therefore,

$$\sum_{i=1}^{n} \left(\frac{Q_i}{T_i}\right) \leq 0. \tag{2.23}$$

This proves the first part of the theorem.

If O is reversible, we shall reverse it and use the same arguments. We arrive at the same inequality, that is,

$$-\sum_{i=1}^{n} \left(\frac{Q_i}{T_i}\right) \leq 0\,. \tag{2.24}$$

Combining this with the previous inequality (which holds good for a reversible O) we get,

$$\sum_{i=1}^{n} \left(\frac{Q_i}{T_i}\right) = 0. \tag{2.25}$$

This completes the proof.

A Corollary follows, viz., for reversible transformation, the integral $\int \frac{dQ}{T}$ is independent of the path, which, in turn, depends only on the initial and the final states of the transformation.

Proof: Let the initial state be A and the final state B.

Let I and II denote two arbitrary reversible paths joining A and B. And, let II' be the reverse of II. Clausius theorem implies that,

$$\int_I \frac{dQ}{T} + \int_{II'} \frac{dQ}{T} = 0$$

But,

$$\int_{II'} \frac{dQ}{T} = -\int_{II} \frac{dQ}{T}$$

Hence,

$$\int_I \frac{dQ}{T} = \int_{II} \frac{dQ}{T}. \tag{2.26}$$

This corollary allows one to define the state function entropy S as follows.

Let fixed state O be a reference state. The entropy S of any state A is defined by,

$$S_A \equiv \int_0^A \frac{dQ}{T}, \tag{2.27}$$

where the path integration is any reversible path joining B to A. Thus, the entropy is defined up to an arbitrary additive constant. In other words, this definition assumes that every equilibrium state A is accessible from any state 0 through a reversible transformation or the surface of the equation of state consists of one continuous sheet. If the surface of the equation of state is two disjoint sheets, the definition only defines S for each sheet up to an additive constant. The absolute value of this constant becomes relevant when we go from one sheet to the next. This is the subject of the third law. It states that at absolute zero, $S = 0$ for all sheets.

The difference between the entropy of two states is completely defined as,

$$S(A) - S(B) = \int_B^A \frac{dQ}{T}, \tag{2.28}$$

where any reversible path joining B to A defines the path of integration.

Therefore, for any infinitesimal reversible transformation, the entropy change is given by,

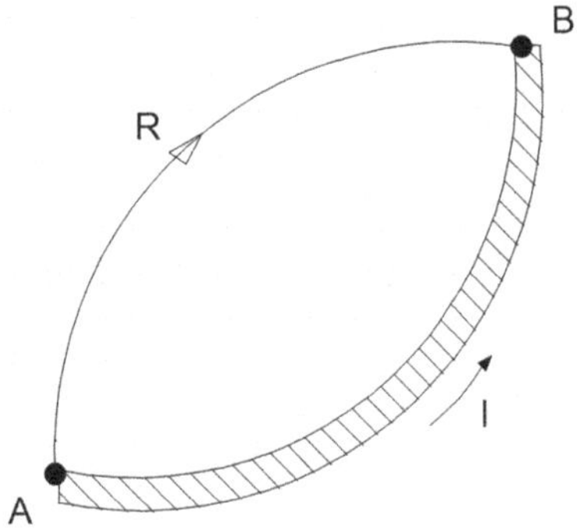

Fig. 2.5 Both reversible path *R* and irreversible path *I* connect states *A* and *B*

$$dS = \frac{dQ}{T}. \tag{2.29}$$

Properties of the entropy:

(a) **For an arbitrary transformation,**

$$\int_A^B \frac{dQ}{T} \leq S(B) - S(A), \tag{2.30}$$

If the transformation is reversible equality holds.

Proof:

Let *R* and *I* denote respectively any reversible and any irreversible path joining *A* to *B*. By the definition of S for path *R* the assertion holds. Let cyclic transformation be made up of I plus the reverse of *R*. It follows from Clausius' theorem that (Fig. 2.5)

$$\int_I \frac{dQ}{T} - \int_R \frac{dQ}{T} \leq 0. \tag{2.31}$$

Or

$$\int_I \frac{dQ}{T} \leq \int_R \frac{dQ}{T} \equiv S_B - S_A. \tag{2.32}$$

(b) **The entropy of a thermally isolated system never decreases.**

Proof: By definition, a thermally isolated system cannot exchange heat with the external world.

Therefore, for any transformation *dQ* = *0*. From the last mentioned property,

$$S(B) - S(A) \geq 0. \tag{2.33}$$

If the transformation is reversible, equality holds.

A consequence: A thermally isolated system is in a state of maximum entropy when it is in a state of equilibrium, consistent with the external constraints.

Entropy: Physical interpretation:

Consider the case where one mole of ideal gas expands isothermally from volume V_1 to V_2 by two routes: (a) reversible isothermal expansion, and (b) irreversible free expansion. The change in entropy of the gas, given the external surroundings, is calculated as follows.

2.3.3 Reversible Isothermal Expansion

The states of the gas (and not its surroundings) are represented in a *P-V* diagram (but not the surroundings) (Fig. 2.6).

$U = U(T)$ for an ideal gas. This is isothermal transformation. Hence, $\Delta U = 0$.

The amount of heat absorbed equals the work done (the shaded area in the *P-V* diagram). Therefore,

$$\Delta Q = RT \log_e \frac{V_2}{V_1} = RTln \frac{V_2}{V_1}. \tag{2.34}$$

Hence,

$$(\Delta S)_{gas} = \int \frac{dQ}{T} = \frac{\Delta Q}{T} = R \log_e \frac{V_2}{V_1} = Rln \frac{V_2}{V_1}. \tag{2.35}$$

The amount of heat ($-\Delta Q$) is supplied by the reservoir.

Hence,

$$(\Delta S)_{reservoir} = -\left(\frac{\Delta Q}{T}\right) = -Rln \frac{V_2}{V_1} \tag{2.36}$$

For the whole arrangement, the change in entropy is equal to 0. The amount of work $W = \Delta Q = RT\ ln \frac{V_2}{V_1}$ is stored in the spring connected to the piston. Reversing the transformation, this can be used to compress the gas.

2.3.4 Free Expansion

As in the reversible isothermal expansion, the initial and the final states remain the same (Fig. 2.7).

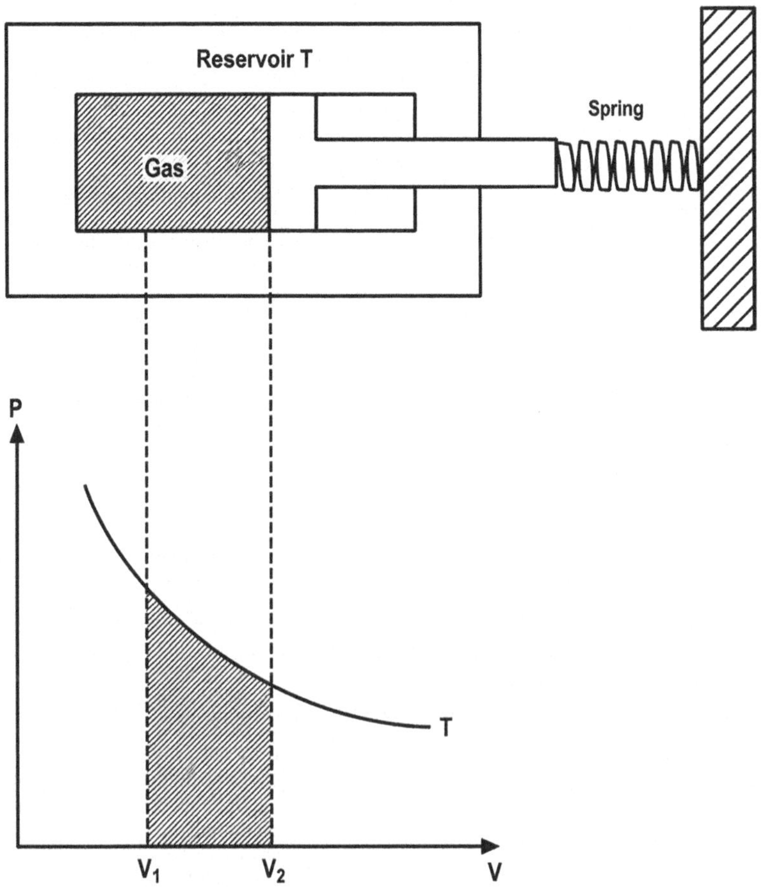

Fig. 2.6 An ideal gas experiencing reversible, isothermal expansion

Since S is a state function, $(\Delta S)_{gas}$ is the same. Or

$$(\Delta S)_{gas} = R\, log_e \frac{V_2}{V_1}$$

No heat is supplied by the reservoir. So,

$$(\Delta S)_{reservoir} = 0$$

Therefore, for the system of (gas + reservoir), increase in entropy is,

$$(\Delta S)_{total} = R\, log_e \frac{V_2}{V_1} \tag{2.37}$$

Compared to previous case, an amount of useful energy

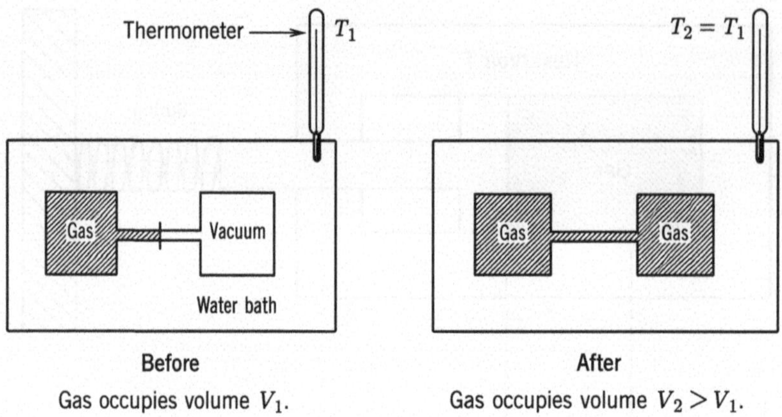

Fig. 2.7 Free-expansion experiment (after Joule)

$W = T(\Delta S)_{total}$ is "wasted" because it could have been extracted by expanding the gas reversibly.

This illustrates that irreversibility is generally "wasteful" and increases in the total entropy of the system. Therefore, the entropy of a state could be viewed as a measure of the unavailability of useful energy.

Another Case: Heat conducted is an irreversible process that increases the total energy.

Suppose heat from a reservoir at temperature T_2 is conducted by a metal bar to one at a temperature T_1 at a rate of Q per second. The net increase in entropy per second of the entire system is,

$$Q\left(\frac{1}{T_1} - \frac{1}{T_2}\right) > 0 \tag{2.38}$$

To transfer heat between the two reservoirs in a reversible manner one has to operate a carnot engine.

Here is a "Thought experiment". The entropy of the entire universe can never decrease as it is an isolated system. We have clear evidence looking around that in the universe most changes are irreversible. Therefore, the entropy of the universe constantly increases. It will lead to a "heat death" of the universe—a state of maximum entropy.

Is this the fate of the universe?

If the Second Law of Thermodynamics is rigorously correct, "yes" will be the answer. But our universe is not such a universe, but this conclusion cannot be reached within thermodynamics. Molecular laws govern our universe. Its invariance under time reversal denies the distinction between the past and the future, which is not correct. Between the past and the future—which is contradicted by experience.

The proper answer to this question is "No", because the second law of thermodynamics is not a rigorous law.

Then a second question arises about the extent to which the second law of thermodynamics is correct. Kinetic theory, which is based on molecular laws, answers that the second law of thermodynamics is correct "on an average". In macroscopic phenomena deviations from this law are so very rare and can be ignored. This aspect is beyond the scope of the present book.

Some consequences follow.

They use the fact that dS is an exact differential.

Note that

$$dQ = C_V dT + \left[\left(\frac{\partial U}{\partial V}\right)_T + P\right] dV$$

Putting $dQ = TdS$, we get

$$dS = \left(\frac{C_V}{T}\right) dT + \frac{1}{T}\left[\left(\frac{\partial U}{\partial V}\right)_T + P\right] dV \tag{2.39}$$

Since dS is an exact differential,

$$\left(\frac{\partial}{\partial V}\right)_T \left(\frac{C_V}{T}\right) = \left(\frac{\partial}{\partial T}\right)_V \left[\frac{1}{T}\left(\frac{\partial U}{\partial V}\right)_T + \frac{P}{T}\right] \tag{2.40}$$

Putting, $C_V = \left(\frac{\partial U}{\partial T}\right)_V$ and carrying out the differentiations on the right-hand side and rearranging, we get

$$\left(\frac{\partial U}{\partial V}\right)_T = T\left(\frac{\partial P}{\partial T}\right)_V - P. \tag{2.41}$$

It is now possible to calculate $\left(\frac{\partial U}{\partial V}\right)_T$ for an ideal gas. Using the equation of state P = $\left(\frac{NkT}{V}\right)$, we obtain $\left(\frac{\partial U}{\partial V}\right)_T = \frac{NKT}{V} - P = 0$; (since for an ideal gas PV = NkT). Hence, U is a function of T only.

This was earlier deduced from Joule's free expansion experiment with the help of the First law. Here, it is seen as a logical consequence of the Second law. Substituting, Eq. (2.41) in Eq. (2.39), yields

$$TdS = C_V dT + T\left(\frac{\partial P}{\partial T}\right)_V dV. \tag{2.42}$$

Going through similar steps with the other form of the dQ equation in which P and T are the independent variables, we obtain,

$$TdS = C_P dT - T\left(\frac{\partial V}{\partial T}\right)_P dP.$$
$$\left\{dQ = \left[\left(\frac{\partial U}{\partial T}\right)_P + P\left(\frac{\partial V}{\partial T}\right)_P\right]dT + \left[\left(\frac{\partial U}{\partial P}\right)_T + P\left(\frac{\partial V}{\partial P}\right)_T\right]dP\right\}. \tag{2.43}$$

The next step is to express Eqs. (2.42) and (2.43) so that only conveniently measurable quantities appear in those equations.

A **Lemma** (i.e., a subsidiary or intermediate theorem in an argument or proof) follows. Let x, y, z be quantities that satisfy the relation,

F (x, y, z) = 0, Let w be a function of any two of x, y, z.

Then,

$$\left(\frac{\partial x}{\partial y}\right)_w \left(\frac{\partial y}{\partial x}\right)_w = \left(\frac{\partial x}{\partial Z}\right)_w. \tag{2.44}$$

$$\left(\frac{\partial \varkappa}{\partial y}\right)_Z = \frac{1}{\left(\frac{\partial y}{\partial x}\right)_Z}. \tag{2.45}$$

$$\left(\frac{\partial x}{\partial y}\right)_z \left(\frac{\partial y}{\partial z}\right)_x \left(\frac{\partial z}{\partial x}\right)_y = -1 \text{ (the so called chain relation)}. \tag{2.46}$$

The proof is straight forward.

Define the following measurable quantities.

$$\alpha \equiv \frac{1}{V}\left(\frac{\partial V}{\partial T}\right)_P \text{(coefficient of thermal expansion)}; \tag{2.47}$$

$$k_T \equiv -\frac{1}{V}\left(\frac{\partial V}{\partial P}\right)_T \text{(isothermal compressibility)}; \tag{2.48}$$

$$k_S = -\frac{1}{V}\left(\frac{\partial V}{\partial P}\right)_S \text{(adiabatic compressibility)}. \tag{2.49}$$

From the Lemma,

$$\left(\frac{\partial P}{\partial T}\right)_V = -\frac{1}{(\partial T/\partial V)_P(\partial V/\partial P)_T} = \frac{\left(\frac{\partial V}{\partial T}\right)_P}{-\left(\frac{\partial V}{\partial P}\right)_T} = \frac{\alpha}{k_T}. \tag{2.50}$$

Equations (2.47) and (2.48) can now be written as,

$$TdS = C_V dT + \frac{\alpha T}{k_T} dV; \tag{2.51}$$

$$TdS = C_P dT - \alpha TUdP. \tag{2.52}$$

These are known as the *TdS* equations.

($C_P - C_V$) for any substance in terms of other experimental quantities can be expressed by equating the right-hand side of Eqs. (2.47) and (2.48). One then obtains,

$$C_V dT + T\left(\frac{\partial P}{\partial T}\right)_V dV = C_P dT - T\left(\frac{\partial V}{\partial T}\right)_P dP. \tag{2.53}$$

Rearranging the above equation

$$(C_P - C_V) = \frac{T\left(\frac{\partial V}{\partial T}\right)_P}{\left(\frac{\partial T}{\partial P}\right)_V} = -T\left[\left(\frac{\partial V}{\partial T}\right)_P\right]^2 \left(\frac{\partial P}{\partial V}\right)_T. \tag{2.54}$$

$$\left(\frac{\partial T}{\partial P}\right)_V \cdot \left(\frac{\partial P}{\partial V}\right)_T \cdot \left(\frac{\partial V}{\partial T}\right)_P = -1. \tag{2.55}$$

i.e.,

$$\left(\frac{\partial T}{\partial P}\right)_V = -\left(\frac{\partial P}{\partial V}\right)_T \left(\frac{\partial V}{\partial T}\right)_P \tag{2.56}$$

Choose P and V as the independent variables. Then

$$dT = \left(\frac{\partial T}{\partial V}\right)_P dV + \left(\frac{\partial T}{\partial P}\right)_V dP. \tag{2.57}$$

Substitution of this expression into the previous equation (the expression for ($C_P - C_V$)) yields,

$$\left[(C_P - C_V)\left(\frac{\partial T}{\partial V}\right)_P - T\left(\frac{\partial P}{\partial T}\right)_V\right] dV + \left[(C_P - C_V)\left(\frac{\partial T}{\partial P}\right)_V - T\left(\frac{\partial V}{\partial T}\right)_P\right] dP = 0.$$

Since dV and dP are independent of each other, the coefficients of dV and dP must vanish separately.

Therefore,

$$(C_P - C_V) = \frac{T\left(\frac{\partial P}{\partial T}\right)_V}{\left(\frac{\partial T}{\partial V}\right)_P} = -T\left[\left(\frac{\partial V}{\partial T}\right)P\right]^2 \cdot \left(\frac{\partial P}{\partial V}\right)_T.$$

(The second equality of the chain relation)

$$C_P - C_V = \frac{TV\alpha^2}{K_T}. \tag{2.58}$$

This shows that *$(C_P - C_V) > 0$*, if *$kT \geq 0$*. From experience, for most part *$kT \geq 0$*. But neither the First nor the Second law of Thermodynamics suggests this.

But it can be proved in statistical mechanics, where the nature of intermolecular forces is made use of and this is referred to as Van Hove's theorem.

Now the ratio $\gamma \equiv \frac{C_P}{C_V}$ will be evaluated.

Equations Eqs. (2.47) and (2.48) are valid for adiabatic transformations for which $dS = 0$.

Therefore,

$$C_V = -T\left(\frac{\partial P}{\partial T}\right)_V\left(\frac{\partial V}{\partial T}\right)_S.$$

But $C_P = T\left(\frac{\partial V}{\partial T}\right)_P\left(\frac{\partial P}{\partial T}\right)_S$. Therefore,

$$\begin{aligned}\gamma \equiv \frac{C_P}{C_V} &= -\frac{\left(\frac{\partial V}{\partial T}\right)_P\left(\frac{\partial P}{\partial T}\right)_S}{\left(\frac{\partial P}{\partial T}\right)_V\left(\frac{\partial V}{\partial T}\right)_S} = \frac{\left(\frac{\partial V}{\partial T}\right)_P}{\left(\frac{\partial P}{\partial T}\right)_V}\left(\frac{\partial P}{\partial V}\right)_S \\ &= \frac{\left(\frac{\partial V}{\partial P}\right)_T}{\left(\frac{\partial V}{\partial P}\right)_S}.\end{aligned} \tag{2.59}$$

Combining (2.58) and (2.59), we can solve for C_P and C_V separately. That is,

$$C_V = \frac{TV\alpha^2 k_S}{(k_T - k_S)k_T}; C_P = \frac{TV\alpha^2}{(k_T - k_S)}.$$

2.3.5 *Thermodynamic Potentials*

Two auxiliary state functions are defined at this stage.

1. Helmholtz free energy, A (simply free energy);
2. Gibbs' thermodynamic potential, G (or Gibbs potential).

Definitions:

$$A = U - TS, \tag{2.60}$$

$$G = A + PV. \tag{2.61}$$

For determining the equilibrium state of a non-isolated system these equations are used.

Physical meaning of free energy, A: The change of free energy in an isothermal transformation is the negative of the maximum possible work done by the system.

Suppose a system undergoes an arbitrary isothermal transformation (from state A to state B).

From second law $\int_A^B \frac{dQ}{T} \leq S(B) - S(A)$. Since T = constant, $\frac{\Delta Q}{T} \leq \Delta S$, where ΔQ is the amount of heat absorbed during the transformation and $\Delta S = S(B) - S(A)$.

From

$$\Delta U = \Delta Q - \Delta W \text{ (I law)}, A = U - TS; W \leq -\Delta U + T\Delta SP = -\left(\frac{\partial A}{\partial V}\right)_T \quad (2.62)$$

Here W is the work done by the system.

$$\text{Since the right} - \text{side} = -\Delta A, \text{we have } W \leq -\Delta A \quad (2.63)$$

If the transformation is reversible, equality will hold.

When $W = 0$, then Eq. (2.63) turns into a useful theorem.

Theorem The Helmholtz free energy never increases for a mechanically isolated system at constant T.

Corollary The state of equilibrium for a mechanically isolated system kept at a constant temperature, is the minimum Helmholtz free energy.

In an infinitesimal reversible transformation,

$$dA = -PdV - SdT. \quad (2.64)$$

Hence,

$$P = -\left(\frac{\partial A}{\partial V}\right)_T; \quad (2.65a)$$

$$S = -\left(\frac{\partial A}{\partial T}\right)_V \quad (2.65b)$$

which are known as Maxwell's relations.

If the function $A(V, T)$ is known, then P and S are calculable using Eqs. (2.65a) and (2.65b).

2.3.6 Principle of Minimization of Free Energy

Let a gas in a cylinder be at constant T. A sliding piston divides the total volume V into V1 and V2, where the pressures are P1 and P2 respectively.

Question: How to find the equilibrium position if the piston is released and allowed to slide freely?

Answer: The position of the piston must minimize the free energy of the total system.

Once equilibrium has been established, then,as the free energy is a minimum a slight change in the piston's position should not change the free energy. That is, $\delta A = 0$. But, A is a function of V_1, V_2 and T. Hence, $0 = \delta A = \left(\frac{\partial A}{\partial V_1}\right)_T \delta V_1 + \left(\frac{\partial A}{\partial V_2}\right)_T \delta V_2$.

Since $V_1 + V_2 = V =$ constant, $\delta V_1 = -\delta V_2$; $0 = \left[\left(\frac{\partial A}{\partial V_1}\right)_T - \left(\frac{\partial A}{\partial V_2}\right)_T\right]\delta V_1$.

As δV_1 is arbitrary, $\left(\frac{\partial A}{\partial V_1}\right)_T = \left(\frac{\partial A}{\partial V_2}\right)_T$

From Eq. (2.65a), $\left(P = -\left(\frac{\partial A}{\partial V}\right)_T\right)$; $P_1 = P_2$. This result is rather obvious.

2.3.7 *Gibbs' Potential*

Theorem The Gibbs' potential never increases at constant temperature (T) and pressure (P).

Corollary At constant T and P, the state of minimum Gibbs' potential represents the equilibrium state.

Proof: If T is constant, $w \leq -\Delta A$ in any transformation.

If P is constant, Then, $w = P\Delta V$. Then, we have

$$P\Delta V + \Delta A \leq 0; \Delta G \leq 0.$$

In an infinitesimal reversible transformation,

$$dG = -SdT + VdP. \tag{2.66}$$

Whence,

$$S = -\left(\frac{\partial G}{\partial T}\right)_P \tag{2.67a}$$

$$V = \left(\frac{\partial G}{\partial P}\right)_T. \tag{2.67b}$$

Two more Maxwell's relations result from the differential changes of the enthalpy: $H = U + PV$; $dH = TdS + VdP$.

Where

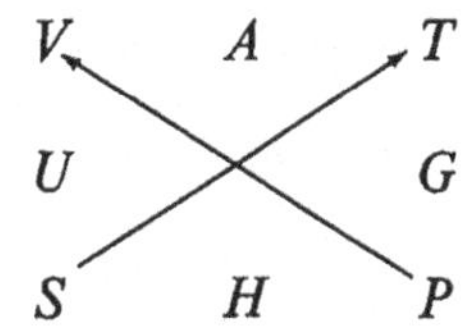

Fig. 2.8 Maxwell relations based on different thermodynamic variables

$$V = \left(\frac{\partial H}{\partial P}\right)_S; \tag{2.68a}$$

and

$$T = \left(\frac{\partial H}{\partial S}\right)_P. \tag{2.68b}$$

Equations (2.65a) and (2.65b) are also Maxwell equations:

$$P = -\left(\frac{\partial A}{\partial V}\right)_T; S = -\left(\frac{\partial A}{\partial T}\right)_V.$$

The other Maxwell's relations are:

$$T = \left(\frac{\partial U}{\partial S}\right)_V \tag{2.69a}$$

$$P = \left(\frac{\partial U}{\partial V}\right)_S. \tag{2.69b}$$

They result from the First law, $dU = -PdV + TdS$.

- The 8 Maxwell relations so far derived are convenient for thermodynamic calculations.
- These 8 Maxwell relations are conveniently summarized by the above diagram (Fig. 2.8).

2.4 The Third Law of Thermodynamics

The second law leads to the definitions presented below. (The entropy of a substance up to an arbitrary additive constant is the first consequence.)

- Definition of entropy depends on the existence of a reversible transformation connecting an arbitrarily chosen reference state O to a chosen state A. Such a reversible transformation exists if both O and A lie on one plane of the equation of state surface.

- Take two different surfaces or meta-stable phases of the same substance. More than one disjointed sheet may make up the equation of state surface. Then, the reversible path may not exist.
- So, the second law does not uniquely define the difference in entropy of 2 states, A and B, if A refers to one substance and B to another.
- Nernst supplied a rule in 1905 for such a determination. Now it is known as the third law of thermodynamics.
- **The Third Law**: At absolute zero the entropy of a system is a universal constant, which is taken as zero for mathematical convenience.
- The generality of this statement is based on (a) that it refers to any system, and (b) that it reveals that S = 0 at T = 0 regardless of other parameters of which S may be a function.

Therefore, the third law **presents the entropy of any state of any system unique**. It states that heat capacity of a system much be zero at absolute zero.

Let R be any reversible path connecting a state of a system at absolute zero to state A, whose entropy is unknown.

We assume that $C_R(T)$ is the heat capacity of the system along path R.

From the Second Law

$$S(A) = \int_0^{T_A} C_R(T) \frac{dT}{T} \text{ (i)}. \tag{2.70}$$

From the Third Law,

$$S(A) \underset{T_A \to 0}{\overset{0}{\longrightarrow}} = 0 \text{ (ii)}. \tag{2.71}$$

Hence

$$C_R \to_{T \to 0} 0 \text{ (iii)}. \tag{2.72}$$

C_R may be C_V or C_P.

Equation (2.71) has been verified for all systems tested so far.

Another consequence of the Third Law that is not so obvious is that at absolute zero the coefficient of thermal expansion of any substance is zero.

The proof is obtained from the *TdS* equations. From them one deduces the following.

$$\left(\frac{\partial S}{\partial T}\right)_P = \frac{C_P}{T} \text{(iv)} \tag{2.73}$$

$$\left(\frac{\partial S}{\partial P}\right)_T = -\left(\frac{\partial V}{\partial T}\right)_P \tag{2.74}$$

Combining the two equations one obtains

$$\left(\frac{\partial C_P}{\partial P}\right)_T = -T\left(\frac{\partial^2 V}{\partial T^2}\right)_P. \tag{2.75}$$

From Eqs. (2.74) and (2.70),

$$V\alpha \equiv \left(\frac{\partial V}{\partial T}\right)_P = -\left(\frac{\partial S}{\partial P}\right)_T = -\frac{\partial}{\partial P}\int_0^T C_P \frac{dT}{T} = -\int_0^T \left(\frac{\partial C_P}{\partial P}\right)_T \frac{dT}{T}. \tag{2.76}$$

The integrations are along a path of constant P.

From Eq. (2.75),

$$V\alpha = \int_0^T \left(\frac{\partial^2 V}{\partial T^2}\right)_P dT = \left(\frac{\partial V}{\partial T}\right)_P - \left[\left(\frac{\partial V}{\partial T}\right)_P\right]_{T=0}. \tag{2.77}$$

Therefore,

$$\alpha \xrightarrow[T\to 0]{} 0. \tag{2.78}$$

Similarly,

$$\left(\frac{\partial P}{\partial T}\right)_V \to_{T\to 0} 0. \tag{2.79}$$

Combined with Eq. (2.78), this implies that $P - T$ diagram of the melting curve has zero tangent at T = 0.

Experimentally it is seen that

$$C_P = T^x\left(a + bT + cT^2 + \cdots\right). \tag{2.80}$$

Here x is a positive constant. a, b, c,…, are functions of P.

Differentiating Eq. (2.80) with respect to P,

$$\left(\frac{\partial C_P}{\partial_P}\right)_T = T^x\left(a' + b'T + c'T^2 + \ldots\right). \tag{2.81}$$

Putting in Eq. (2.76),

$$V\alpha = -\int_0^T dT\left(a'T^{x-1} + b'T^x + \dots\cdot\right) = -T^x\left(\frac{a'}{x} + \frac{b'T}{x+1} + \frac{c'T^2}{x+2} + \dots\right).$$

Therefore,

$$\left(\frac{V\alpha}{C_P}\right)_{\overrightarrow{T\to 0}} \text{ is a finite constant.} \tag{2.82}$$

An important consequence of this derivation: A system cannot be cooled to absolute zero by changing the thermodynamic parameters by finite amounts. For instance, from one of the TdS equations it follows that through an adiabatic change of pressure dP, the temperature changes by

$$dT = \left(\frac{V\alpha}{C_P}\right) TdP. \tag{2.83}$$

From Eq. (2.82), to change T by a finite amount, the change in P needed is unbounded as T → 0. An alternative form of the Third Law is to state that Absolute Zero is unattainable. This is independent of the second law, which merely points out that no Carnot engine with its lower reservoir at at absolute zero can exist[2].

An Application of the Third Law of Thermodynamics:

$$\text{Helmholtz free energy } A = U - TS. \tag{2.84}$$

$$\text{From the third law, } A = U - T\int_0^T \frac{C_V}{T'}dT'. \tag{2.85}$$

On the right hand side of Eq. (2.85) the integration is over a path of constant volume. No arbitrary constants except the one contained in U are present.

As

$$U = \int_0^T C_V dT' + \text{constant} \tag{2.86}$$

both U and A up to the same arbitrary additive constant can be determined from measurements of C_V.

The following is a practical use of these ideas, viz., the determination of the melting point of solid quartz. The low temperature stable phase at low temperatures

[2]From second law dQ = TdS. So, at absolute zero dQ = 0. In other words, at absolute zero transformation is adiabatic. Therefore, there is no way to reversibly heat the system to a higher temperature. Or, no Carnot engine with its lower temperature reservoir at absolute zero can exist. The abstract notions receive meaning through Quantum Statistical Mechanics.

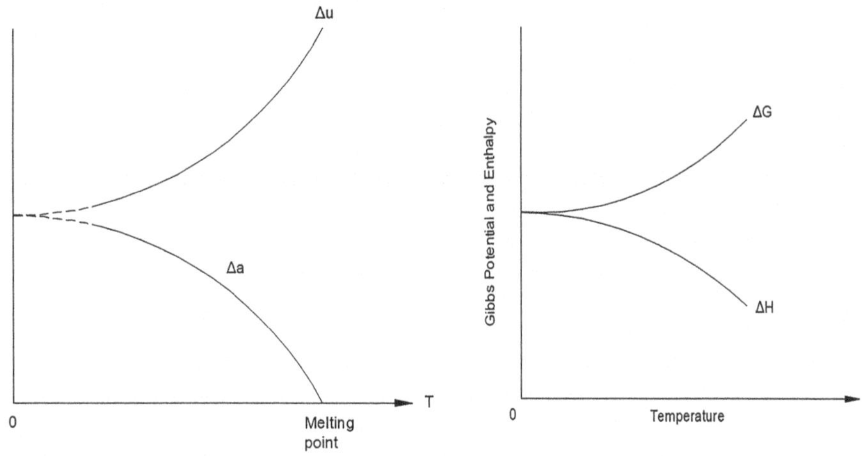

Fig. 2.9 An application of the third law in melting point determination

of quartz is a crystalline solid. Liquid phase (glass) can be supercooled in metastable equilibrium well below the melting point. So, it is not easy to measure directly the melting point of quartz. But it can be determined indirectly through Eq. (2.85).

If specific heat at constant volume,C_V, of both solid and liquid quartz is measured through a range of temperatures, and ΔC_V, which s a function of T, denotes their difference, then, the difference in internal energy/unit mass of the two phases is determined by numerical integration at constant V, of ΔC_V, as follows.

$$\Delta U = \int \Delta C_V dT'. \tag{2.87}$$

Using Eq. (2.86), the difference in the Helmholtz free energy per unit mass of the two phases is given by,

$$\Delta A = \Delta U - T \int_0^T \frac{\Delta c_V}{T'} dT'. \tag{2.88}$$

ΔU and ΔA as a function of T at constant volume are potted accordingly (Fig. 2.9).

The melting point is when $\Delta A = 0$ as the condition of phase equilibrium at fixed T and V is when the free energies per unit mass are equal. In practice, $\Delta A = 0$ is obtained by direct integration up to that point or by extrapolation.

Degree of Irreversibility: When the enery of a system undergoing a process energy is degraded, that process is irreversible. The degree of degradation differs from one process to another. So, a quantitative measure to define the degree of irreversibility (extent of degradation) is necessary.

Mathematically, an increase in the entropy of the system, ΔS, is the measure of the degree of irreversibility: $\Delta S = S(A) - S(B) = 0$, if it is a reversible transformation.

Therefore, There is no Entropy creation in a reversible process. It is just transferred from one part of the system to the other.

Criterion for equilibrium: If a system in equilibrium will continue to remain in equilibrium.

A non-equilibrium state will spontaneously move to its equilibrium state. This spontaneous process is irreversible and when a system moves from its initial non-equilibrium state to its final equilibrium state, the system entropy increases. This attainment of the equilibrium state coincides with when the entropy reaches a maximum value. Then, the system attains equilibrium. Therefore, for knowing the equilibrium state entropy can be a criterion..

When the internal energy of an isolated system is constant, U, and constant volume, V, equilibrium is attained when the entropy is a maximum. This is consistent with U and V remaining constant. Let the constant volume chemical reaction,

$$A + B = C + D$$

occur inside an adiabatic enclosure.

If the entropy increases, the reaction will proceed from left to right or vice versa (Fig. 2.10).

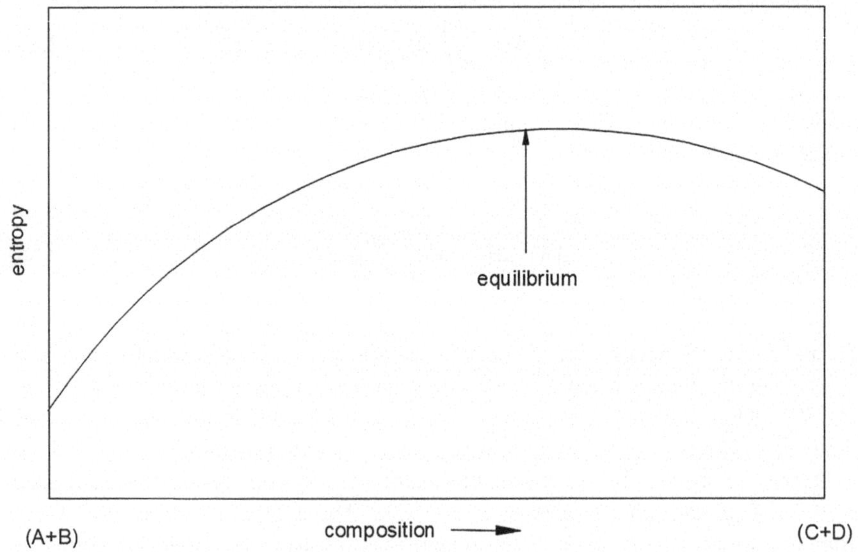

Fig. 2.10 In a closed system at constant volume and internal energy, the reaction $A + B = C + D$ is presented as a function of the reactant present

The equilibrium state of the system is reached when the entropy is a maximum. Evidently the entropy decreases on either side of the maximum. So, such movements will not be spontaneously favored.

A combined statement of the First and the Second Laws is as follows.

From the first law, In a closed system for an incremental change,

$$dU = \delta Q - \delta w. \tag{2.89}$$

For a reversible process, the second law leads to

$$dS = \frac{\delta Q}{T} or\ \delta Q = TdS.$$

Taken together, from the two laws one gets,

$$dU = TdS - PdV. \tag{2.90}$$

However, there are two restrictions. (1) the system is closed, i.e., does not exchange matter with the surroundings during the process. (2) The only work done by the system is because of the change in volume.

Equation (2.90) relates the dependent variable, U, to the independent variables, S and V, i.e.,

$$\mathrm{U} = \mathrm{U}\ (\mathrm{S}, \mathrm{V}).$$

The total differential of this function is,

$$dU = \left(\frac{\partial U}{\partial S}\right)_V dS + \left(\frac{\partial U}{\partial V}\right)_S dV. \tag{2.91}$$

Comparing Eqs. (2.90) and (291), $T \equiv \left(\frac{\partial U}{\partial S}\right)_V$.

$$P = -\left(\frac{\partial U}{\partial V}\right),$$

when $S = S(U, V)$ or

$$dS = \left(\frac{\partial S}{\partial U}\right)_V dU + \left(\frac{\partial S}{\partial V}\right)_U dV. \tag{2.92}$$

From Eq. (2.90), $dS = \frac{dU}{T} + \frac{PdV}{T}$.

Comparing with Eq. (2.92), one obtains $\left(\frac{\partial S}{\partial V}\right)_U = \frac{P}{T}$.

In a system in which internal energy and volume are constant and entropy is maximized equilibrium occurs. Equilibrium occurs in a system of constant entropy and volume, if the internal energy is minimized.

Considerable difficulty is encountered in a real system in arranging the state of the system so that, simultaneously, the required entropy and the required volume is a minimum.

Equations for a closed system are as follows.

We know from Eq. (2.90) $dU = TdS - PdV$.

$$H = U + PV \text{ and so, } dH = TdS + VdP. \tag{2.93}$$

$$A = U - TS \text{ and so, } dA = -SdT - PdV. \tag{2.94}$$

$$G = H - TS \text{ and so, } dG = -SdT + VdP. \tag{2.95}$$

$$\begin{aligned}(dA &= TdS - PdV - SdT - TdS,\\ G &= A + PV = U - TS + PV = H - TS,\\ G &= H - TS \text{ and so } dG = -SdT + VdP.)\end{aligned}$$

2.4.1 Gibbs Free Energy, G

It is known that $G = H - TS = U + PV - TS$.

If the of state of a system changes from 1 to 2,

$$\begin{aligned}(G_2 - G_1) &= (H_2 - H_1) - (T_2S_2 - T_1S_1),\\ &= (U_2 - U_1) + (P_2V_2 - P_1V_1) - (T_2S_2 - T_1S_1).\end{aligned}$$

In a closed system, the first law states that $(U_2 - U_1) = Q - W$,

$$(G_2 - G_1) = Q - W + (P_2V_2 - P_1V_1) - (T_2S_2 - T_1S_1).$$

For an isothermal process, $T_1 = T_2 = T$.

Also, let $P_2 = P_1 = P$. Then,

$$(G_2 - G_1) = Q - W + P(V_2 - V_1) - T(S_2 - S_1).$$

According to during a process the first law the system has a total work of W done on or by it. If additional chemical or electrical work, apart from the work of expansion against an external pressure, id done then these work terms are also included in W.

Therefore, $W = W' + P(V_2 - V_1)$. Here P $(V_2 - V_1)$ is the P-V work done at constant P by changing the volume. W' is the sum of all non- $P - V$ work done. Using these values,

$$(G_2 - G_1) = Q - w' - T(S_2 - S_1).$$

It is known that $Q \leq T(S_2 - S_1)$.

Then, $w' \leq -(G_2 - G_1)$.

For equality, one can write $-W' = (G_2 - G_1) + T\Delta S$.

During a process at constant temperature and pressure no work other than P-V work is done. That is, $w' = 0$. $(G_2 - G_1) + T\Delta S_{irr} = 0$. Such a process occurs spontaneously, with an increase in entropy and the Gibbs free energy decreases.

At thermodynamic equilibrium $dS_{irr} = 0$ is a criterion. Then in a process at cnonstant temperature and pressure at equilibrium an increment of an isothermal, isobaric process occurs at equilibrium $dGs = 0$. If Gibbs free energy can decrease or remain constant and G can be a minimum consistent with the P and T values, and equilibrium is attained when G attains the minimum value consistent with the values of P and T then the system will be in equilibrium. This criterion of equilibrium has practical value and is used in different fields.

Variations in the composition and size of the system could be present. But so far only close systems of fixed size and fixed composition have been studied. It should be noted that there are two independent variables. When they are fixed, the state of the system is uniquely fixed.

If the size and the composition change during a process, TWO independent variables are not sufficient to define the state of a system. For example, a process at constant T and constant P are equivalent, when G is a minimum.

If the number of moles of different species can vary during a chemical reaction, i.e., the composition is a variable, in isothermal and isobaric processes the minimization of G would occur only when the system has a unique composition. For instance, in a gaseous system at constant P and T of CO, CO_2, H_2, and H_2O G would be a minimum if the reaction $CO + H_2O = CC_2 + H_2$ is in equilibrium. As G is an extensive property, the number of moles within a system should be specified.

Therefore

$$\mathrm{G} = \mathrm{G}\left(\mathrm{T}, \mathrm{P}, n_i, n_j, n_k, \ldots\right), \tag{2.96}$$

where n_i, n_j,n_k, ... are the number of moles of species i, j, k, ... present in a system. When the independent variables are fixed, the state of the system is fixed only when all the independent variables are fixed.

Differentiating the equation

$$\boldsymbol{d}\mathrm{G} = \left((\partial G/\partial T)_{P,i,j,k,\ldots} \boldsymbol{d}\mathrm{T} + (\partial G/\partial P)_{T,i,j,k,\ldots} \boldsymbol{d}\mathrm{P} + (\partial G/\partial n_i)_{T,P,,n_j,n_k\ldots\ldots} \mathrm{d}n_i + \left(\partial G/\partial n_j\right)_{P,T,n_i,n_k\ldots\ldots} \mathrm{d}n_j + \ldots \right. \tag{2.97}$$

If the number of moles of all the individual species remains constant during the process, the equation reduces to one of constant volume and composition of the species present. Therefore,

$$\left((\partial G/\partial T)_{P,n_i,n_j,n_k}\right) = -S + (\partial G/\partial P)_{T,n_i,n_j,n_k,\ldots} = V.$$

Substitution in (2.96) gives,

$$dG = -S.dT + V.dP + \sum_i^k (\partial G/\partial n_i)_{T,P,n_j,,\ldots k}\, dn_i \tag{2.98}$$

The Chemical Potential:

The term $\left(\frac{\partial G}{\partial n_i}\right)_{T,P,n_j\ldots}$ is reerred to as the chemical potential of the species "i" and is represented as $\left(\frac{\partial G}{\partial n_i}\right)_{T,P,n_j,\ldots} = \mu_i$. Therefore, at constant P and T chemical potential of "i" in a homogeneous phase represents the rate of increase of G with n_i when the species "i" is added to the system and the number of moles of all the other species remains constant. Alternatively, in large enough systems at constant T and P the addition of 1 mole of i does not noticeably change the composition of the system. Then this addition causes an increase in the Gibbs free energy μ_i of the system.

An alternative form of Eq. (2.98) is,

$$dG = -SdT + VdP + \sum_1^k \mu_i dn_i \tag{2.99}$$

when G is a function of *T*, *P* and composition.

Thus, Eq. (2.99) can be applied to *open systems* which exchange matter and heat with their surroundings and to closed systems which undergo changes in composition caused by chemical reactions.

The equations derived earlier, viz.,

$$dU = TdS - PdV, dH = TdS + VdP \text{ and } dA = -SdT - PdV,$$

can be applied to open systems by including the dependences on the composition.

i.e.,

$$dU = TdS - PdV + \sum_{i=1}^k \left(\frac{\partial U}{\partial n_i}\right)_{S,U,n_j,\ldots} dn_i. \tag{2.100}$$

$$dH = TdS + VdP + \sum_{i=1}^k \left(\frac{\partial H}{\partial n_i}\right)_{S,P,n_j} dn_i. \tag{2.101}$$

$$dA = -SdT - PdV + \sum_{i=1}^k \left(\frac{\partial H}{\partial n_i}\right)_{T,V,n_j} dn_i. \tag{2.102}$$

From Eqs. (2.99), (2.100), (2.101) and (2.102),

$$\left(\frac{\partial G}{\partial n_i}\right)_{T,P,n_j....}=\mu_i=\left(\frac{\partial U}{\partial n_i}\right)_{S,V,n_j....}=\left(\frac{\partial H}{\partial n_i}\right)_{S,P,n_j...}=\left(\frac{\partial A}{\partial n_i}\right)_{T,V,n_j....}$$

The complete set is to be given by Eqs. (2.99)–(2.102).

Although all these equations are basic in nature, Eq. (2.99) is called the fundamental equation because of its practical use.

From the First Law, $dU = \delta Q - \delta w$. From Eq. (2.100), a reversible chemical reaction in a closed system implies a process involving a reversible change in composition,

$$\delta Q = TdS \text{ and } \delta W = PdV + \Sigma\mu_i dn_i.$$

$\Sigma\mu_i dn_i$ is the chemical work done by the system which was represented as w' earlier. (i.e., *Total work* = (*P* – *V work*) + *chemical work*).

From Eqs. (2.100)–(2.102) and (2.98) one obtains,

$$T=\left(\frac{\partial U}{\partial S}\right)_{V,comp}=\left(\frac{\partial H}{\partial S}\right)_{P,comp}. \quad (2.103)$$

$$P=-\left(\frac{\partial U}{\partial V}\right)_{S,comp}=-\left(\frac{\partial A}{\partial V}\right)_{T,comp}. \quad (2.104)$$

$$V=\left(\frac{\partial H}{\partial P}\right)_{S,comp}=\left(\frac{\partial G}{\partial P}\right)_{T,comp}. \quad (2.105)$$

$$S=-\left(\frac{\partial A}{\partial T}\right)_{V,comp}=-\left(\frac{\partial G}{\partial T}\right)_{P,comp}. \quad (2.106)$$

Earlier the transformation formula (also known as the upstairs–Downstairs–Inside–Out Formula) was prsented. From differential Calculus,

$$\left(\frac{\partial x}{\partial y}\right)_z\left(\frac{\partial y}{\partial z}\right)_x\left(\frac{\partial z}{\partial x}\right)_y=-1 \quad (2.107)$$

This relation can invoke any three state functions.

Gibbs-Helmholtz equation:

We know that $G = H - TS$, $dG = -SdT + VdP$. Or $\left(\frac{\partial G}{\partial T}\right)_P = -S$.

Therefore, at constant P and composition, $G = H + T\left(\frac{\partial G}{\partial T}\right)$, or $GdT = HdT + TdG$. Dividing T^2,

$$\frac{TdG - GdT}{T^2}=\frac{-HdT}{T^2} \quad (2.108)$$

From differential calculus,

$$d\left(\frac{x}{y}\right) = (ydx - xdy)/y^2,$$

$$\frac{d(G/T)}{dT} = \frac{-H}{T^2}.$$

This is known as the Gibbs–Helmholtz equation. It is applies to a closed system of fixed composition.

For a change of state at constant P of a closed system of fixed composition, the relation of the change in G gets changed as

$$\frac{d(\Delta G/T)}{dT} = \frac{-\Delta H}{T^2}. \tag{2.108a}$$

This equation is useful in experimental thermodynamics as it permits ΔH, the heat of a reaction, to be obtained by measuring the variation of ΔG, the free energy change for the reaction, with T. Alternatively, it allows ΔG to be obtained by measuring ΔH. This is used in situations like phase composition, phase transformation relations etc.

Relation between A and U:

$$A = U - TS = U + T\left(\frac{\partial A}{\partial T}\right)_V. \tag{2.109}$$

Similarly,

$$\frac{d(A/T)}{dT} = -\frac{U}{T^2}. \tag{2.110}$$

This equation is only applicable to a closed system of constant volume and fixed composition. For a change of state under these constraints,

$$\frac{d(\Delta A/T)}{dT} = -\frac{\Delta U}{T^2}. \tag{2.111}$$

2.5 Statistical Interpretation of Entropy

The First Law was easily accepted as the significance of internal energy was well known. As entropy was not easy to comprehend, the Second Law was accepted slowly.

As a perpetual motion machine was not invented, the second law is valid. It has not yet been disproved within the scope of classical thermodynamics. Physical interpretation of entropy results from quantum theory and statistical mechanics.

2.5.1 Entropy and Disorder on an Atomic Scale

Gibbs stated that the Entropy is a measure of its "degree of mixed–up-ness" at the atomic or molecular scale. That is, when entropy is high, the particles of a system are more mixed up, e.g., the Entropy of the gaseous state is greater than that of the liquid state, which, in turn, is greater than that of the solid state.

At the macro-level, a latent heat of melting, q, is absorbed in the solid to liquid transformation. That is, entropy increases by q/T_m. During constant pressure melting, $q = \Delta H$ or $\Delta S_{melting} = \frac{\Delta H_{melting}}{T_m}$.

If the degree of order of a combined system (substance + heat reservoir) remains unchaged because of a change of phase, then that happens at the equilibrium melting or freezing temperature of a substance. At that temperature the solid and the liquid are in equilibrium with one another. So, at that temperature the phase change occurs reversibly.

2.6 The Concept on a Microstate

Statistical mechanics assumes that the equilibrium of a state of a system is just the most probable of all possible states. The subject determines the criteria governing the properties of this most probable state.

Quantum Theory proposes that a particle confined to move within a fixed volume, has its energy quantized, i.e., the particle has only a certain discrete allowed value of energy. The other energy levels are forbidden to it. The energy becomes continuous only when no restriction is placed on the position of the particle.

The effect of quantization of energy is shown as follows. Let in a perfect crystal all the lattice sites be occupied by identical particles. The particle characteristics and crystal structure determine the quantization of the allowed energy levels. In increasing energy levels the (allowed) energy levels are represented as ϵ_0, ϵ_1, ϵ_2.... .

$$n! = n(n-1)(n-2)......1; 0! = 1. \tag{2.112}$$

$$P(n, r) = \frac{n!}{(n-r)!}. \tag{2.113}$$

$$c_x^n = \binom{n}{k} = \frac{n!}{k!(n-k)!}. \tag{2.114}$$

Let U be the energy contained by a crystal of n particles. Then, two questions arise.

Question 1: Given that the total energy has to remain U, in how many ways can the n particles be distributed?

Question 2: Which is the most probable distribution?

Each distinguishable arrangement is called a complexion, or a microstate. All the microstates together form a single macrostate.

The most probable microstate is determined as follows.

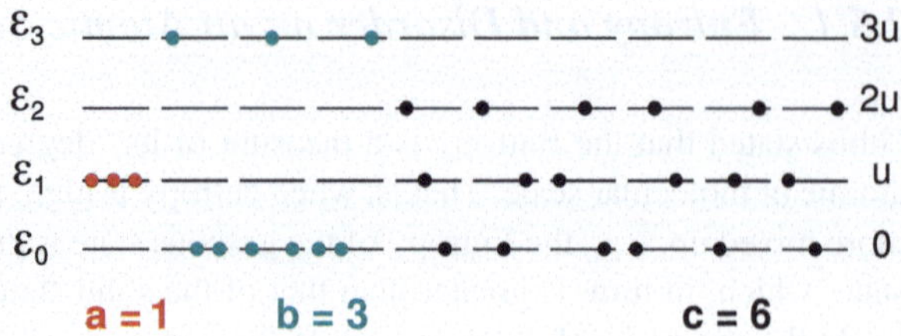

Fig. 2.11 Complexions or microstates within distributions of particles among energy levels. The system is of constant energy

The Macrostate in classical thermodynamics is fixed when the values of the independent variables are fixed. For example, in a system of fixed composition, U, V and n (the number of atoms/molecules) should be defined because the constancy of volume is required to determine the quantization of the energy levels.

With respect to the microstates within a macrostructure, it is assumed that each of the possible microstates is equally probable. That is, it is equal to $\left(\frac{1}{x}\right)$, where x is the number of possible microstates.

When $x = 10$. These 10 microstates occur in 3 distributions. One arrangement for distribution a; 3 for distribution b and 6 for distribution c (Fig. 2.11).

In the system the distribution c is "most probable" because the probability of distribution a is (1/10), of b is (3/10) and of c is (6/10).

The two ways of viewing the physical significance of these probabilities:

(a) If an instantaneous observation of the system were possible, the probability of observing the distribution c is (6/10).
(b) Over an infinite time interval, the system changes rapidly from one microstate to another. The fraction of the time spent in all the arrangements of distribution c would be (6/10).

The total energy of the system and the number of particles which it contains increases. Therefore, for fixed U, V, n the number of distinguishable arrangements (microstates) increases. These microstates still correspond to a single macrostate. In engineering systems at least one mole of atoms is considered. That is, n = 1 mole of a system $=6.023 \times 10^{23}$ atoms/molecules. $(6.023 \cdot 10^{23})$, the so-called Avagadro number.

The number of arrangements within the most probable distribution is much larger than in all the other distributions.

Ω is the number of arrangement within a given distribution.

Let the particles remain at the energy levels so that n_0 are in level ϵ_0 , n_1 in level ϵ_1, n_2 in level ϵ_2, …, n_r in the highest level of occupancy ϵ_r. Then,

$$\Omega = \frac{n!}{n_0!n_1!n_2!\ldots.n_r!} = \frac{n!}{\prod_{i=0}^{i=r} n_i!}. \tag{2.115}$$

Therefore, $\Omega\ (distribution\ a) = \frac{3!}{3!0!0!} = 1.$

$$\Omega\ (distribution\ b) = \frac{3!}{2!1!0!} = 3.$$

$$\Omega\ (distribution\ c) = \frac{3!}{1!1!1!} = 6.$$

The most probable distribution maximizes the value of Ω.

When the values of n_i are large, Stirling's approximation can be used, i.e., $lnX! = XlnX - X$.

Taking log in Eq. (2.115),

$$ln\,\Omega = nlnn - n\sum_{i=0}^{i=r}(n_i ln\, n_i - n_i). \tag{2.116}$$

For a macrosystem U = constant, n = constant,

$$U = constant = n_0\varepsilon_0 + n_1\varepsilon_1 + n_2\varepsilon_2 + \cdots n_r\varepsilon_r = \sum\nolimits_{i=0}^{i=r} n_i\varepsilon_i. \tag{2.117}$$

and

$$n = constant = n_0 + n_1 + n_2 + \cdots n_r = \sum\nolimits_{i=0}^{r} n_i. \tag{2.118}$$

From Eqs. (2.116) and (2.118),

$$\delta U = \sum_i \epsilon_i \delta n_i = 0 \tag{2.119}$$

$$\delta n = \sum_i \delta n_i = 0 \tag{2.120}$$

Also, from Eq. (2.115),

$$\begin{aligned}\delta \ln \Omega &= -\sum\left(\delta n_i \ln n_i + \frac{n_i \delta n_i}{n_i} - \delta n_i\right)\\ &= -\sum(\delta n_i \ln n_i).\end{aligned} \tag{2.121}$$

For maximum possible value of Ω, a small rearrangement of particles among the energy levels will not change the value of Ω (or $ln\Omega$). For a set of n_i's when Ω has its maximum value, then

$$\delta ln\Omega = -\sum(\delta n_i\, ln\, n_i) = 0. \tag{2.122}$$

Therefore, when Ω has the maximum value for a given microstate, the Eqs. (2.115), (2.116) and (2.120) should be satisfied simultaneously.

Using the method of undetermined multipliers the set of values of the most probable distribution is obtained. See below.

Equation (2.119) is multiplied by a constant β, which has the dimensions of reciprocal energy, to give

$$\Sigma \beta \epsilon_i \delta n_i = 0. \tag{2.123}$$

Equation (2.120) is multiplied by a dimensionless constant α, to give,

$$\Sigma \alpha \delta n_i = 0. \tag{2.124}$$

Equations (2.121) and (2.124) are added to give,

$$\sum_{i=0}^{\infty} (ln\, n_i + \alpha + \beta \epsilon_i) \delta n_i = 0 \tag{2.125}$$

If the bracketed terms are equal to zero, i.e., $lnn_i + \alpha + \beta \epsilon_i = 0$, this is possible.

Or

$$n_i = e^{-\alpha} e^{-\beta \epsilon_i}. \tag{2.126}$$

Summing over all r energy levels, $\sum_{i=0}^{i=r} n_i = n = e^{-\alpha} \sum_{i=0}^{i=\infty} e^{-\beta \epsilon_i}$

$$\sum e^{-\beta \epsilon_i} = e^{-\beta \epsilon_0} + e^{-\beta \epsilon_1} + e^{-\beta \epsilon_1} + \ldots e^{-\beta \epsilon_r}.$$

The magnitude of β is determined by the quantization of the energy—the so-called Partition Function, P_f. That is,

$$e^{-\alpha} = \frac{n}{P_f} \; and \; n_i = \frac{ne^{-\beta \epsilon_i}}{P_f}. \tag{2.126a}$$

Therefore, the most probable distribution is one in which the occupancy of the energy-level decreases increases exponentially. The shape of this distribution is schematically presented in Fig. 2.12.

The shape of the exponential curve above is decided by β, which is reciprocally related to the absolute temperature.

$$B = \frac{1}{kT}$$

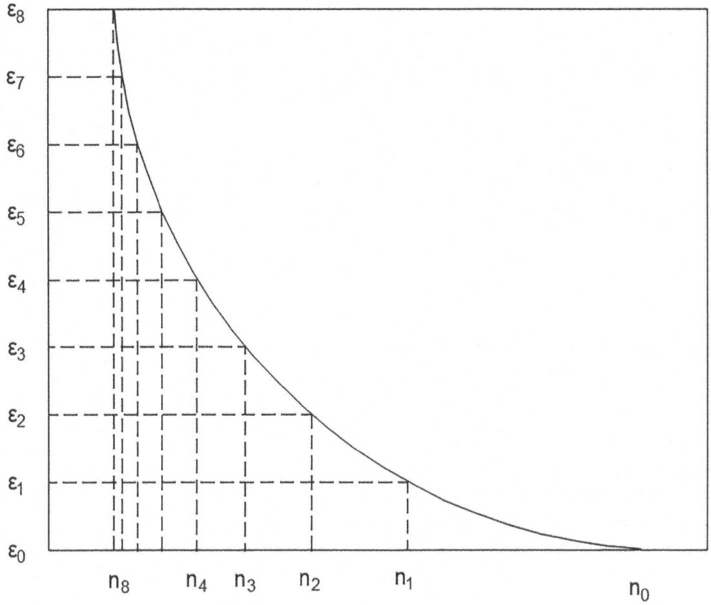

Fig. 2.12 The most probable (exponential decrease) distribution of quantized energy levels of particles (schematic)

$$k = Boltzmann's\ constant.$$

$$k = \frac{R}{N_0} = \frac{8.344}{6.0232 \times 10^{23}} = 1.38054 \times 10^{-23}\ J/K$$

where N_0 is the Avagadro number.

Nature of the exponential distribution of particles Fig. 2.12 is decided by temperature.

The values of *U, V and n* determine the macrosystem. The dependent variable (T) is then fixed.

From Fig. 2.13, it follows that if T increases, β will decrease or vice versa. Then, the shape of the figure will change (see below).

A rise in T causes the upper energy levels and they get relatively more populated. This amounts to an increase in the average energy of particles, *(U/n)*. This amounts to an increase in U when V and n are constant.

If n in a system is very large, the number of arrangements within the most probable distribution, Ω_{max}, is the only term which contributes significantly to the total number of arrangements, Ω_{total}. In other words, Ω_{max} is significantly larger than the sum of all the other arrangements. So, for a large number of particles, $\Omega_{total} \approx \Omega_{max}$.

As $\beta = \left(\frac{1}{kT}\right)$, Eq. (2.116) becomes, $\ln \Omega_{total} = \ln \Omega_{max} = n \ln n - \sum n_i \ln n_i$, in which values of n_i are given by Eq. (2.126a).

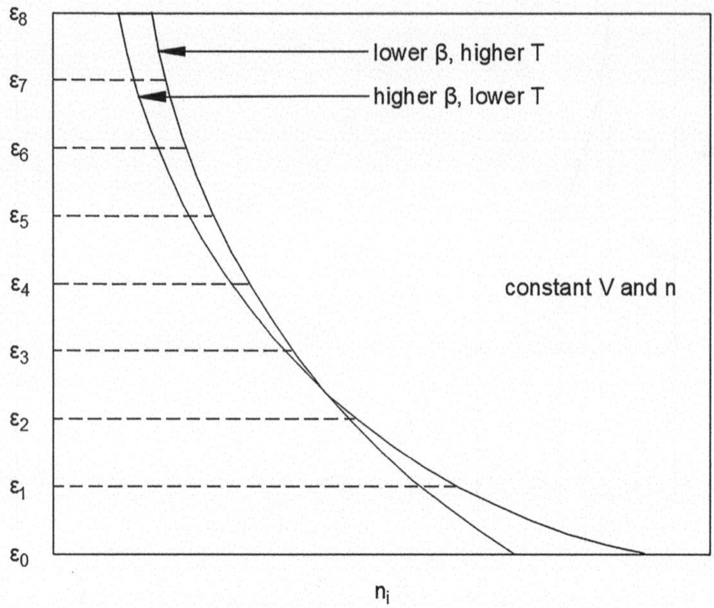

Fig. 2.13 The temperature dependence of the most probable distribution of particles among the quantized energy levels in a system of constant volume (a closed system)

$$\begin{aligned} \ln \Omega_{total} &= n \ln n - \sum \frac{n}{P_f} e^{-\epsilon_i/kT} \ln\left(\frac{n}{P_f} e^{-\epsilon_i/kT}\right), \\ &= n \ln n - \frac{n}{P_f} \Sigma\left[e^{-\epsilon_i/kT}\left(\ln n - \ln P_f - \frac{\epsilon_i}{kT}\right)\right], \\ &= n \ln n - \frac{n}{P_f}(\ln n - \ln P_f) \sum e^{-\epsilon_i/kT} + \frac{n}{P_f kT} \Sigma\, \epsilon_i e^{-\epsilon_i/kT}. \end{aligned}$$

But

$$U = \sum n_i \epsilon_i = \sum \frac{n}{P_f} \epsilon_i e^{-\epsilon_i/kT} = \frac{n}{P_f} \Sigma \epsilon_i e^{-\epsilon_i/kT}$$

$$\sum \epsilon_i e^{-\epsilon_i/kT} = \frac{U P_f}{n} \quad \ln \Omega = n \ln P_f + \frac{U}{kT} \tag{2.128}$$

2.7 Thermal Equilibrium and the Boltzmann Equation

Let a system of particles be in thermal equilibrium with a heat bath. This combined system has fixed values of *U, V, n,* where,

$$U = U_{particle\ system} + U_{heat\ bath};$$
$$V = V_{particle\ system} + V_{heat\ bath};$$

n = number of particles in the system + the heat bath of fixed size.

If particles system and heat bath are in thermal equilibrium, small exchanges of energy between them are possible. For constant U, V, n, Eq. (2.128) gives,

$\delta ln\Omega = \frac{\delta U}{kT}$ (P_f is dependent only on ε_i and T).

At constant total volume exchange,

$\delta U = \delta q$, i.e., the energy exchange is a heat exchange. Thus,

$$\delta ln\Omega = \frac{\delta q}{kT}. \quad (2.129)$$

For constant T (reversible) exchange, $\frac{\delta q}{T} = \delta S$.

Therefore,

$$\delta S = k\ \delta ln\Omega.$$

As S and Ω are state functions, on integration the equation becomes

$$S = k\ ln\ \Omega. \quad (2.130)$$

Equation (2.130) is called Boltzmann's equation—a quantitative relationship between the entropy and its "Degree of mixed-up-ness". Ω is the number of ways in which the system energy is distributed among the particles.

Ω_{max}, is the (the maximum) value of the most probable state for constant U, V *and* n values. As this is the equilibrium state, S is a maximum. In other words, the Boltzmann equation points out a physical meaning to entropy.

Heat capacity: We have

$$C_V = \left(\frac{\partial V}{\partial T}\right)_V. \quad (2.131)$$

$$C_P = \left(\frac{\partial H}{\partial T}\right)_P. \quad (2.132)$$

Notation: When E is an extensive property per mole of the system, E' is the same property of n moles of the entire system.

Then, $E' = nE$.

Then, Eqs. (2.131) and (2.132) become

$$dU' = nC_V dT\ or dU = C_V dT. \quad (2.133)$$

$$dH' = nC_P dT\ or dH = C_P dT. \quad (2.134)$$

C_P and C_V are respectively constant pressure and constant volume molar heat capacities.

Integrating Eq. (2.134),

$$\Delta H = H(T_2, P) - H(T_1, P) = \int_{T1}^{T2} C_P dT. \tag{2.135}$$

Variation of C_P with T should be known for knowing the temperature dependence of enthalpy and its S dependence.

Similarly, by integrating Eq. (2.133), the temperature dependence of internal energy can be known. One must also know the C_V variation with T.

2.8 Theoretical Calculation of the Heat Capacity

An empirical rule was given in 1819 by Dulong and Petit, viz., "3R (24.9 J/K) is the value of the molar heat capacity of all solids".

On the other hand, in 1865 Kopp presented a rule that "at ordinary temperature" the sum of the molar heat capacity of the constituent elements in a solid chemical compound equals the molar heat capacity of the chemical compound. Subsequent experiments revealed that the molar heat capacity of all elements was close to 3R at room temperature. But it increases with increasing T. In addition, the values can be significantly lower than 3R at low temperatures. See Fig. 2.14 pertaining to Pb, Cu, Si, and diamond, which shows the variations of C_V with T.

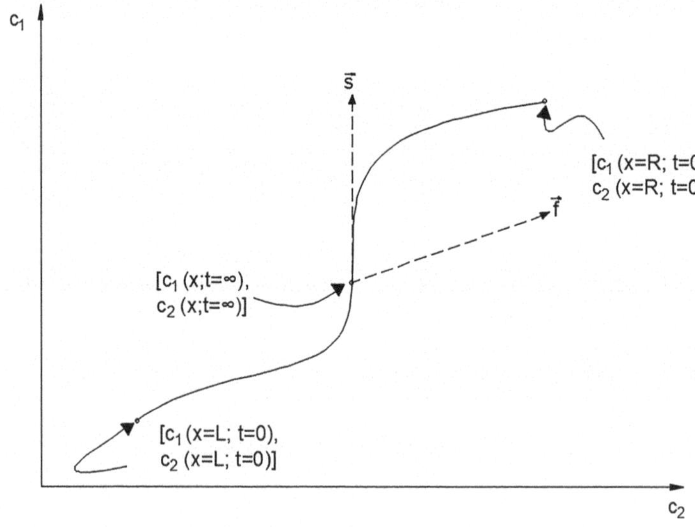

Fig. 2.14 variations of C_V with T for Pb, Cu, Si, and diamond [1]

The prediction of the temperature dependence of the heat capacity of a solid element was a major success story of quantum theory. Einstein did the first calculation in 1907, considering a crystal containing *n* atoms, each behaving like a harmonic oscillator vibrating about the lattice point independently. As the behavior of each oscillator is independent, a single fixed frequency, ν, can be assigned to each oscillator. A system of oscillators is known as an Einstein Crystal.

According to quantum theory, the energy of the ith energy level of a harmonic oscillator is given by,

$$\epsilon_i = \left(i + \frac{1}{2}\right) h\nu. \tag{2.136}$$

i is an integer with values ranging from 0 to ∞; h = Planck's constant of action; ν is the thermal vibration frequency. Every oscillator has 3 degrees of freedom (*x, y, z* directions). Therefore, Einstein crystal ≡ 3n linear harmonic oscillators.

Or,

$$U' = 3\sum n_i \epsilon_i, \tag{2.137}$$

where n_i = number of atoms in the ith energy level.

Substituting Eq. (2.136) and $n_i = \frac{ne^{-\beta\epsilon_i}}{P_f}$ in Eq. (2.137),

$$U' = 3\Sigma\left(i + ½\right)h\nu\left(\frac{ne^{-h\nu\left(i+\frac{1}{2}\right)/kT}}{\sum e^{-h\nu\left(i+\frac{1}{2}\right)/kT}}\right).$$

$$= 3nh\nu\left(\frac{\sum ie^{-h\nu\left(i+\frac{1}{2}\right)/kT}}{\Sigma e^{-h\nu\left(i+\frac{1}{2}\right)/kT}} + \frac{\frac{1}{2}\sum e^{-h\nu\left(i+\frac{1}{2}\right)/kT}}{\Sigma e^{-h\nu\left(i+\frac{1}{2}\right)/kT}}\right).$$

$$= 3nh\nu\left(\frac{\sum ie^{-h\nu_i/kT}}{\sum e^{-h\nu_i/kT}} + \frac{1}{2}\right) = \frac{3}{2}nh\nu\left(1 + \frac{2\Sigma ie^{-h\nu_i/kT}}{\sum e^{-n\nu_i/kT}}\right).$$

Taking

$$\sum ie^{-h\nu_i/kT} = \sum ix^i$$

where $x^i = e^{-h\nu_i/kT}$, gives

$x(1 + 2x + 3x^2 + \ldots) = \frac{x}{(1-x)^2}$, and $\sum e^{-h\gamma_i/KT} = \sum x^i = 1 + x + x^2 + \ldots = \frac{1}{(1-x)}$.

Therefore,

$$U' = \frac{3}{2}nh\nu\left(1+\frac{2x}{1+x}\right) = \frac{3}{2}nh\nu\left(1+\frac{2e^{h\nu_i/kT}}{1-e^{h\nu_i/kT}}\right)$$
$$= \frac{3}{2}nh\nu + \frac{3nh\nu}{(e^{h\nu_i/kT}-1)}. \tag{2.138}$$

Equation (2.138) quantitatively tells us how the variation energy of the system depends on the temperature. By definition, differentiation of Eq. (2.138) with respect to temperature at constant V gives C_V, the heat capacity at constant volume.

When volume is constant, quantization of the energy levels results.

$$C_V = \left(\frac{\partial U'}{\partial T}\right)_v = 3nh\nu\left(e^{h\nu/kT}-1\right)^{-2}\frac{h\nu}{kT^2}e^{h\nu/kT} = 3nk\left(\frac{h\nu}{kT}\right)^2\frac{e^{h\nu/kT}}{\left(e^{h\nu/kT}-1\right)^2}.$$

Defining $\frac{h\nu}{k} = \theta_E$, when θ_E is the Einstein characteristic temperature and taking $n =$ Avagadro number, results in constant volume molar heat capacity of the crystal as,

$$C_V = 3R\left(\frac{\theta_E}{T}\right)\frac{e^{\theta_E/T}}{(e^{\theta_E/T}-1)^2}. \tag{2.139}$$

As $\frac{T}{\theta_E}$ (and hence T) increases, $C_V \rightarrow 3R$, which is according to Dulong and Petit's law. As $T \rightarrow 0$, $C_V \rightarrow 0$. This is in agreement with experimental results.

Curve fitting eq. (2.138) to experimentally measured heat capacity data the Einstein equation the actual temperature θ_E, *and* ν can be obtained.

The Einstein equation represents experimental heat capacities at high temperatures fairly well. But at lower temperatures the theoretical values approach zero more quickly than the experimental values. This is because the oscillators vibrate with different frequencies.

Assuming that the range of frequencies of the oscillators is the same as that of elastic frequencies of oscillators in a continuous solid, in 1912 Debye proposed his Theory of Heat Capacity with a constraint that the range of frequencies of vibration available to the oscillators also are the same. Interatomic distances in the solid decide the lower limit of these vibrations. That is, if the wavelength equals the interatomic distance, then neighboring atoms would also be in the same phase of vibration. Then, the atoms will not vibrate relative to each other. Theoretically, the shortest allowable wavelength is twice the interatomic distance. In such a case, neighboring atoms vibrate in opposition to one another.

If minimum wavelength, λ_{min}, is 5×10^{-8} cm, and the wave velocity v, in the solid is 5×10^5 cm.s^{-1}, the maximum frequency of vibration of an oscillator is $\nu_{max} = \frac{v}{\lambda_{min}} = \frac{5\times10^5}{5\times10^{-8}} = 10^{13}$s. The assumption of Debye was that the number of vibrations/unit volume per unit frequency range increasing parabolically with increasing frequency in the allowed range $\theta \leq \nu \leq \nu_{max}$ decided the frequency distribution.

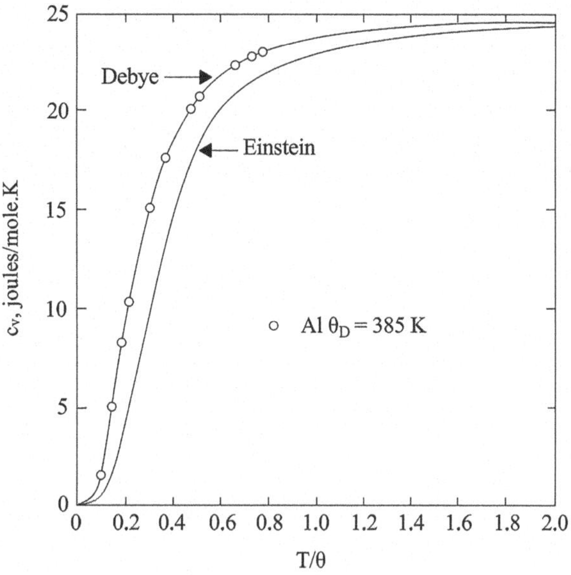

Fig. 2.15 Debye heat capacity, the Einstein heat capacity, and the experimental heat capacity of aluminum are compared [1]

Integrating the Einstein equation over this range of frequencies one gets,

$$C_V = \frac{9nh^3}{k^2\theta^3}\int_0^{\nu_D}\nu^2\left(\frac{h\nu}{kT}\right)^2\frac{e^{h\nu/kT}}{\left(1-e^{h\nu/kT}\right)^2}d\nu \text{ and representing, } x = \left(\frac{h\nu}{kT}\right)$$
$$C_V = 9R\left(\frac{T}{\theta_D}\right)^3\int_0^{\theta_D/T}\frac{x^4e^{-x}}{(1-e^{-x})^2}dx, \tag{2.140}$$

where ν_D (the Debye frequency) $= \nu_{max} = \frac{h\,\nu_D}{k}$ is the characteristic Debye temperature of the solid.

Debye's equation gives an excellent fit with the experimental data at the lower temperatures (Fig. 2.15).

The integral in Eq. (2.140) from 0 to ∞ gives a value 25.98.

That is, at very low T's, Eq. (2.140) leads to,

$$C_V = 9R \times 25.98\left(\frac{T}{\theta_D}\right)^3 = 1943\left(\frac{T}{\theta}\right)^3 \tag{2.138}$$

which is termed as Debye's T^3 law for low temperature heat capacities.

Debye's theory does not consider the electronic contribution to C_V. Since $C_V = \left(\frac{\partial U}{\partial T}\right)_V$, there is an electronic contribution to heat capacity in any temperature range in which the energy of the electrons changes with T. According to electron gas theory of metals, electronic contribution to specific heat is α T. Therefore, when T is large, electronic contribution is large. Therefore, at high temperatures

$C_V = 24.94 + bT$ *J/k*.mole, where bT is the electronic contribution. As the number of electrons/ atom present in the electron cloud is not known, b cannot be calculated theoretically. In addition, the theoretical approach ignores anharmonicity of the lattice vibrations at elevated temperatures. As the accuracy of theoretical calculations is uncertain, in practice the variation of C_P with T is measured first and the the analytical relationship is developed.

Problems

1. Helium gas in a closed system is compressed adiabatically, reducing its volume by 25%. Calculate the ratio of the final pressure to the initial pressure

Note: Helium is a monatomic gas and is considered to be the closest to an ideal gas

Solution:

- Final volume $V_f = 0.75V_i$ (Volume is reduced by 25%)
- Helium is a monatomic gas, so $\gamma = 5/3 = 1.66$

$$\frac{P_f}{P_i} = \left(\frac{V_i}{V_f}\right)^{\gamma}$$

$$\frac{P_f}{P_i} = \left(\frac{1}{0.75}\right)^{1.66}$$

$$\frac{P_f}{P_i} = 2.15$$

Therefore, the ratio of the final pressure to the initial pressure is **2.15**

2. A heat engine has two heat reservoirs with a constant heat capacity C_p. Let the initial temperature of the reservoirs be T_1 and T_2 respectively. Assume that the reservoirs remain isobaric and undergo no phase change. Derive the expression for maximum work extractable from the system

Solution:

$\mathbf{W_{max} = Q_{in} - Q_{out}}$

Final equilibrium temperature, $T_f = (T_1 + T_2)/2$

Heat **lost** by the **hot** reservoir: $Q_{lost} = C_p (T_1 - T_f)$

Heat **gained** by the **cold** reservoir: $Q_{gained} = C_p (T_f - T_2)$

$\mathbf{Q_{lost} = Q_{gained}}$

$$W_{max} = C_p(T_1 - T_f) + C_p(T_f - T_2)$$

$$\textit{substituting } T_f = \frac{T_1 + T_2}{2}$$

$$W_{max} = C_p\left(T_1 - \frac{T_1 + T_2}{2}\right) + C_p\left(\frac{T_1 + T_2}{2} - T_2\right)$$

$$W_{max} = C_p\left(\frac{2T_1 - (T_1 + T_2)}{2}\right) + C_p\left(\frac{T_1 + T_2 - 2T_2}{2}\right)$$

$$W_{max} = C_p\left(\frac{T_1 - T_2}{2}\right) + C_p\left(\frac{T_1 - T_2}{2}\right)$$

$$W_{max} = C_p(T_1 - T_2)$$

The expression for maximum work extractable from the system is $\boldsymbol{C_p (T_1 - T_2)}$

3. A system comprising 100 g argon gas at 1 atm pressure and zero deg. Celsius temperature is heated isobarically. If the heat added to the system is 35 kJ and the resultant work done by the system due to reversible expansion is 9.5 kJ, calculate the initial and final values of P, V and T. Also, find the change in internal energy for the process. Assume the gas to be ideal

Solution:

$m = 100 \text{ g} = 0.1 \text{ kg}$

$P_i = 1 \text{ atm} = 101.325 \text{ kPa}$

$T_i = 273 \text{ K}$

(continued)

	Q = 35 kJ W = 9.5 kJ M = 40 g/mol $C_v = 1.5\ R = 12.5$ J/mol.K $C_p = 2.5\ R = 20.8$ J/mol.K No. of moles = m/M = 100/40 = **2.5 moles** • **Using Ideal Gas Eqn:** $V_i = \frac{nRT_i}{P_i}$ $V_i = \frac{2.5 \times 8.314 \times 273}{101.325}$ $V_i = 0.056m^3$ • **Final Internal Energy Change:** $\Delta U = Q - W$ $\Delta U = 35 - 9.5 =$ **25.5 kJ** • **Final Temperature:** $\Delta T = \frac{\Delta U}{nC_v}$ $\Delta T = \frac{25500}{2.5 \times 12.5} = 81.6K$ $T_f = T_i + \Delta T = 273 + 81.6 = 354.6K$ • **Final volume:** $V_f = \frac{nRT_f}{P_f}$ $V_f = \frac{2.5 \times 8.314 \times 354.6}{101.325}$ $V_f = \mathbf{0.073m^3}$
4.	1 mole of helium gas is compressed from 10 to 5 L at a constant temperature of 300 K. Calculate the amount of work done assuming helium to be ideal and compare it with the value of work done calculated by considering the true behaviour of helium Note: Non-ideal gases are better described by Van der Waals equation of state **Solution:** Number of moles: n = 1 mole Initial volume: $V_i = 10$ L = 0.01 m^3 Final volume: $V_f = 5$ L = 0.005 m^3 Temperature: T = 300 K Universal gas constant: $R = 8.314$ J/mol. K 1. Assuming helium to be ideal: $W = -nRTln\left(\frac{V_f}{V_i}\right)$ $W = -(1)(8.314)(300)\ ln\left(\frac{0.005}{0.01}\right) = 1728J$ 2. Assuming helium to be non-ideal: $\left(P + \frac{a}{V^2}\right)(V - b) = RT$ a = 0.03412 L^2. bar/mol^2 (Van der Waals constant for helium) b = 0.0237 L/mol R = 0.08314 L·bar/mol·K T = 300 K $W = nRT\ ln\left(\frac{Vf-b}{Vi-b}\right) - an\left(\frac{1}{V_f} - \frac{1}{V_i}\right)$ Substituting value:

(continued)

	$$W=(1\times 0.08314\times 300)\ln\left(\frac{5-0.0237}{10-0.0237}\right)-0.03412\times 1\left(\frac{1}{5}-\frac{1}{10}\right)$$ $$=24.942\ln(0.4976)-0.03412\times(0.2-0.1)$$ $$=(24.942\times 0.6995)-(0.03412\times 0.1)$$ $$\approx -17.46-0.00341$$ $$\approx -17.46\text{ J}$$ • **Work done assuming helium as an ideal gas: 1728 J** • **Work done considering Van der Waals equation (real gas behavior): −17.46 J** This shows that **the ideal gas assumption significantly overestimates the work done** compared to a real gas model
5.	A system comprising 5 moles of an ideal monatomic gas at an initial temperature of 600 K is supplied with 10,000 J of heat. The gas does 11,000 J of work as a consequence. Calculate the final temperature of the gas **Solution:** n = 5 moles $T_i = 600$ K Q = 10,000 J W = 11,000 J $C_v = 1.5\text{ R} = 12.471$ J/mol.K $\Delta U = Q - W$ $U = 10000 - 11000$ $\mathbf{\Delta U = -1000\ J}$ $\Delta U = nC_v\Delta T$ $-1000 = 5\times 12.471\times(T_f-600)$ $-1000 = 62.355\times(T_f-600)$ $T_f-600=\frac{-1000}{62.355}$ $T_f-600\approx -16.04$ $T_f\approx 583.96$ K The final temperature of the gas is **583.96 K**
6.	A solid tank comprising 1 mole of helium is supplied with 500 J of heat. Calculate the change in internal energy and temperature if the gas is assumed to do no work and the work done on the gas is zero **Solution:** n = 1 mole W = 0 J Q = 500 J $C_v = 1.5\text{ R} = 12.471$ J/mol.K $\Delta U = Q - W$ $U = 500 - 0$ $\mathbf{\Delta U = 500\ J}$ $\Delta U = nC_v\Delta T$ $500 = 1\times 12.471\times\Delta T$ $=\frac{500}{12.471}$ ≈ 40.1 K **Change in internal energy:** $\Delta U = 500$ J **Change in temperature:** $\Delta T\approx 40.1$ K

(continued)

7.	A steam generator produces 1620 cm^3 of steam from 1 L of water. Compute the change in internal energy during the phase transformation. The latent heat of vaporization of water is 2.26×10^6 J/kg **Solution:** $V_{water} = 1\text{ L} = 0.001\ \mathbf{m^3}$ $V_{steam} = 1620\ cm^3 = \mathbf{0.00162\ m^3}$ $\rho_{water} = 1000\ kg/m^3$ $m = \rho \times V_{water} = 1000 \times 0.001 = 1$ kg $L(water) = 2.26 \times 10^6\ J/kg$ $P = 101.325$ kPa $\mathbf{\Delta U = Q - W}$ $\mathbf{Q = mL}$ $Q = (1) \times (2.26 \times 10^6) = \mathbf{2.26 \times 10^6 J}$ $\mathbf{W = P\Delta V}$ $W = 101.325 \times 10^3 \times (0.00162 - 0.001)$ $W = 101325 \times 0.00062$ $\mathbf{W = 62.8\ J}$ $\Delta U = Q - W = 2.26 \times 10^6 J - 62.8$ J $\mathbf{\Delta U = 2.25994 \times 10^6 J}$ Change in internal energy during the phase transformation is **2.25994 × 10⁶ J**
8.	Hydrogen gas in a system is compressed isothermally such that its volume is halved. If the initial volume is 1 L at normal temperature and pressure, calculate the change in heat and the internal energy. **Solution:** $V_i = 1\text{ L} = \mathbf{1 \times 10^{-3}\ m^3}$ $V_f = 0.5\text{ L} = \mathbf{5.0 \times 10^{-4}\ m^3}$ $P_i = 101.325$ kPa $\Delta U = 0$ $W = -nRT\ \ln\left(\frac{V_f}{V_i}\right)$ $\Delta U = Q + W$ For an isothermal process, $\Delta U = 0$ so $\mathbf{Q = -W}$ $n = \frac{P_i V_i}{RT}$ $n = \frac{(1.01325 \times 10^5)(1.0 \times 10^{-3})}{8.314 \times 298}$ $n = \frac{(101.325 \times 10^5)(1.0 \times 10^{-3})}{2477.572}$ ≈ 0.041 moles $W = -nRT \ln\left(\frac{V_f}{V_i}\right)$ $W = -(0.041)(8.314)(298) \ln\left(\frac{0.5}{1}\right)$ $= -(0.041)(2477.572)(-0.693)$ $= (0.041)(1717.45)$ ≈ 70.42J $\mathbf{Q = -70.42J}$

(continued)

9\. A sealed steel tank with a volume of 2 m^3 is fitted with a paddle wheel and contains 5 kg of carbon monoxide. The paddle wheel adds 10 W of energy every hour to the CO gas. As a result, the specific internal energy of CO is increased by 8 kJ/kg in 1 h. Assuming that there is no change in kinetic and potential energies, calculate the specific volume of the gas at the final state, energy transfer by work and energy transfer by heat transfer

Solution:
Volume V = 2 m^3
Mass of carbon monoxide, m = 5 kg
Power added by paddle wheel, W = 10 W
Duration, t = 1 h = 3600 s
Increase in specific internal energy, Δu = 8 kJ/kg
Final specific volume, v = V/m = 2/5 = **0.4 m^3/kg**
Energy transfer by work = W × t = 10 W × 3600 s = **36 kJ**
For a closed system, ΔU = Q − W = 5 kg × 8 kJ/kg = **40 kJ**
$Q = \Delta U + W$
Q = 40 + 36 = **76 kJ**

10\. Determine the change in specific enthalpy of an ideal gas in a system if the initial and final temperatures are 500 K and 1000 K respectively

Solution:

$$\mathbf{\Delta H = C_p \Delta T}$$

$$C_p \approx 1.04\ \mathrm{kJ/kg};$$

$$\mathbf{\Delta T = 500\ K}$$

$$\Delta H = 1040 \times 500$$

$$\mathbf{\Delta H = 520\ kJ/kg}$$

11\. Two storage tanks A and B are connected by a valve. Tank A contains 1 kg of CO at 65 °C at a pressure of 0.6 bar. Tank B contains 8 kg of CO at room temperature (25 °C) at a pressure of 1.1 bar. The gases in both tanks are allowed to mix and thermally equilibrate by opening the valve. Once equilibrium is achieved the temperature becomes 40 °C in both the tanks. Calculate the final equilibrium pressure and the total heat transfer considering C_p as constant

Solution:

For **Tank A**:

- Mass of CO = 1 kg
- Temperature = 65 °C = 338 K
- Pressure = 0.6 bar
- Volume to be determined

For **Tank B**:

- Mass of CO = 8 kg
- Temperature = 25 °C = 298 K
- Pressure = 1.1 bar
- Volume to be determined

Final temperature, T_f = 40 °C = 313 K
Final pressure to be determined)
Volume remains constant (no expansion work).
PV = mRT

$$V_A = \frac{m_A\, RT_A}{P_A}$$

$$= \frac{(1)(0.2968)\ (338)}{0.6}$$

$$= 167.1\ \mathrm{L}$$

$$= 0.1671\ \mathrm{m^3}$$

(continued)

$$V_B = \frac{m_B RT_B}{P_B}$$
$$= \frac{8 \times 0.2968 \times 298}{P_B}$$
$$= 643.5 \text{ L}$$
$$= 0.6435 \text{ m}^3$$

$V_T = V_A + V_B = 0.1671 + 0.6435 = 0.8106 \text{ m}^3$

Total mass of CO:

$m_T = m_A + m_B = 1 + 8 = 9$

Apply the ideal gas law in equilibrium,

$$P_f V_f = mRT_f$$
$$P_f = \frac{9 \times 0.2968 \times 313}{0.8106}$$
$$= 1.02 \text{ bar}$$

$Q = \Delta U = mCv\Delta T$

Total change in internal energy:

$$Q = m_A C_v (T_f - T_A) + m_B C_v (T_f - T_B)$$
$$= (1)\,(0.7432)\,(313 - 338) + (8)(0.7432)(313 - 298)$$
$$= (1)\,(0.7432)\,(-25) + (8)\,(0.7432)\,(15)$$
$$= -18.58 + 89.18$$
$$= 70.6 \text{ KJ}$$

Final Equilibrium Pressure: $P_f = 1.02$ bar

Total Heat Transfer: $Q = 70.6$ kJ

12. In a piston-cylinder setup, air is compressed polytropically from 1 atm pressure and 20 °C to 7 atm pressure. Calculate the work done and the heat transfer by using the ideal gas model with a constant specific heat and a 'n' value of 1.4.

Solution:

- Initial Pressure $P_1 = 1$ atm $= 101.325$ kPa
- Initial temperature: $T_1 = 20$ °C $= 293$ K
- Final pressure: P2 $= 7$ atm $= 708.925$ kPa
- Polytropic index: $n = 1.4$
- Gas constant for air: $R = 0.287$ kJ/kg·K
- Specific heat capacities:
  - $C_p = 1.005$ kJ/kg·K
  - $C_v = 0.718$ kJ/kg·K

$$T_2 = T_1 \left(\frac{P_2}{P_1}\right)^{\frac{n-1}{n}}$$
$$= 293 \times \left(\frac{7}{1}\right)^{\frac{1.4-1}{1.4}}$$
$$= 293 \times \left(\frac{7}{1}\right)^{\frac{0.4}{1.4}}$$
$$= 293 \times 2.219$$
$$= 650.3 \text{ K}$$

(continued)

$$W = R\left(\frac{T_2 - T_1}{1 - n}\right)$$
$$= 0.287\left(\frac{650.3 - 293}{1 - 1.4}\right)$$
$$= 0.287\left(\frac{357.3}{-0.4}\right)$$
$$= \frac{102.6}{-0.4}$$
$$= 256.5\ \text{KJ/Kg}$$
$$Q = \Delta U + W$$
$$\Delta U = C_v(T_2 - T_1)$$
$$= (0.718)\,(650.3 - 293)$$
$$= 265.5\ \text{KJ/Kg}$$
$$Q = 256.5 + (-256.5) = 0$$

Work done: $W = -256.5$ kJ/kg (**work done on the system**)
Heat transfer: $Q = 0$ kJ/kg (**adiabatic process**)

13. An open glass contains 200 mL of water and 100 g of ice cubes. The initial temperature of the water and ice is 25 °C and 0 °C respectively. The mixture reaches equilibrium after all the ice cubes melt. Assuming no heat transfer between the contents of the glass and the outside environment, calculate the final equilibrium temperature. The enthalpy of fusion for ice to water is 333.5 kJ/kg

Solution:
Mass of water: **0.2 kg**
Mass of ice: **0.1 kg**
Initial temperature of water: **25 °C**
Initial temperature of ice: **0 °C**
Latent heat of fusion of ice: **333.5 kJ/kg**
Specific heat of water: **4.18 kJ/kg·K**
Energy required to melt ice:
$Q_{melt} = mi\ hf = 0.1 \times 333.5 = 33.35$ KJ
Energy lost by water while cooling to T_f:
$Q_{water} = m_w C_w(25 - T_f) = 0.2 \times 4.18 \times (25 - T_f) = 0.836\ (25 - T_f)$
Energy gained by melted ice to reach T_f:
$Q_{melted\ ice} = m_i C_w(T_f - 0) = 0.1 \times 4.18 \times T_f = 0.418T_f$
Energy balance equation:
$$Q_{water} = Q_{melt} + Q_{melted\ ice}$$
$$0.836\ (25 - T_f) = 33.35 + 0.418\ T_f$$
Solving for T_f:
$$20.9 - 0.836\ T_f = 33.35 + 0.418\ T_f$$
$$-12.45 = .254\ T_f$$
$$T_f = -9.93\,^\circ\text{C}$$
Since $T_f < 0$ °C, **not all ice melts**. The final temperature remains **0 °C**, with some ice left

14. A system consists of 5 moles of argon gas at a temperature of 300 K and a pressure of 1 atm. Calculate the total energy of the system. Also, determine the entropy, enthalpy, Helmholtz free energy and Gibbs free energy of the system

Solution:
Number of moles: $n = 5$ moles
Temperature: $T = 300$ K
Pressure: $P = 1$ atm $=$ **101.325 kPa**

(continued)

Gas constant: R = 8.314 J/mol·K

Calculate internal Enegy (U)

$$\begin{aligned} U &= \frac{3}{2}\text{nRT} \\ &= \frac{3}{2}(5)(8.314)(300) \\ &= 1869.15\ \text{J} \end{aligned}$$

Calculate enthalpy (H):

$$H = \frac{5}{2}nRT$$

$$\begin{aligned} H &= \frac{5}{2}(5)(8.314)(300) \\ &= 3115.25\ \text{J} \end{aligned}$$

Calculate entropy (S):

$$\begin{aligned} S &= nC_p \ln\left(\frac{T}{T_0}\right) - nR\ln\left(\frac{P}{P_0}\right) \\ &= (5)\ (20.785)\ \ln\left(\frac{300}{298}\right) - (5)(8.314)\ln\left(\frac{1}{1}\right) \\ &= (5)\ (20.785)\ (0.00667) \\ &= 0.694\ \text{J/K} \end{aligned}$$

Calculate Helmholtz free energy (A)

$$\begin{aligned} A &= U - TS \\ &= 1869.15 - (300 \times 0.694) \\ &= 1660.95\ \text{J} \end{aligned}$$

Calculate Gibbs free Energy (G)

$$\begin{aligned} G &= H - TS \\ &= 3115.25 - (300 \times 0.694) = 2907.05\ \text{J} \end{aligned}$$

15. Two copper blocks with initial temperatures of 290 K and 320 K and a heat capacity of 200 J/K are brought into contact. Assuming that there is no heat loss to the surrounding environment from the blocks, compute the change in entropy that occurs as the blocks are brought in contact and verify if the ΔS is in line with the second law

Solution:

- **Block 1:**
  - Initial temperature, $T_1 = 290$ K
  - Heat capacity, C = 200 J/K
- **Block 2:**
  - Initial temperature, $T_2 = 320$ K
  - Heat capacity, C = 200 J/K

Determine final equilibrium temperature (T_f)

$$\begin{aligned} T_1 &= \left(\frac{CT_1 + CT_2}{C + C}\right) \\ &= \left(\frac{200 \times 290 + 200(320)}{400}\right) = 305\text{K} \end{aligned}$$

Calculate entropy change for block 1

$$\Delta S_1 = 200\ \ln\left(\frac{305}{290}\right)$$

$$200 \times 0.0505 = 10.10\ \text{J/K}$$

Calculate entropy change for block 2

(continued)

	$\Delta S_2 = 200 \ln\left(\frac{305}{320}\right)$ $= 200 \times (-0.0481) = -9.62\ J/K$ Calculate Total entropy change $\Delta S_{total} = \Delta S_1 + \Delta S_2$ $\Delta S_{total} = 10.10 – 9.62 = 0.48\ J/K$ Since $\mathbf{\Delta S_{total} > 0}$, the process **follows the second law of thermodynamics**
16.	A researcher for an experimental work had to find out the specific heat capacity of a liquid. He decided to follow the 'drop calorimetry' method due to the unavailability of an electric calorimeter at his lab. He took 1 kg of the liquid in an isothermal container equipped with a temperature monitor and noted the temperature of the liquid to be 23.5 °C. He then heated an aluminium block weighing 100 g to 100 °C by immersing it in boiling water and quickly transferred the block into the isothermal container, immersing it completely in the liquid. After thermal equilibrium is reached, he noted that the liquid's final temperature is 28.3 °C. From this data determine the specific heat capacity of the liquid Note: Molar mass of aluminium, M = 26.9815 g/mol **Solution:** • **Liquid:** – Mass = 1 kg = 1000 g – Initial temperature T_i = 23.5 °C – Final temperature T_f = 28.3 °C • **Aluminium block:** – Mass = 100 g – Initial temperature T_i = 100 °C – Final temperature T_f = 28.3 °C – Specific heat capacity of Al = 0.9 J/g·K $m_{Al}C_{Al}(T_{Al,i} – T_f) = m_l C_l (T_f – T_{l,i})$ $(100)(0.9)(100 - 28.3) = (1000)(C_t)(28.3 - 23.5)$ $(100)(0.9)(71.7) = (1000)(Ct)(4.8)$ $645.3 = 4800 C_l$ $C_l = \left(\frac{645.3}{4800}\right)$ The specific heat capacity of the liquid is found to be **134.4 J/kg·K**
17.	A piston-cylinder system contains 20 L of steam at a pressure of 150 kPa at 170 °C. The piston is moved to compress the steam to a pressure of 650 kPa. This process is polytropic with a polytrophic exponent of 1. Evaluate if this process violates the second law **Solution:** • V_1 = 20 L = 0.02 m^3,P_1 = 150 kPa, T_1 = 170 °C • P_2 = 650 kPa, Polytropic exponent—n = 1 (isothermal process). Since PV^n = constant and n = 1, the process is isothermal and the temperature remains constant during compression $\Delta S = R \ln\left(\frac{P_1}{P_2}\right)$ $R = 0.4615\ KJ/Kg.K$ (for steam) $\Delta S = 0.4615 \ln\left(\frac{150}{650}\right)$ $= 0.4615 \times (-1.4676)$ $= -0.677\ KJ/Kg.K$ ΔS is ***negative***, meaning steam entropy decreases.Heat is rejected, causing the entropy of

(continued)

	surroundings to increase.Total entropy change (system + surroundings) is not negative Therefore, the process **does not violate** the **second law of thermodynamics**
18.	Helium gas in a system with 300 kPa pressure at 350 °C is isothermally compressed to 400 kPa. Calculate the heat transfer per mass using the first law and compare it with the heat transfer per mass calculated using the second law. Assume the gas to be ideal **Solution:** • $P_1 = 300$ kPa, T = 350°C = 623 K (Isothermal process) • P2 = 400 kPa • Gas constant for helium: R = 2.0769 kJ/kg·K First law calculation: For an isothermal process: Q = W $q = RT \ln \left(\frac{P_2}{P_1}\right)$ $= (2.0769)(623) \ln \left(\frac{400}{300}\right)$ $= (2.0769)(623)(0.2877)$ $= 371.7\ \text{KJ/kg}$ Second law calculation: Entropy change : $\Delta S = RT \ln \left(\frac{P_1}{P_2}\right)$ $\Delta S = 2.0769 \ln (0.75)$ $= 2.0769 \times (-0.2877)$ $= -0.5977\ \text{KJ/Kg.K}$ Heat Transfer : $q = T\,\Delta S$ $q = (623)(-0.5977)$ $= -372\ \text{KJ/Kg}$ **Comparison:** First law: q = 371.7 kJ/kg Second law: q = −372 kJ/kg • **Same magnitude, opposite sign** (heat is rejected) – Heat transfer per mass = **−372 kJ/kg** (heat rejected) – **First and second law results match**
19.	An ideal gas at 350 K has a volume of 10 L at a pressure of 12 atm. Calculate: (a) The final volume of the system, (b) The work done by the system, (c) The heat entering or leaving the system, (d) The change in the internal energy, and (e) The change in the enthalpy When the gas undergoes a reversible **isothermal expansion** to a pressure of 8 atm and reversible **adiabatic expansion** to a pressure of 8 atm **Solution** Data: $T_1 = 350$ K; $V_1 = 10$ L; $P_1 = 12$ atm; $P_2 = 8$ atm; $C_v = 1.5R$; R = 8.3144 J/mol · K PV = nRT; $n = \left(\frac{P_1 V_1}{RT_1}\right) = \left(\frac{12 \times 10}{0.08206 \times 350}\right) = 4.16$ moles **1. Isothermal expansion:**

(continued)

	$P_1V_1 = P_2V_2$ $V_2 = \left(\frac{P_1V_1}{P_2}\right)$ $= \left(\frac{12 \times 10}{8}\right)$ $= 15$ L $W = nRT \ln\left(\frac{V_2}{V_1}\right)$ $= 4.16 \times 8.3144 \times 350 \times \ln\left(\frac{15}{10}\right)$ $= 4206.8 \times 0.405$ $= 1706.8$ J Since it is isothermal, $\mathbf{\Delta U = 0}$ and $\mathbf{Q = W = 1706.8\ J}$. Also, $\mathbf{\Delta H = 0}$ **2. Adiabatic expansion:** $P_1V_1^{\gamma} = P_2V_2^{\gamma}$ $V_3 = V_1\left(\frac{P_1}{P_2}\right)^{\frac{1}{\gamma}}$ $V_3 = 10 \times \left(\frac{12}{8}\right)^{\frac{3}{5}}$ $V_3 = 10 \times 1.26$ $= 12.6$ L $T_3 = T_1\left(\frac{V_1}{V_3}\right)^{\gamma - 1}$ $= 350 \times \left(\frac{10}{12.6}\right)^{\frac{2}{3}}$ $= 350 \times 0.86$ $= 301$ K $W = nc_vR\ (T3 - T1)$ $= 4.16 \times 1.5 \times 8.3144 \times (301 - 350)$ $= -2420.5$ J $\Delta H = nc_pR\ (T_3 - T_1)$ $C_p = c_v + R = 1.5R + R = 2.5R$ $\Delta H = 4.16 \times 2.5 \times 8.3144 \times (301 - 250)$ $= -4.34.2$ J
20.	One mole of a **monatomic ideal gas**, initially at T1 = 273 K and P1 = 1 atm, undergoes the following two reversible processes: (a) **Compression at constant pressure** until the volume is **halved** (b) **Cooling at constant volume** until the pressure is **halved** **Solution:** $V_1 = RT_1/P_1 = (0.08206)(273)/1$ **V1 = 22.4 L** **Compression at constant pressure** $V_2 = \frac{V_1}{2} = \frac{22.4}{2} = 11.2$ L $T_2 = \frac{P_2V_2}{R} = \frac{(1)(11.2)}{0.08206} = 136.5$ K **Cooling at constant volume**

(continued)

	$P_3 = \frac{P_2}{2} = \frac{1}{2} = 0.5$ atm $T_3 = \frac{P_3 V_2}{R} = \frac{(0.5)(11.2)}{0.08206} = 68.25$ K
21	A monatomic ideal gas is initially at P = 2 atm, V = 1.5 liters, and T = 400 K. The gas undergoes the following three reversible processes: **Cooling at constant pressure**: The gas is cooled at constant pressure to a volume V. Calculate the value of *VVV* and the total work done on or by the gas **Solution:** • **Initial pressure**: P = 2 atm • **Initial volume**: $V_1 = 1.5$ L • **Initial temperature**: $T_1 = 400$ K • **Final temperature**: Assume $\mathbf{T_2 = 300\ K}$ (since the gas is cooled) • **Gas constant**: R = 0.08206 L·atm/(mol·K) **PV = nRT** $n = \frac{PV}{R}$ $= \frac{(2)(1.5)}{(0.08206)(400)}$ $= \frac{3}{(32.824)} = 0.0914$ moles $\frac{V_2}{V_1} = \frac{T_2}{T_1}$ $V_2 = V_1 \times \frac{T_2}{T_1}$ $= 1.5 \times \frac{300}{400}$ $= 1.5 \times 0.75$ $= 1.125$ L $W = P(V_2 - V_1)$ $= 2 \times (1.125–1.5)$ $= 2 \times (-0.375)$ $= -0.75 \times 101.3$ $= -76.0$ J **Negative means work is done on the gas**
22.	Two moles of a monatomic ideal gas are initially at a pressure of 1 atm and a temperature of 300 K. The gas undergoes a reversible expansion, absorbing 34,166 J of heat while performing 1216 J of work on its surroundings. Determine the final temperature of the gas **Solution:** • **Number of moles:** n = 2 • **Initial temperature:** T1 = 300 K • **Heat added:** Q = 34,166 J • **Work done by gas:** W = 1216 J • $C_v = 1.5\ R = 1.5 \times 8.314 = 12.471$ J/mol/ $\mathbf{\Delta U = Q - W = n\ C_v dT}$

(continued)

<table>
<tr><td></td><td>
$$\Delta U = Q - W = 34166 - 1216$$
$$\Delta U = \mathbf{32950\ J}$$
$$\Delta U = nCv\ (T_2 - T_1)$$
$$32,950 = 2 \times 12.471 \times (T_2 - 300)$$
$$32,950 = 24.942\ (T_2 - 300)$$
$$(T_2 - 300) = \frac{32,950}{24.942}$$
$$= 1620.9\ K$$
</td></tr>
<tr><td>23.</td><td>
One mole of an ideal gas is initially at 273 K and 1 atm pressure. When 3000 J of heat is added at constant pressure, the gas expands, performing 832 J of work. Determine:
(a) The final temperature and pressure of the gas,
(b) The changes in internal energy (ΔU) and enthalpy (ΔH),
(c) The values of C_V and C_P for the gas. Assume the process is reversible
Solution:
• Number of moles: n = 1
• Initial temperature: T1 = 273 K
• Initial pressure: P1 = 1 atm
• Heat added at constant pressure: Q = 3000 J
• Work done by the gas: W = 832 J
• Universal gas constant: R = 8.314 J/mol·K
• Process is reversible and at constant pressure
$\Delta U = Q - W = n\ C_v\ dT$
For a constant pressure process, $Q = n\ C_P\ dT$

$$\Delta T = Q/(n\ C_p)$$
$$\Delta T = \frac{3000}{1 \times \frac{7}{2} \times 8.314}$$
$$= \frac{3000}{29.099}$$
$$= 103.1\ K$$
$$T_2 = T_1 + \Delta T = 273 + 103.1 = 376.1\ K$$

Internal energy ΔU:

$$\Delta U = nc_v\Delta T$$
$$= 1 \times \frac{5}{2} \times 8.314 \times 103.1$$
$$= 2143.7\ J$$

Enthalpy change ΔH:

$$\Delta H = nc_p\Delta T$$
$$= 1 \times \frac{7}{2} \times 8.314 \times 103.1$$
$$= 3000\ J\ (\text{as expected since } Q = \Delta H \text{ at constant pressure})$$
$$C_V = 2.5\ R = 2.5 \times 8.314 = 20.785\ J/mol\ K$$
$$C_P = 3.5\ R = 3.5 \times 8.314 = 29.099\ J/molK$$
$$C_P - C_V = R = 8.314\ J/mol.K(\text{which is correct})$$
</td></tr>
<tr><td>24.</td><td>
A gas sample, initially at 1 atm and 25 °C, undergoes the following reversible cyclic process:
First cycle
1. Isothermal expansion until the pressure reduces to 0.6 atm
2. Isobaric expansion until the temperature increases to 120 °C
3. Isothermal compression back to 1 atm
</td></tr>
</table>

(continued)

4. Isobaric compression back to 25 °C

Second cycle

1. Isobaric expansion until the temperature reaches 120 °C
2. A decrease in pressure at constant volume to P atm
3. Isobaric compression at P atm until the volume reaches 26 L
4. An increase in pressure at constant volume back to 1 atm

Determine the value of P such that the work done on the gas in the first cycle equals the work done by the gas in the second cycle

Solution:

- **Isothermal expansion (1 → 2)**
- **Isobaric expansion (2 → 3)**
- **Isothermal compression (3 → 4)**
- **Isobaric compression (4 → 1)**

Isothermal expansion (1–2)

Formula for isothermal work:

$W = nRT \ln \frac{V_2}{V_1}$

Since PV = nRT,we can express volume interms of pressure,

$\frac{V_2}{V_1} = \frac{P_1}{P_2}$

Given that pressure drops from 1 atm to 0.6 atm,

$$W = nRT \ln \frac{P_1}{P_2}$$

$$= 1 \times 8.314 \times 298 \ln \left(\frac{1}{0.6}\right)$$

$$= 1266.8 \text{ J}$$

Isobaric expansion (2–3)

$W = nc_p \Delta T$

For an ideal monoatomic gas:

$$Cp = \frac{5}{2} R$$

$$W_{23} = nCp\,(T_3 - T_2)$$

Given $T_3 = 120\,°C = 393$ K

$$W_{23} = (1) \times \left(\frac{5}{2} \times 8.314\right) \times (393 - 298)$$

$$= (1) \times 20.785 \times 95$$

$$= 1974.6 \text{ J}$$

Step 1.3: Isothermal compression (3 → 4)

Since the gas returns to its initial pressure, the work done is equal in magnitude but opposite in sign to the **isothermal expansion**:

$\mathbf{W_{34} = -W_{12} = -1266.8\ J}$

Step 1.4: Isobaric compression (4 → 1)

Since the gas returns to its initial state, the work done here is the **negative of the isobaric expansion**:

$\mathbf{W_{41} = -W_{23} = -1974.6\ J}$

Total work done in first cycle

$W_{cycle\ 1} = W_{12} + W_{23} + W_{34} + W_{41} = 1266.8 + 1974.6 - 1266.8 - 1974.6$

$$\mathbf{W_{cycle\ 1} = 0}$$

Step 2: Work done in the second cycle

The **second cycle** consists of:

1. **Isobaric expansion (A → B)**
2. **Pressure decrease at constant volume (B → C)**

(continued)

3. **Isobaric compression (C → D)**
4. **Pressure increase at constant volume (D → A)**

Step 2.1: Isobaric expansion (A → B)

Since it follows the same path as W23 in the first cycle,

$\mathbf{W_{AB} = 1974.6\ J}$

Step 2.2: Constant volume pressure drop (B → C)

No work is done during this step since: $\mathbf{W_{BC} = 0}$

Step 2.3: Isobaric compression (C → D)

Since the gas is compressed **at P atm** to a final volume of **24.5 L**, the work done is:

$$\mathbf{W_{CD} = -P\Delta V}$$

$$V_B = V_A \times \frac{T_B}{T_A}$$

$$V_B = 24.5 \times \frac{393}{298} = 32.3 \text{ litres}$$

Step 2.4: Constant volume pressure increase (D → A)

Since the process is at constant volume, **no work is done**: $\mathbf{W_{DA} = 0}$

Total work done in second cycle

$$\mathbf{Wcycle2 = W_{AB} + W_{BC} + W_{CD} + W_{DA}}$$

$$= 1974.6 + 0 - P(32.3 - 24.5) + 0$$

$$= 1974.6 - P(7.8)$$

For the cycles to have equal work

$$W_{cycle1} = W_{cycle2}$$

$$0 = 1974.6 - P(7.8)$$

$$P = \frac{1974.6}{7.8} = 243.8 \text{ atm}$$

The required pressure P is **253.8 atm** such that the work done on the gas in the first cycle equals the work done by the gas in the second cycle

---

25. A heat engine absorbs heat from a hot reservoir at 500 K and rejects heat to a cold reservoir at 300 K. The engine operates in a Carnot cycle. Calculate the efficiency of the engine

Solution:

$$\eta = 1 - \frac{T_C}{T_H}$$

Where:

$T_H = 500$ K (hot reservoir temperature)

TC = 300K (Cold reservoir temperature)

Substitute the values:

$$\eta = 1 - \frac{300}{500} = 1 - 0.6 = 0.4$$

Thus the efficiency of the engine is 40%

---

26 A closed system contains 2 moles of gas at an initial temperature of 300 K and pressure of 1 atm. If the gas undergoes an isothermal expansion to twice its volume, calculate the work done by the gas

Solution:

$$W = nRT\ \ln\left(\frac{V_f}{V_i}\right)$$

$$W = 2 \times 8.314 \times 300 \times \ln(2)$$

$$W \approx 2 \times 8.314 \times 300 \times 0.693 \approx 3452.6\text{J}$$

Thus, the work done by the gas is approximately **3452.6 J**

(continued)

27.	A system absorbs 150 J of heat while doing 50 J of work. Calculate the change in internal energy of the system **Solution:** Using the first law of thermodynamics: $\Delta U = Q - W$ • $Q = 150\,J$ • $W = 50\,J$ Thus, the change in internal energy is **100 J**
28.	Two systems, A and B, are in thermal contact with each other, and a third system, C, is in thermal contact with system B. If system A is in thermal equilibrium with system B, and system B is in thermal equilibrium with system C, what can be said about the relationship between systems A and C? **Solution:** According to the zeroth law of thermodynamics, if two systems are each in thermal equilibrium with a third system, they must be in thermal equilibrium with each other. Therefore, if system A is in thermal equilibrium with system B, and system B is in thermal equilibrium with system C, system A will also be in thermal equilibrium with system C
29.	If system A has a temperature of 300 K and system B has a temperature of 300 K, and they are in thermal contact, can we conclude that they are in thermal equilibrium according to the zeroth law? **Solution:** Yes, according to the zeroth law of thermodynamics, if two systems (A and B) are in thermal contact and their temperatures are equal (300 K), they will be in thermal equilibrium. Therefore, since the temperatures of both systems are the same, they are in thermal equilibrium
30.	A gas expands from an initial volume of 2 L to a final volume of 5 L at a constant pressure of 1 atm. How much work is done by the gas if the expansion is reversible? **Solution:** The work done by the gas during an isobaric process (constant pressure) is given by: $W = P\Delta V$ Where: $P = 1\ \text{atm} = 1.013 \times 10^5 \text{Pa}$ $\Delta V = V_f - V_i$ $= 5L–2L$ $= 3L$ $= 3 \times 10^{-3} m^3$ Thus: $W = 1.013 \times 10^5 \times 3 \times 10^{-3} = 303.9\ J$ Therefore, the work done by the gas is **303.9 J**
31.	A system absorbs 500 J of heat and does 200 J of work. Using the first law of thermodynamics, calculate the change in internal energy **Solution:** According to the first law of thermodynamics: $\Delta U = Q - W$ Where: • $Q = 500\,J$ • $W = 200\,J$ Substitute the values: $\Delta U = 500 - 200 = 300\ J$ Thus, the change in internal energy is **300 J**

(continued)

32. At 0 K, the entropy of a perfect crystal is zero. If a system undergoes a reversible process at constant temperature, how does the entropy of the system change?

Solution:

According to the **third law of thermodynamics**, the entropy of a perfect crystal at absolute zero (0 K) is zero. During a reversible process at constant temperature, the change in entropy is:

$\Delta S = Q/T$

Since the process is reversible and at constant temperature, the change in entropy depends on the heat transferred. For a perfect crystal at 0 K, if the system undergoes a reversible process, the entropy change is calculated based on the heat transfer

33. The number of microstates Ω of a system is 10^5. Calculate the entropy of the system using the statistical interpretation

Solution:

The statistical definition of entropy is given by:

$S = k_B \ln \Omega$

Where:

$k_B = 1.38 \times 10^{-23}$ J/K (Boltzmann Constant)

$\Omega = 10^5$

Substitute the values:

$$S = 1.38 \times 10^{-23} \ln(10^5)$$
$$= 1.38 \times 10^{-23} \times 11.5129$$
$$= 1.59 \times 10^{-22}\,\mathrm{J/K}$$

Thus the entropy is 1.59×10^{-22} J/K

34. For a system, the number of microstates is $\Omega = 10^6$. Calculate the change in entropy when the number of microstates increases from 10^4 to 10^6

Solution:

Using the statistical definition of entropy:

$$\Delta S = k_B \ \ln \left(\frac{\Omega_2}{\Omega_1}\right)$$

Where,

$\Omega_1 = 10^4$

$\Omega_2 = 10^6$

$k_B = 1.38 \times 10^{-23}$ J/K

Substitute the values,

$$\Delta S = 1.38 \times 10^{-23} \ \ln \left(\frac{10^6}{10^4}\right)$$
$$= 1.38 \times 10^{-23} \ln (100)$$
$$= 6.36 \times 10^{-23}\ \mathrm{J/K}$$

35. A system has $\Omega = 10^5$\omega $= 10^5$ accessible microstates at a temperature of 400 K. Calculate the entropy of the system

Solution:

Using the formula for entropy:

$S = kB \times \ln \Omega$

$$S = 1.38 \times 10^{-23} \ \ln (10^5)$$
$$= 1.38 \times 10^{-23} \times 11.5129$$
$$= 1.59 \times 10^{-22}\ \mathrm{J/K}$$

Reference

1. Gaskell, D. R. (2009). *Introduction to the thermodynamics of materials*. Taylor and Francis.

Chapter 3
Thermodynamics of Reactions

Contents

Behavior of solutions: At sufficiently high temperatures and adequately low pressures, the atoms or molecules in a gas mixture interact negligibly and the interactions can be neglected. Then, the mixture can be viewed as an ideal gas mixture. At high temperatures and low pressures, all gas mixtures exhibit the same mixing properties. In condensed solutions, strong interactions exist among atoms, molecules, or ions. These interactions have a significant influence on the behavior of a solution. Atomic size, electronegativity, electron-to-atom ratio, etc., determine the extent of solubility of a component in a solution and whether components can chemically react to form new species.

K. A. Padmanabhan et al., *Thermodynamics and Kinetics of Materials*, Engineering Materials, https://doi.org/10.1007/978-3-032-14527-7_3

3.1 Solution Thermodynamics

This is concerned with the relationships among vapor pressure, temperature, and composition of a solution.

Raoult's and Henry's laws: If A, a pure liquid, is kept inside a closed, evacuated vessel at temperature T, it will evaporate until the pressure in the vessel attains the saturated vapor pressure of liquid A, P_A^0 at T. A dynamic equilibrium is then established, i.e., evaporation rate of liquid A = condensation rate of vapor, B. Evaporation rate, $r_{e(A)}$, is determined by the bond energy between atoms of A at the liquid surface. Interatomic forces ensure that each surface atom is located near the potential energy well bottom.

For an atom to leave the liquid surface to enter the vapor phase, it should possess an activation energy, E*. Intrinsic evaporation rate, $r_{e(A)}$, is decided by E* (potential well depth) and temperature T. Condensation rate, $r_{C(A)}\alpha$, is the number of atoms in the vapor phase that strike the liquid surface in unit time.For given T, $r_{C(A)}$ is proportional to the pressure of the vapor.

Therefore, $r_{c(A)} = \left(k_1 P_A^o\right)$. At equilibrium,

$$r_{eq\ (A)} = \left(k_1 P_A^o\right) \tag{3.1}$$

Figure 3.1 illustrates why the saturated vapor pressures of liquids are exponential functions of T. The surface atom energies are quantized and the surface atom distribution among the available quantized levels is given by $n_i = \left[\frac{n\left(\frac{E_i}{kT}\right)}{P_f}\right]$ where $\left(\frac{n_i}{n}\right)$ is the fraction of atoms in the E_i^{th} energy level, P_f is the partition function,

$$P_f = \sum_0^\infty exp\left(\frac{-E_i}{kT}\right). \tag{3.2}$$

If the quantized levels are closely spaced,

$\mathrm{P_f} = \int_0^\infty \exp\left(\frac{-E_i}{kT}\right) dE = kT_1$. This is the average energy /atom at temperature, $\mathrm{T_1}$.

Therefore, the fractions of atoms at the surface which have energy greater than E*, the activation energy for evaporation, is

$$\frac{n_i}{n} = \left(\frac{1}{kT}\right) \int_{E^*}^\infty \exp\left(\frac{-E_i}{kT}\right) dE = \exp\left(\frac{-E^*}{kT}\right). \tag{3.3}$$

From Eq. (3.1), $r_{e(A)}\alpha \left(\frac{n^*}{n}\right)$. Therefore, the rate of evaporation increases exponentially with increasing temperature and decreases exponentially with increasing E* value.

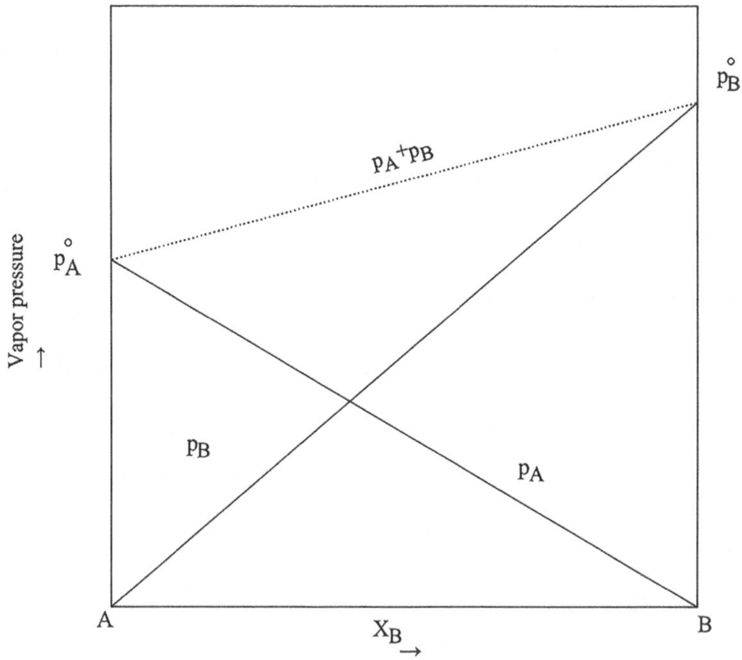

Fig. 3.1 The vapor pressures in a binary Raoultian solution due to the components [1]

Similarly,

$$r_{e(B)} = k'P_B^0. \tag{3.4}$$

In these equations, k_1, k' are constants of proportionality.

Add a small quantity of liquid B to liquid A. If X_A is the mole fraction of A in the solution and the atomic diameters of A and B are similar, then, assuming the surface composition to be the same as that of the bulk liquid, the surface sites fraction occupied by A atoms is X_A. As A can only evaporate from surface sites occupied by A atoms, the evaporation rate of A is decreased by X_A. At equilibrium, as evaporation and condensation rates are equal, the equilibrium vapor pressure of A exerted by the A-B solution is decreased from P_A^0 to P_A, where,

$$r_{e(A)}X_A = kP_A. \tag{3.5}$$

Similarly, for liquid B containing a small amount of A,

$$r_{e(B)}X_B = k'P_B. \tag{3.6}$$

From Eqs. (3.1) and (3.5),

$$P_A = X_A P_A^0. \tag{3.7}$$

From Eqs. (3.4) and (3.6),

$$P_B = X_B p_B^0. \tag{3.8}$$

Equations (3.7) and (3.8) express Raoult's law, "the vapor pressure exerted by a component "i" in a solution is equal to the product of the mole fraction of i in the solution and the saturated vapor pressure of pure liquid i at the temperature of the solution", see Fig. 3.1 above. The components of such a solution exhibit Raoultian behavior.

Equations (3.5) and (3.6) imply that the magnitudes of the A-A, B-B, and A-B bond energies in the solution are identical. Then, the potential energy well depth of a surface atom is independent of the type of atoms.

Let A-B bond energy be considerably more negative than that of the A-A and the B-B bonds. Also, let the solution of A in B to be sufficiently dilute so that every A atom on the surface of the liquid is surrounded by only B atoms. For this case, the surface A atoms are each located in a deeper potential energy well than are the A atoms at the surface of pure A. Then, A atoms must be lifted from deeper wells. Here, the intrinsic rate of evaporation of A is decreased from $r_{e(A)}$ *to* $r'_{e(A)}$. Equilibrium between the condensed solution and the vapor phase occurs when

$$r'_{e(A)} X_A = k P_A. \tag{3.9}$$

From Eqs. (3.5) and (3.9),

$$P_A = \frac{r'_{e(A)}}{r_{e(A)}} X_A P_A^0. \tag{3.10}$$

As $r'_{e(A)} < r_{e(A)}$, P_A in Eq. (3.10) is smaller than P_A in Eq. (3.7).

Equation (3.10) can be rewritten as,

$$P_A = k_A X_A. \tag{3.11}$$

If the mole fraction of A in solution A-B increases, the probability of all A atoms on the surface of the liquid being surrounded only by B atoms decreases.

Then, the depth of the potential energy well decreases and $r'_{e(A)}$ value increases (Fig. 3.2).

Beyond some critical mole fraction of A, $r'_{e(A)}$ varies with composition. Then, Eq. (3.11) is not valid for this solution. So, depending on temperature T, Eq. (3.11) is valid only over an initial range of concentration of A in B of the solution and it also depends on the relative magnitudes of A-A, B-B and A-B bond energies. Similar arguments apply for dilute solutions of B in A.

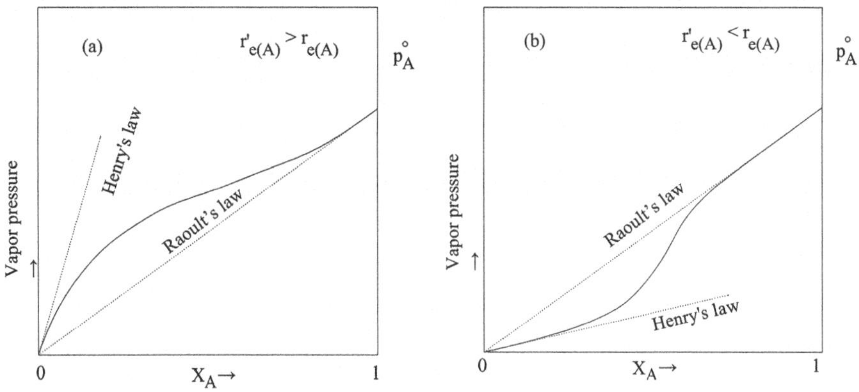

Fig. 3.2 (**a**) Positive deviation behavior of vapor pressure of a component in a binary solution, (**b**) when negative deviation from Raoult's law is present [1]

Therefore,

$$P_B = k_B X_B. \tag{3.12}$$

Equations (3.11) and (3.12) are referred to as Henry's law. The solutions that obey Eqs. (3.11) and (3.12) are said to exhibit Henrian behavior.

If the A-B bond energy is more negative than the A-A and B-B bond energies, then, as $r'_{e(A)} < r_{e(A)}$, the Henry's law line lies below the Raoult's law line. If the A-B bond energy is less negative than the A-B and B-B bond energies, $r'_{e(A)} > r_{e(A)}$. Therefore, Henry's law line lies above the Raoult's law line.

3.2 The Thermodynamic Activity of a Component in a Solution

Fugacity of a real gas is a concept used to compute chemical equilibrium accurately by replacing the mechanical partial pressure. It is equal to the pressure of an ideal gas that has the same temperature and molar Gibbs free energy as the real gas.

The thermodynamic activity of a component in any state at a temperature T is the ratio of the fugacity of that substance in that state to its fugacity in its standard state. It is used to multiply the pressure or partial pressure for an ideal gas in equations to generate equations applicable to the real gas of a substance in that state. That is, for a species or substance i, the activity of i,

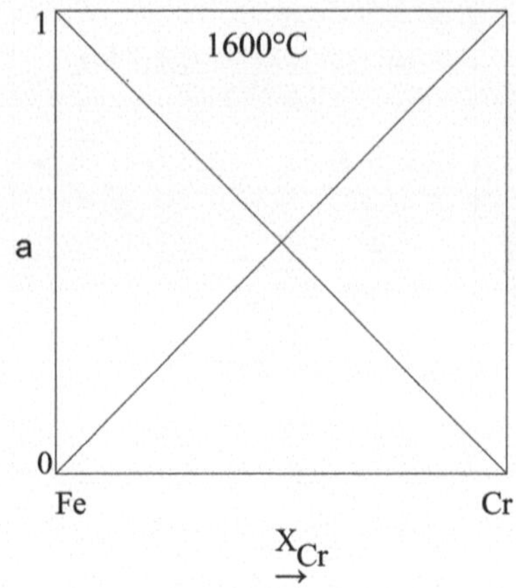

Fig. 3.3 Activities in the binary system iron-chromium at 1600 °C [1]

$$a_i = \frac{f_i}{f_i^0}. \tag{3.13}$$

In a condensed solution, f_i is the fugacity of the component i in the solution at temperature T. f_0 is the fugacity of pure i in the standard state at temperature T.

If a vapor in equilibrium with a condensed solution is ideal, then $f_i = P_i$. Then,

$$a_i = \frac{P_i}{P_i^0}. \tag{3.14}$$

That is, the activity of i in solution, with respect to pure i, is the ratio of the partial pressure of i exerted by the solution to the saturated vapor pressure of pure i at the same T.

If component i exhibits Raoultian behaviour, then

Equations (3.5) and (3.14) give

$$a_i = X_i, \tag{3.15}$$

Raoult's law can be expressed as shown here. Figure 3.3 illustrates the Raoultian behavior in a binary solution where the activities of both components are considered.

In this way, the definition of activity normalizes the vapor pressure-composition relationship (with respect to the saturated vapor pressure in the standard state). The composition range where Henry's law is obeyed by solute I is similar.

Equations (3.11) and (3.14) give

$$a_i = k_i X_i. \tag{3.16}$$

Henry's law can be expressed in this manner.

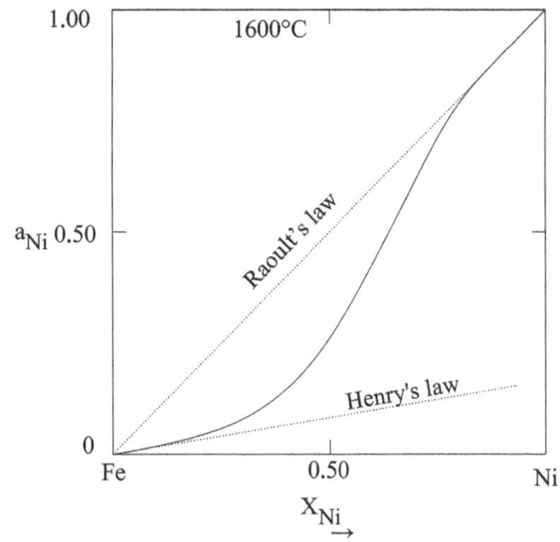

Fig. 3.4 The activity of nickel in the system iron-nickel at 1600 °C [1]

Henrian behavior is displayed in Fig. 3.4, using the activity of a component of a binary solution.

3.3 The Gibbs-Duhem Equation

Extensive thermodynamic properties of only one component of a binary (or multicomponent) solution are amenable to experimental measurement in many cases. Then, the corresponding properties of the other components can be estimated from a relationship that exists between the properties of the components, the so-called Gibbs-Duhem equation.

The value of an extensive thermodynamic property of a solution depends on T, P, and the number of moles of the components in the solution. If Q is an extensive molar property, then, $Q^{'} = Q^{'}(T, P, n_i, \eta_j, n_k, \cdots)$. At constant T and P,

$$dQ' = \left(\frac{\partial Q'}{\partial n_i}\right)_{T,P,n_j,n_k\ldots.} dn_i + \left(\frac{\partial Q'}{\partial n_j}\right)_{T,P,n_i,n_k\ldots..} dn_j + \left(\frac{\partial Q'}{\partial n_k}\right)_{T,P,Pn_i,n_j,n_l\ldots} dn_k + \cdots \tag{3.17}$$

Assume

$$\overline{Q_i} = \left(\frac{\partial Q'}{\partial n_i}\right)_{T,P,\eta_j,n_k}.$$

Then, from Eq. (3.17)

$$dQ' = \overline{Q}_i dn_i + \overline{Q}_j dn_j + \overline{Q}_k dn_k + \cdots. \tag{3.18}$$

$\overline{Q}_i$ is the increase in the value of Q' for a mixture or solution. When 1 mole of i is added to a large quantity of the solution at constant T and P, the value of Q′ for the solution increases to $\overline{Q}_i$. The quantity of solution should be large enough so that the addition of 1 mole of i to the solution does not change the composition measurably.)

Then,

$$Q' = n_i\overline{Q}_i + n_j\overline{Q}_j + n_k\overline{Q}_k + \cdots \tag{3.19}$$

or

$$dQ' = n_i\overline{dQ_i} + n_j\overline{dQ_j} + n_k\overline{dQ_k} + \ldots + \overline{Q}_i dn_i + \overline{Q}_j dn_j + \overline{Q}_k dn_k + \cdots. \tag{3.20}$$

From Eqs. (3.18) and (3.20), at constant T and P,

$$n_i\overline{dQ_i} + n_j\overline{dQ_j} + n_k\overline{dQ_k} + \ldots. = 0.$$

Or,

$$\Sigma n_i\overline{dQ_i} = 0. \tag{3.21}$$

Dividing Eq. (3.21) by n, the total number of moles of all the components of the solution, gives

$$\sum X_i\overline{dQ_i} = 0. \tag{3.22}$$

Equations (3.21) and (3.22) are equivalent. They are known as the **Gibbs-Duhem equation**.

3.4 The Gibbs Free Energy of Formation of a Solution

Gibbs free energy is an extensive property, e.g., see eq. (3.21), valid for fixed T and P.

$$G' = n_A\overline{G}_A + n_B\overline{G}_B \text{ (additive linear approximation.)} \tag{3.23}$$

where $\overline{G}_A and\ \overline{G}_B$ are, respectively, the partial molar Gibbs free energies of A and B in the solution. Dividing by $(n_A + n_B)$,

$$G = X_A\overline{G}_A + X_B\overline{G}_B. \tag{3.24}$$

$$dG = X_A d\overline{G}_A + X_B d\overline{G}_B + \overline{G}_A dX_A + \overline{G}_B dX_B. \tag{3.25}$$

From the Gibbs-Duhem equation, $X_A d\overline{G}_A + X_B d\overline{G}_B = 0.$
Therefore,

$$dG = \overline{G}_B dX_A + \overline{G}_B dX_B. \tag{3.26}$$

Or

$$\frac{dG}{dX_A} = \overline{G}_A - \overline{G}_B. \tag{3.27}$$

Equation (3.27) is multiplied by

$$X_B = > X_B\frac{dG}{dX_A} = X_B\overline{G}_A - X_B\overline{G}_B. \tag{3.28}$$

Adding Eqs. (3.28) and (3.24), one gets,

$$G + X_B\frac{dG}{dX_A} = G_A(X_A + X_B)\ as\ (X_A + X_B) = 1. \tag{3.29}$$

Or

$$\overline{G}_A = G + X_B\frac{dG}{dX_A}.$$

Similarly,

$$\overline{G}_B = G + X_A\frac{dG}{dX_B}. \tag{3.30}$$

Formation of a solute changes the Gibbs free energy of a condensed solution at T and a lower equilibrium pressure P_i, which will be a function of the pure component i at T and the equilibrium vapor pressure P_i^o.

Examine the following 3-step process:

(a) The evaporation of 1 mole of pure condensed i to vapor i at P_i^o .
(b) A decrease in the pressure of 1 mole of vapor i from $P_i^o\ to\ P_i$.
(c) The condensation of 1 mole of vapor i from the pressure P_i to a condensed solution.

The difference between the molar free energy of i in the solution and the molar free energy of pure i is given by $G_{(a)} + \Delta G_{(b)} + \Delta G_{(c)}$. However, as steps (a) and (c) are condensed at equilibrium, both $\Delta G_{(a)} and\ \Delta G_{(c)}$ are zero.

The overall change in Gibbs free energy which accompanies the isothermal 3-step process is thus,

$\Delta G_{(b)} = RTln\left(\frac{P^i}{P^i_0}\right)$ $[dG = -SdT + VdP + \sum \mu_i dn_i]$; T and $n_i{}'S$ are constants.

$$dG = VdP.$$

For 1 mole of an ideal gas, $dG = \frac{RT}{P} dP; \Delta G_{(b)} = RT \ln\left(\frac{P_i}{P_i^0}\right)$.

From eq. (3.14), $\Delta G_{(b)} = G_{i(solution)} - G_{i(pure)} = RTlna_i$.

If ΔG_i^M = partial molar Gibbs free energy of the solution i,

$$\Delta G_i^M = \overline{G}_i - G_i^o = RT \ln a_i. \tag{3.31}$$

If n_A moles of A and n_B moles of B are mixed to form a solution at constant T and P, the Gibbs free energy before mixing $= n_A G_A^o + n_B G_B^o$ and after mixing $= n_A \overline{G}_A + n_B \overline{G}_B$.

The change in the Gibbs free energy caused by the mixing process, $\Delta G'^M$ is given by,

$$\Delta G'^M = \left(n_A \overline{G}_A + n_B \overline{G}_B\right) - \left(n_A G_A^o + n_B G_B^o\right) = n_A\left(\overline{G}_A - G_A^o\right) + n_B\left(G_B - G_B^o\right).$$

Using Eq. (3.31),

$$\Delta G'^M = n_A \Delta \overline{G}_A + n_B \ln a_B. \tag{3.32}$$

Or

$$\Delta G'^M = RT(\eta_A \ln a_A + n_B \ln a_B). \tag{3.33}$$

For 1 mole of solution, Eqs. (3.32) and (3.33) will transform, respectively to

$$\Delta G^M = X_A \Delta G_{n'}^M + X_B \Delta \overline{G}_B \tag{3.34}$$

(Fig. 3.5) and

$$\Delta G^M = RT(X_A ln\, a_A + X_B\, ln\, a_B). \tag{3.35}$$

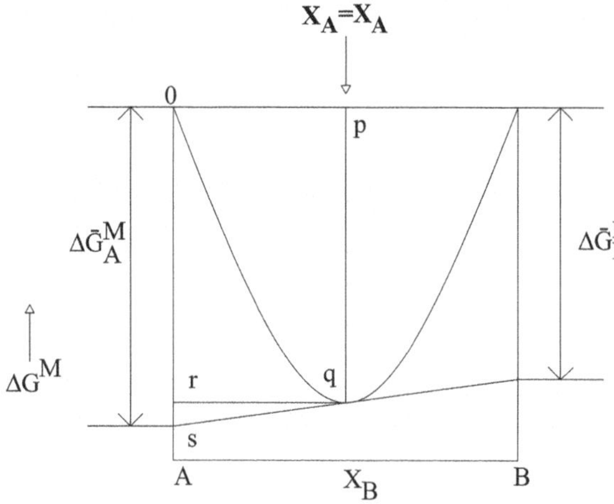

Fig. 3.5 Compositional variation of the molar Gibbs free energy of formation of a binary solution [1]

3.4.1 *Non-ideal Solutions*

When the activities of the components of a solution do not equal their mole fractions then it is known as a non-ideal solution.

In this case, the activity coefficient, γ_i, of a component of a solution is

$$\gamma_i = \frac{a_i}{X_i}. \tag{3.36}$$

γ_i can be greater than or less than unity. $\gamma = 1$ is Raoultian behavior.

If $\gamma_i > 1$, component i exhibits a positive deviation from Raoultian ideal behavior.

If $\gamma_i < 1$, a negative deviation from Raoult's law is found.

A temperature increase of a non-ideal solution decreases the degree of deviation from ideal behavior. That is, if $\gamma_i > 1$, then an increase in temperature decreases γ_i towards unity. If $\gamma_i < 1$, with increasing T, γ_i increases towards unity.

If ΔH^M = molar heat of formation of a solution, ΔH^M is positive.

When $\gamma_i > 1$, on increasing T, γ_i tends to unity (i.e., decreases). That is, the process of mixing is endothermic. When $\gamma_i < 1$, ΔH^M is negative and the mixing process is exothermic.

Mixing is exothermic when the A-B bond energy is more negative than both A-A and B-B bond energies. Then it leads to "ordering" in the solution, in which A atoms attempt to have only B atoms as the nearest neighbors and vice versa. That is an exothermic process, which points to a tendency towards compound formation between two components. Conversely, endothermic mixing means that the A-B bond energy is less negative than that of A-A and B-B bonds. This results in phase separation or "clustering" in the solution, i.e., A atoms tend to coordinate with A and B with B. In both cases, the equilibrium configuration of the solution is a compromise between the enthalpy factors, determined by the relative magnitude of the bond energies, which attempt to either completely order or completely unmix the solution, and the entropy factor, which attempts to maximize the randomness of mixing in the solution.

Under the category of Regular Solutions, two classes of solutions have been identified.

1. Ideal or Raoultian solutions in which $a_i = X_i$, i.e., $\Delta\overline{H}_i^M = 0, \Delta\overline{V}_i^M = 0$ *and* $\Delta\overline{S}_i^M = -R\ln X_i$.
2. Non-ideal solutions in which $a_i \neq X_i$, $\Delta\overline{H}_i^M \neq 0$.

Another category constitutes the non-ideal solutions. The simplest mathematical formulation is that of "regular solution behavior". It has been suggested that at constant T, activity coefficients $\gamma_A$and γ_B of the components of a binary solution could be represented by a power series of the form,

$$\ln\gamma_A = \alpha_1 X_B + \frac{1}{2}\alpha_2 X_B^2 + \frac{1}{3}\alpha_3 X_B^3 + \cdots \tag{3.37}$$

and

$$\ln\gamma_B = \beta_1 X_A + \frac{1}{2}\beta_2 X_A^2 + \frac{1}{3}\beta_3 X_A^3 + \cdots$$

Using the Gibbs-Duhem equation,

$$X_A d\ln\gamma_A = -X_B d\ln\gamma_B. \tag{3.38}$$

It has been proved that if these equations are to be valid over the entire range of composition of solutions,

$$\alpha_1 = \beta_1 = 0.$$

Proof Both sides of Eq. (3.38) are expressed as power series of X_A and X_B. The coefficients are then equated. By comparing coefficients of the power series, it can be shown that if the activity coefficient variations can be represented by the quadratic terms alone, then,

$$\alpha_2 = P_2. \tag{3.39}$$

In 1929 Hildebrand, using an equation due to Van Laar, which is based on the Van der Waals equation of state for mixtures, showed that if the value of the Van der Waals' "b" is the value of "b" for both the components, then, in the binary A-B solution,

$$RT \ln \gamma_B = \alpha' X_A^2 \text{ and } RT \ln \gamma_B = \alpha' X_A^2. \tag{3.40}$$

When Eq. (3.40) is obeyed it is referred to as a "Regular Solution".
Using Gibbs-Duhem equation, it has been shown in addition that

$$ln \gamma_A = -X_B X_A \alpha_B \int_{X_A=1}^{X_A=X_A} \alpha_B dX_A \ln \gamma_A$$

And so, if the value of α of one component, say B, is independent of composition, then,

$$\ln \gamma_A = -X_A X_B \alpha_B - \alpha_B (x_A - 1) = -X_A X_B + \alpha_B X_B = \alpha_B X_B (1 - X_A) = \alpha_B X_B^2.$$

From Eq. (3.40), $\alpha_A = \frac{\alpha'}{RT'}$ $\alpha_A = \alpha_B = \alpha$ (follows from Gibbs-Duhem equation) or for a regular solution

$$\alpha = \frac{\alpha'}{RT}. \tag{3.41}$$

So, a regular solution is one in which,

$$\Delta \overline{H}_i^M \neq O, \ \ \Delta \overline{S}_i^M = \Delta \overline{S}_i^{M,id} = -R \ln X_i.$$

3.4.2 *The Concept of Excess Function*

$$G = G^{\text{id}} + G^{XS} \tag{3.42}$$

where G = molar Gibbs free energy of the solution, G^{id} = molar Gibbs free energy of the solution, when it is ideal, G^{XS} = excess molar free energy of the solution

$$\Delta G^M = \Delta G^{M,\text{id}} + G^{XS}. \tag{3.43}$$

For any solution $\Delta G^M = \Delta H^M - \Delta \Delta^M$ and for an ideal solution, $\Delta G^{M,\ id} = -T\Delta S^{M,\ id}$. Then,

$$G^{XS} = \Delta G^M - \Delta G^{M,id} = \Delta H^M - T\left(\Delta S^M - \Delta S^{M,id}\right). \tag{3.44}$$

For a regular solution, $\Delta S^M = \Delta S^{M,\ id}$

Thus,

$$G^{XS} = \Delta H^{M}. \tag{3.45}$$

Now, $\Delta G^{M} = RT(X_A \ln a_A + X_B \ln a_B) = RT(X_A \ln X_A + X_B \ln X_B) + RT(-(X_A \ln X_A + X_B \ln X_B) + RT(X_A \ln \gamma_A + X_B \ln \gamma_B)$and as

$$\Delta G^{M,id} = RT(X_A \ln X_A + X_B \ln X_B), G^{XS} = RT(X_A \ln \gamma_A + X_B \ln \gamma_B) \tag{3.46a}$$

$$= X_A \overline{G}_A^{XS} + X_B \overline{G}_B^{XS}. \tag{3.46b}$$

For a regular solution,

$$\ln \gamma_A = \alpha X_B^2;\ \ln \gamma_B = \alpha X_A^2.$$

Taken together with Eq. (3.46a),

$$G^{XS} = RT\alpha X_A X_B. \tag{3.47}$$

or as $\alpha = \frac{\alpha'}{RT}, G^{XS} = \alpha' X_A X_B$, which is independent of T.

Also, from $\left(\frac{\partial G^{XS}}{\partial T}\right)_{P,comp} = -S^{XS}$ and as S^{XS} for a regular solution is zero, G^{XS} and hence ΔS^{M} are independent of T.

From Eqs. (3.46a), (3.46b) and (3.47) at any given composition,

$$\overline{G}_A^{XS} = RT,\ \ln \gamma_{A(T_1)} = RT_2 \ln \gamma_{A(T_2)} = \alpha' X_B^2.$$

So, for a regular solution,

$$\frac{\ln \gamma_A\ at\ T_2}{\ln \gamma_A\ at\ T_1} = \frac{T_1}{T_2}. \tag{3.48}$$

This equation is of great practical value as it changes activity data of regular solution at one temperature to that at another.

3.5 A Statistical Model of Solutions

The Statistical Mixing Model applied to two species of equal molar volume and which do not change in volume when mixed helps in interpreting the regular solution behavior.

Interatomic forces exist only between neighboring atoms n the pure state and in solution. Therefore, the solution energy equals the sum of interatomic bond energies.

Let 1 mole of a mixed crystal contain N_A atoms of A and N_B atoms of B so that,

$$X_A = \frac{N_A}{N_A + N_B} = \frac{N_A}{N_0} \text{ and } X_B = \frac{N_B}{N_0}$$

where $N_0$is the Avagadro number.

Three types of interatomic bonds are prrsent in a mixed crystal or solid solution:

1. Each A-A bond has an energy E_{AA}.
2. Each B-B bond has an energy E_{BB}.
3. Each A-B bond has an energy E_{AB}.

When the atoms are infinitely apart, relative energy is zero. So, all the bond energies, E_{AA}, E_{BB} and E_{AB} negative.

In a crystal Z is the coordination number of an atom.

If P_{AA} is the number of A-A bonds, P_{BB} of the B-B bonds and P_{AB} of the A-B bonds, the energy of the solution,

$$E = P_{AA}E_{AA} + P_{BB}E_{BB} + P_{AB}E_{AB}. \tag{3.49}$$

Therefore, the calculation of P_{AA}, P_{BB}, P_{AB} will allow the calculation of E. As each A-A bond involves 2 atoms, the total number of atoms equals number of A-B bonds +2xthe number of A-A bonds.

Thus,

$$N_A Z = P_{AB} + 2P_{AA} \; or \; P_{AA} = \frac{N_A Z}{2} - \frac{P_{AB}}{2}. \tag{3.50}$$

Likewise,

$$N_B Z = P_{AB} + 2P_{BB} \; or \; P_{BB} = \frac{N_B Z}{2} - \frac{P_{AB}}{2}. \tag{3.51}$$

Substituting Eqs. (3.50) and (3.51) in Eq. (3.49),

$$E = \left(\frac{N_A Z}{2} - \frac{P_{AB}}{2}\right)E_{AA} + \left(\frac{N_B Z}{2} - \frac{P_{AB}}{2}\right)E_{BB} + P_{AB}E_{AB} = \frac{1}{2}ZN_A E_{AA}$$
$$+ \frac{1}{2}ZN_B E_{BB} + P_{AB}\left[E_{AB} - \frac{1}{2}(E_{AA} + E_{BB})\right]. \tag{3.52}$$

(Z is the total number of lattice sites avaiable.)

The energies of the pure components before mixing: N_A atoms be present in pure A. Then, Z x the number of A-A bonds = the number of atoms x the number of bonds per atom, i.e., $P_{AA} = \frac{1}{2}N_A Z$. Similarly, for N_B atoms in pure B, $P_{BB} = \frac{1}{2}N_B Z$.

Thus, ΔE^M, (the energy of mixing/solution) – (the energy of unmixed components)

$$= P_{AB}\left[E_{AB} - \frac{1}{2}(E_{AA} + E_{BB})\right].$$

For the mixing process,

$$\Delta H^M = \Delta E^M - P\Delta V^M.$$

It is stipulated that $\Delta V^M = 0$. Then,

$$\Delta H^M = \Delta E^M = P_{AB}\left[E_{AB} - \frac{1}{2}(E_{AA} + E_{BB})\right]. \tag{3.53}$$

That is, when E_{AA}, E_{AB}, E_{BB} are known, ΔH^M depends on P_{AB}.

For an ideal solution,

$$\Delta H^M = 0; E_{AB} = \frac{(E_{AA} + E_{BB})}{2}. \tag{3.54}$$

If $|E_{AB}| > \left|\frac{1}{2}(E_{AA} + E_{BB})\right|$,

there is a negative deviation from Raoultian ideal behavior.

If $|E_{AB}| < \left|\frac{1}{2}(E_{AA} + E_{BB})\right|$, which is a positive quantity, it suggests a positive deviation from Raoultian ideal behavior.

If $\Delta H^M = 0$, then the mixing of N_A atoms of A with N_B atoms of B is random. Then,

$$\Delta S^M = \Delta S^{M,id} = -R(X_A \ln X_A + X_B \ln X_B).$$

Even when $|\Delta H^M| \leq RT$, the mixing of atoms can be assumed to be random.

Then P_{AB} calculation: 2 neighboring lattice sites are present in a crystal labelled 1 and 2. The probability of an A atom occupying site 1 is,

$$\frac{\textit{The no.of A atoms in the crystal}}{\textit{The no.of lattice sites in the crystal}} = \frac{N_A}{N_0} = X_A.$$

Then, the probability that site 2 is occupancy by a B atom is X_B.

$X_A X_B$ will be the probability that site 1 is occupied by an A atom and site 2 by a B atom simultaneously.

If site 1 is occupied by a B atom and site 2 simultaneously by an A atom, that probability also is $X_A X_B$.

So, the probability that the neighboring sites contain an A-A pair is X_A^2. The neighboring sites contain a B-B pair is X_B^2. The probability that neighboring sites contain an A-B pair or an A-A pair or a B-B pair is $X_A^2 + 2X_A X_B + X_B^2 = (X_A + X_B)^2 = 1$

A mole of a crystal contains ½ Z number of pairs of lattice sites. Then the number of A-B pairs = the number of pair of sites x, the probability of an A-B pair.

i.e.,

$$\frac{1}{2}ZN_0 \times 2X_AX_B = ZN_0X_AX_B. \tag{3.55}$$

Similarly, $P_{AA} = \frac{1}{2}ZN_0 \times X_A^2 = \frac{1}{2}ZN_0X_A^2$. $P_{BB} = \frac{1}{2}ZN_0X_B^2$

Substituting Eq. (3.55) in Eq. (3.51),

$$\Delta H^M = ZN_0X_AX_B\left[E_{AB} - \frac{1}{2}(E_{AA} + E_{BB})\right].$$

If $\Omega = zN_0\left[E_{AB} - \frac{1}{2}(E_{AA} + E_{BB})\right]$,

$$\Delta H^M = \Omega X_AX_B. \tag{3.56}$$

And so, ΔH^M is a parabolic function of composition. Random mixing has been assumed. So, the statistical model is the same as the regular solution model. That is,

$$\Delta H^M = G^{XS} = \Omega X_AX_B = RT\alpha X_AX_B. \tag{3.57}$$

Hence,

$$\alpha = \frac{\Omega}{RT}. \tag{3.58}$$

From the tangent method,

$$\Delta \overline{H}_A^M = \Delta H^M + X_B\frac{\partial \Delta H^M}{\partial X_A}.$$

(This is derived from $G + X_B\frac{dG}{dX_A} = \overline{G_A}(X_A + X_B).Or,\ \overline{G_A} = G + X_B\frac{dG}{dX_A}$ and $\overline{G_B} = G + X_A\frac{dG}{dX_B}$.)

From eq. (3.57), $\frac{\partial \Delta H^M}{\partial X_A} = \Omega(X_B - X_A)$

Thus,

$$\overline{H}_A^M = \Omega X_AX_B + X_B\Omega(X_B - X_A) = \Omega X_B^2, \tag{3.59a}$$

and

$$\Delta \overline{H}_A^M = \Omega X_A^2. \tag{3.59b}$$

As mixing is random,

$$\Delta\overline{S}_A^M = -R\ln X_B \quad ;\Delta\overline{S}_B^M = -R\ln X_B$$

and hence,

$$\Delta\overline{G}_A^M = \Delta\overline{H}_A^M - T\Delta\overline{S}_A^M = \Omega X_B^2 + RT\ln X_A. \tag{3.60}$$

But

$$\Delta\overline{G}_A^M = RT\ln a_A = RT\ln\gamma_A + RT\ln X_A. \tag{3.61}$$

Comparing with Eq. (3.60),

$$\ln\gamma_A = \frac{\Omega}{RT}X_B^2 = \alpha X_B^2. \tag{3.62}$$

Therefore, γ depends on Ω. The values of E_{AA}, E_{BB}, E_{AB} determine the value of Ω.

If Ω is negative, then $\gamma_A < 1$; if Ω is positive, then $\gamma_A > 1$.

Henry's law requires that γ_A and $\ln\gamma_A$ should approach a constant value as $X_A \rightarrow 1$. Thus, as $X_A \rightarrow 1$, $\ln\gamma_A \rightarrow \ln\gamma_A^0 = \frac{\Omega}{RT}$, which is the limiting value asymptotically approached.

In view of the relationship between Henry's and Raoult's laws, Raoult's law is approached asymptotically by component, i, as $Xi \rightarrow 1$. As Ω increases Ω (Z_0 [E_{AB} − (1/2) (E_{AA} + E_{BB})] the statistical model is no longer realistic. That is, if E_{AB} is much greater or less than the average of (E_{AA} + E_{BB}). In such a situation, random mixing cannot be assumed.

Equilibrium configuration of a solution at constant T and P is reached when G (G = H − TS) is a minimum. G is measured relative to the unmixed components.

G is minimized as a compromise between H minimization of H and S maximization. If $|E_{AB}| > \left|\frac{1}{2}(E_{AA} + E_{BB})\right|$, then the minimization of H is equivalent to the maximization of the number of A-B pairs, i.e., a complete ordering in the solution. S maximization suggests total random mixing.

In other words, G maximization amounts to the combined effect of P_{AB} maximization, which increases with an increasingly negative value of Ω and random mixing. The tendency towards this increases with increasing T. Thus, Ω and T values are critical.

When Ω is strongly negative and T is not too high, then the P_{AB} value will be greater than that for random mixing. Then, assuming random mixing is not proper.

If $|E_{AB}| < \left|\frac{1}{2}(E_{AA} + E_{BB})\right|$, then H minimization amounts to the same as the minimization of the number of A-B pairs, which implies complete clustering in the solution.

G is minimized when as a compromise is reached between the minimization of P_{AB} (increases with increasingly positive values of Ω) and random mixing. When Ω is strongly positive and the temperature is not very high, then the P_{AB} value will be less than that for random mixing. In such a case, assuming random mixing is incorrect.

And so, for the statistical/ regular solution model to be valid, the equilibrium solution configuration should not be too distant from random mixing.

The entropy contribution to G is T dependent. Therefore,

1. For all values of Ω, nearly random mixing occurs as the temperature increases, and
2. For a given temperature,when Ω value is small, nearly random mixing is present.

Change of Standard State, Phase Relations and Phase Rule, its applications, Free Energy—Composition Diagrams for Binary Alloy Systems, Determination of Liquidus, Solidus and Solvus Lines

Let A and B mix to form 1 mole of a solution. Then,

$$\Delta G^M = RT(X_A \ln a_A + X_B \ln a_B)$$

where ΔG^M *is the free energy of mixing.*

G changes can be measured. Therefore, G of pure, unmixed components is taken as zero. In an ideal solution, i.e., if $a_i = X_i$, the molar Gibbs free energy of mixing is $\Delta G^{M,\ id} = RT(X_A \ln X_A + X_B \ln X_B)$,

as

$$\Delta H^{M,id} = 0, \Delta G^{M,id} = -T\Delta S^{M,id}.$$

Therefore, the $\Delta G^{M,id}$ variation with composition depends on temperature, T. Then, with the variations of activity with composition generates the tangential intercepts with the compositional variation.

As $X_i \to 0$, $a_i \to 0$ and the tangential intercept $\Delta \overline{G}_i^m = RT \ln a_i \to -\infty$, all G of mixing curves have vertical tangents at the extremities. As it is a logarithmic function, the mixing entropy curve also has at the extremities vertical tangents.

Free Energy, G, of formation of regular solutions:

The non-zero heat of mixing causes the deviation of $\Delta G^M$from $\Delta G^{M,\ id}$. And

$$\Delta G^M - \Delta G^{M,id} = G^{XS} = RT\alpha X_A X_B = \Omega X_A X_B = \Delta H^M.$$

ΔH^M is positive. If Ω and α are positive, ΔH^M is positive.

Two separate solutions at T and P in equilibrium can coexist if

$$\overline{G}_A(in\ solution\ m) = \overline{G}_A(in\ solution\ q) \tag{3.63}$$

and

$$\overline{G}_B(in\ solution\ m) = \overline{G}_B(in\ solution\ q). \tag{3.64}$$

Subtracting G_A^0 from both sides of Eq. (3.1)

$$RT \ln a_{A(in\ solution\ m)} = RT \ln a_{A(in\ solution\ q)}$$

Or

$$G_{A(in\ solution\ m)} = a_{A(in\ solution\ q)}. \tag{3.65}$$

Similarly,

$$G_{B(in\ solution\ m)} = a_{B(in\ solution\ q)}. \tag{3.66}$$

Equations (3.65) and (3.66) state that at constant T and P two solutions (phases) can coexist if

As $\Delta\overline{G}_A^M(in\ m) = \Delta\overline{G}_A^M(in\ q)$,

$\Delta\overline{G}_B^M(in\ m) = \Delta\overline{G}_B^M(in\ q)$. Then at point m the tangent of the curve is present, which is also the tangent to the curve at point q. This double tangent position defines on the Gibbs free energy of mixing the positions of points m and q.

If $\alpha = \Omega/RT$ is sufficiently positive at temperature T, tendency towards clustering of atoms causes phase separation. When B is initially added to A, a homogeneous phase is formed. Saturation of Phase I saturation with B happens at the composition m. Phase II (II solution) of composition q (Phase II saturated with A) appears with a further addition of B. Continued addition of B increases the ratio of phase II to phase I. Finally, the overall composition of the system reaches q, at which point phase I disappears.

Between compositions q and B a homogeneous solution (phase II) is present (Fig. 3.6).

Gibbs free energy of phase I supersaturated with B is represented by mn. Gibbs free energy of phase II supersaturated with B is denoted by curve qp. The equilibrium states of the system are described by line AmqB. Physical significance can be attributed only to this line. In the compositional space G-T-P this line is the section of the system that is at constant pressure and temperature.

3.5.1 Criteria for Phase Stability in Regular Solutions

At a given temperature, say T, a homogeneous solution is stable over the entire range of composition if α is below a critical value. Above this phase separation occurs.

Consider

$$\alpha < \alpha_{critical}; \alpha = \alpha_{critical}; \alpha > \alpha_{icritical}.$$

The critical value of α is decided by that composition at which both $\frac{\partial^2 \Delta G^m}{\partial X_B^2}$ and $\frac{\partial^3 \Delta G^m}{\partial X_B^3}$ are simultaneously equal to zero. Then immiscibility becomes imminent.

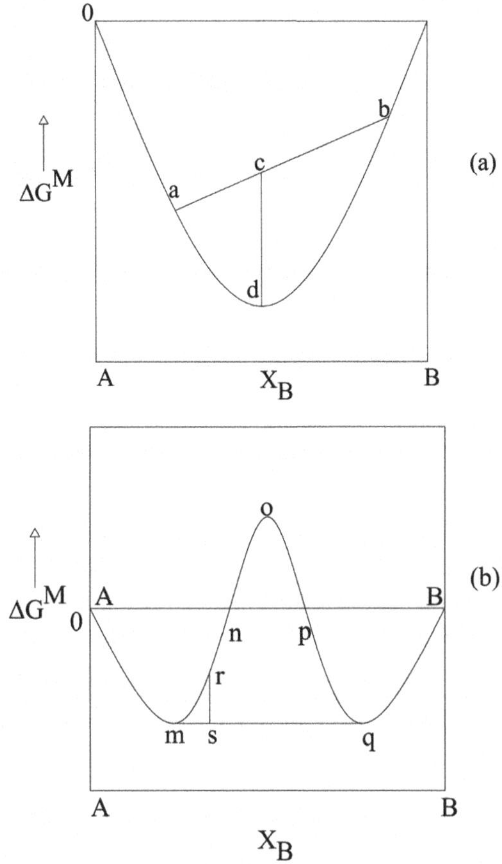

Fig. 3.6 (**a**) The molar Gibbs free energies of completely miscible mixtures of binary components. (**b**) The molar Gibbs free energies of mixing of binary components when they display a miscibility gap [1]

For a regular solution,

$$\Delta G^M = RT(X_A \ln X_A + X_B \ln X_B) + RT\alpha X_A X_B$$

$$\frac{\partial \Delta G^M}{\partial X_B} = RT\left[\ln \frac{X_B}{X_A} + \alpha(X_A - X_B)\right]$$

$$\frac{\partial^2 \Delta G^M}{\partial X_B^2} = RT\left(\frac{1}{X_A} + \frac{1}{X_B} - 2\alpha\right)$$

$$\text{and } \frac{\partial^3 \Delta G^M}{\partial X_B^3} = RT\left(\frac{1}{X_A^2} - \frac{1}{X_B^2}\right).$$

$$\frac{\partial^3 \Delta G^M}{\partial X_B^3} = 0; X_A = X_B = 0.5.$$

Then, $\frac{\partial^2 \Delta G^M}{\partial X_B^2} = 0\ at\ X_A = 0.\ \alpha = 2$ is the critical value of α. Above this phase separation is present.

In any regular system with a positive value of Ω, as α is inversely related to T, a critical temperature is present above which α < 2 and below which α > 2.

$$T_{cr} = \Omega/2\mathrm{R} \tag{3.67}$$

is the critical temperature.

If the heat of mixing and the critical temperatures are known, it is easy to generate the ΔG^M vs. composition plots.

- Gibbs free energy of mixing is convex downwards at all compositions at high enough temperatures.
- A range of composition centered on $X_B = 0.5$ exists over which the Gibbs free energy is convex upwards when the temperature is decreasing.
- Due to the log terms, the tangents to the curve at $X_A = 1$ and $X_B = 1$ are vertical (Fig. 3.7).

As $\Delta \overline{G}_B^M = RT \ln \alpha_B$, the intercepts with the $X_B = 1$ axis reveal the activity.

At $X_B = 0.5$ the activity displays a horizontal inflection.

Proof

$$\Delta \overline{G}_B^M = \Delta \overline{G} + X_A \left(\frac{\partial \Delta G^M}{\partial X_B} \right) = RT \ln a_B$$

Thus,

$$\partial \Delta \overline{G}_B^M = X_A \frac{\partial^2 \Delta G^M}{\partial X_B^2} = \frac{RT}{a_B} \frac{\partial a_B}{\partial X_B}, \tag{3.68}$$

and

$$\frac{\partial^2 \Delta \overline{G}_B^M}{\partial X_B^2} = X_A \frac{\partial^3 \Delta G^M}{\partial X_B^3} - \left(\frac{\partial^2 \Delta G^M}{\partial X_B^2} \right) = \frac{RT \partial^2 a_B}{a_B \partial X_B^2} - \frac{RT}{a_B^2} \left(\frac{\partial a_B}{\partial X_B} \right); (\because X_A = (1 - X_B)). \tag{3.69}$$

The second and the third derivatives of ΔG^M with respect to X_B are zero at T_{cr} and $X_B = 0.5$.

From the first and second derivatives of a_B with respect to X_B are zero (Eqs. (3.68) and (3.69)), a horizontal inflexion point on the activity curve at $X_B = 0.5\ and\ T = T_{cr}$ is produced.

The activity curve has a maximum and a minimum at T < T_{cr}. They are present at $\frac{\partial^2 \Delta G^M}{\partial X^{B^2}}$ and hence $\frac{\partial a_B}{\partial X_B}$. These are, for example, points n and P in Fig. 3.6b indicative of a spinodal composition.

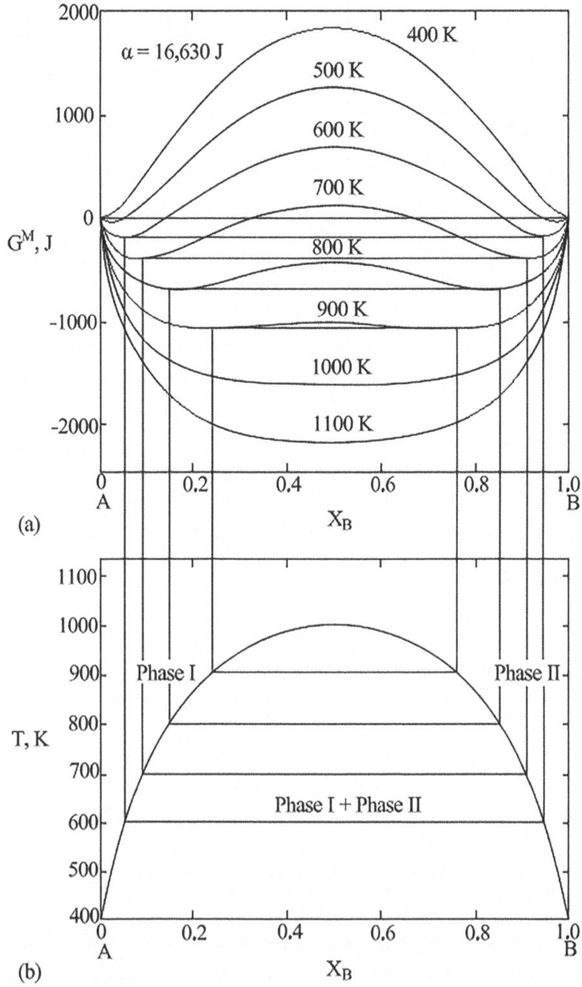

Fig. 3.7 (**a**) The molar Gibbs free energy of mixing in a binary regular solution vs. T for which $\Omega = 16{,}630$ joules. (**b**) Trace of the loci of the double tangent points of Fig. 3.6a [1]

$\left(\frac{\partial a_B}{\partial X_B}\right)$ is positive when there is stability. If this condition is violated in any range, because of these equations, in that range the curve has no physical meaning. The activity is assumed to be constant in that range.

3.6 (Gibbs) Phase Rule

In Fig. 3.8 in 3 areas only one phase is stable. P and T can be independently varied within this range with no upset to the one phase equilibrium.

That is, the equilibrium has TWO degrees of freedom (independent variables).

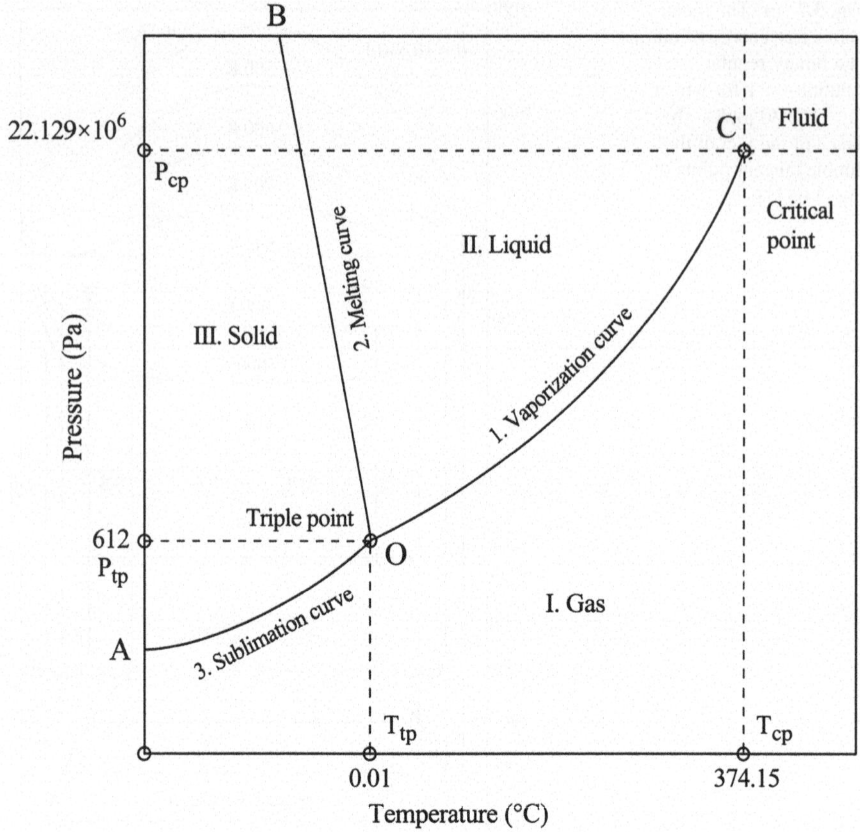

Fig. 3.8 Phase diagram of water—a schematic [1]

The two phases coexist along OA, OB, OC. Hence they have one degree of freedom (P or T). At the triple point, ZERO degree of freedom exists.

Therefore, $F = C - P + 2$, which is called Gibbs Phase Rule.

F = degrees of freedom; C = number of components; P = number of phases.

3.6.1 Phase Diagrams, Gibbs Free Energy and Thermodynamic Activity

Complete solid solubility of components A and B: Their crystal structure should be the same, atomic sizes should be comparable and electronegativities and valences should be similar.

If one of these conditions is not met, a miscibility gap in the solid state would be found.

In Fig. 3.9 the phase diagram A-B, a Eutectic - is displayed.

A and B of different crystal structures display two terminal solid solutions α and β.

Figure 3.9b presents a molar Gibbs free energy of mixing at T_1.

If the molar Gibbs free energy of pure liquid A and pure solid B are "a" and "c", they are located at $\Delta G^M = 0$.

The curve, Fig. 3.9 aeg (I), is the Gibbs free energy of mixing of solid B to form a homogeneous α solid solution. This has the same crystal structure as A. If the species has the same crystal structure as A, this curve would intersect $X_B = 1$ axis at the molar Gibbs free energy.

Similarly, the curve dh (II) represents the Gibbs free energy of mixing of solid B and solid A to form a homogeneous solid solution of crystal structure same as that of B. This curve intersects that $X_A = 1$ axis at that molar Gibbs free energy which A would have if it had the same crystal structure as B.

The molar Gibbs free energy of mixing of liquid A and liquid B to form a homogeneous liquid solution is given by the curve bfc (III).

Everywhere curve III lies below curve II. At T_1 β solutions are unstable. The α solidus composition at T_1 is identified using the double tangents to Curves I and III as e and the liquidus composition as f.

At T_1 representing A as the standard state of the solid and that of B as liquid, activity-composition relationships of the components are drawn—see Fig. 3.9c.

The assumptions: Liquid solutions exhibit Raoultian ideal behavior. Solid solutions show deviations from Raoult's law that are positive.

The horizontal intersection with the liquidus at T_1 is on the A-rich side—see the figure.

The length of AB increases as T goes below T_1. The length of cd (Fig. 3.10) decreases till at $T = T_{m(B)}$. The points c and d coincide at $\Delta G^M = 0$.

The point c (liquid B) lies above point d at $T_2 < T_{m(B)}$. Curve II is partially below curve III (Fig. 3.10). Two tangents, one to the curves I and III, which defines the compositions of the solidus of α and its conjugate liquidus, and the other to the curves II and III can be drawn which defines the compositions of the solidus of β and its conjugate liquidus.

The activity-composition curves at T_2 are also shown. Here, for both the components, the solid is the standard state.

The two liquidus compositions, m and n in Fig. 3.11b, approach each other with decreasing T, and at a unique temperature T_E, the eutectic temperature, they coincide. In other words, the triple tangent to the three curves are formed by the merger of the two double tangents. Doubly saturated eutectic liquid coexists in equilibrium with α and β solid solutions at compositions between the minima in curves I and II, A.

From the phase rule, one degree of freedom is present in this 3-phase equilibrium. This is useful in specifying the pressure of the system.

Therefore, at constant pressure, an example of an invariant is this 3-phase equilibrium.

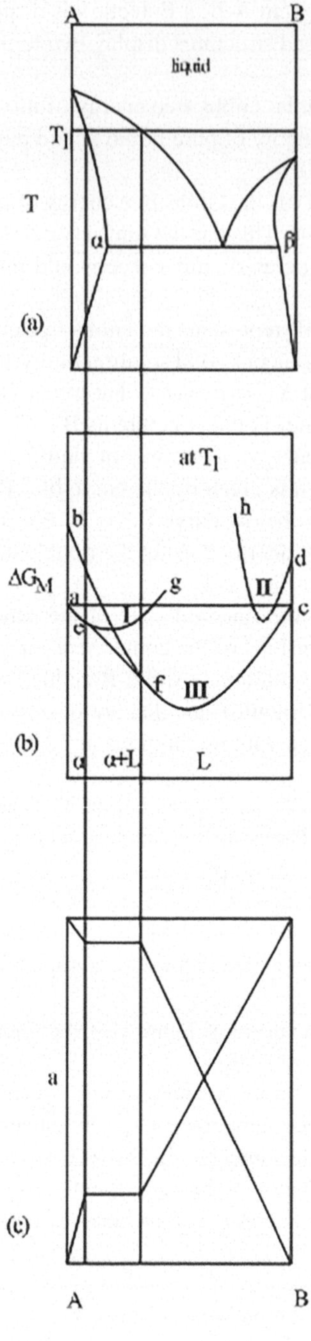

Fig. 3.9 Temperature vs. molar Gibbs free energies of mixing and the activities of the components of the system A–B [1]

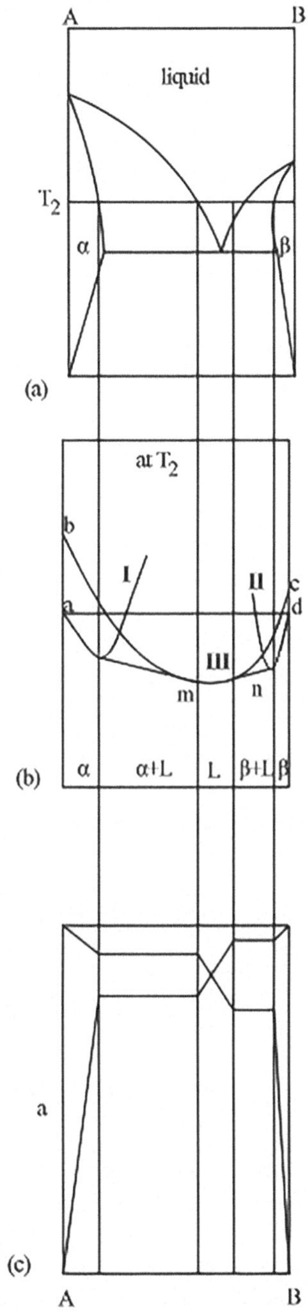

Fig. 3.10 System A-B: Molar Gibbs free energies of mixing and the activities of the components vs. Temperature [1]

Fig. 3.11 Components of the system *A-B*: The temperature effect on the molar Gibbs free energies of mixing and the activities [1]

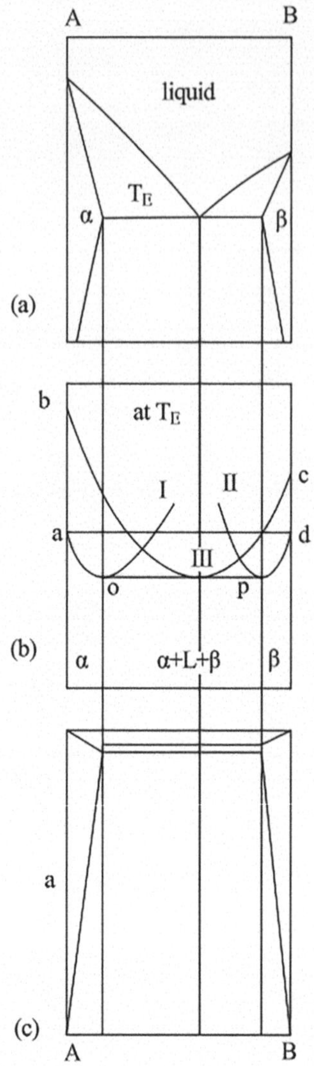

Curves I and II lie below the double tangent to curve III at $T_3 < T_E$. So, the liquid phase is unstable. To a reasonable approximation, A and B are insoluble in one another in solid state if the ranges of solid solubility in α and β phases are immeasurably small.

All Gibbs free energy of mixing curves have vertical tangents at their extremities. Then, a pure substance presents an infinite chemical sink to another substance. That is, an absolutely pure substance cannot be obtained.

A small solid solubility range can be neglected, e.g., Fig. 3.12. Then, curves I and II of Fig. 3.9b (i.e., the Gibbs free energy curves for the formation of α and β) merge with the vertical axes (i.e., α at $X_A = 1$ axis and β at $X_B = 1$ axis).

Fig. 3.12 Activities and molar Gibbs free energy of mixing of a binary eutectic system exhibiting total liquid miscibility and virtually full solid immiscibility [1]

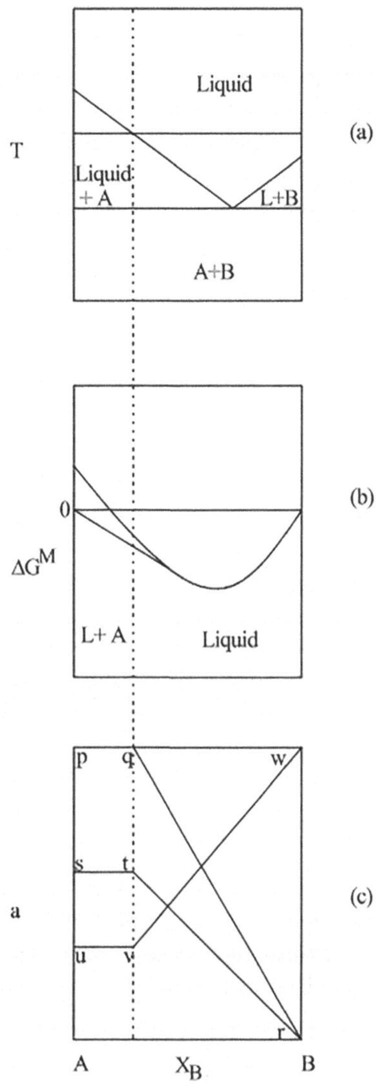

3.6.2 *Gibbs Free Energy of Formation of Liquid Solutions in the Solution A-B at T*

A single tangent could be drawn from the point $X_A = 1$ – pure solid A – to reduce the "double tangents" to the α and the liquid solutions. This assumes that the activity–composition curves pertain to ideal liquid solutions. In Fig. 3.12, pqr is the activity of A for pure solid A at p. The activity "s" is that of pure liquid A vis-a-vis solid A at p. The activity *a* is *str* with respect to liquid A. The latter has unit activity at s. The activity of B with respect to liquid B, is *Auvw*, At w it has unit activity.

Here, in the liquid state, complete miscibility is present. But in the solid state, nearly complete immiscibility exists. The liquidus curves reveal the activity variations in the components of the liquid solutions.

The composition between pure A and the liquidus composition at any T could be practically pure solid A in equilibrium with a liquid solution of the liquidus composition. Thus, at T,

$G^0_{A(s)} = \overline{G}^A_{(s)} = G^0_{A(s)} + RT \ln a_A$ in which a_A is with respect to liquid A as the standard state.

$$\Delta G^0_{m(A)} = -RT \ln a_A. \tag{3.70}$$

If the liquid solutions are Raoultian,

$$\Delta G^0_{m(A)} = -RT \ln X_A. \tag{3.71}$$

Application of Eq. (3.71) **to the calculation of the liquidus lines:**

The system Cd-Bi, to cite an example, Fig. 3.13,

Cd practically does not dissolve in solid Bi. Bi has a maximum solubility in solid Cd at the eutectic temperature of 419 K is 2.75 mole %.

The Bi liquidus is deduced from Eq. (3.71).

$$\Delta G^0_{m(Bi)} = -RT \ln X_{Bi(liquidus)}.$$
$$\Delta H^0_{m(Bi)} = 10,900 \text{ J at T}_{\text{m (Bi)}} = 544 \text{ K}.$$
$$\Delta S^0_{m(B_i)} = \frac{10,900}{545} = 20\ J/K \text{ at } 544 \text{ K}.$$

The temperature variation of molar constant pressure heat capacities of solid and liquid Bi is as follows.

$$\Delta C_{p,Bi(s)} = 18.8 + 22.6 \times 10^{-3} T\ J/s.$$
$$\Delta C_{p,Bi(l)} = 20 + 6.15 \times 10^{-3} T + 21.1 \times 10^5 T^{-2}\ J/K.$$

Thus, $\Delta C_{p,\ Bi} = C_{\beta Bi(l)} - C_{\beta Bi(s)} = 1.2 - 16 \cdot 45 \times 10^{-3} T + 21.1 \times 10^5 T^{-2}\ J/K$.And

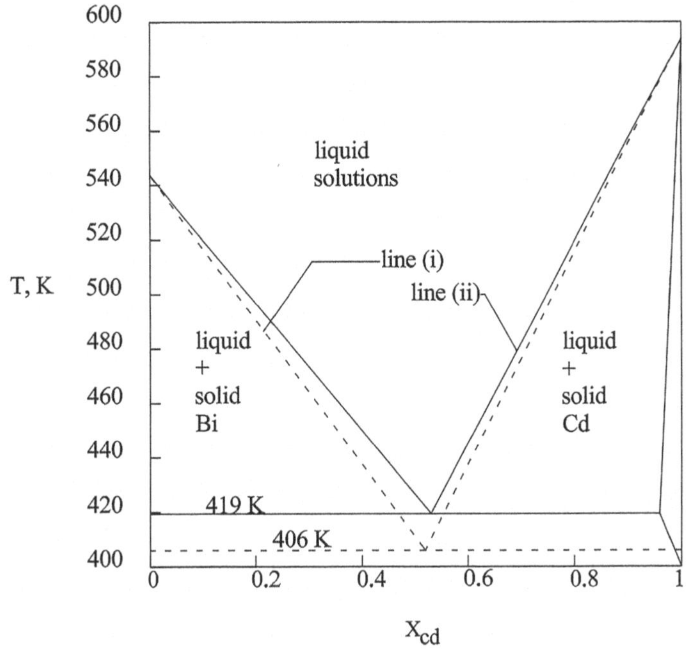

Fig. 3.13 The Bi–Cd phase diagram. The broken and full lines are respectively the calculated and the measured liquidus lines. Assumption: Ideal mixing is present. No solid solution is present [1]

$$\begin{aligned}\Delta G^0_{m(Bi),544} &= \Delta H^0_{m(Bi)544} + \int_{544}^{T} \Delta C_{p,Bi} dT - T\left(\Delta S^0_{m,Bi,455} + \int_{544}^{T} \frac{\Delta C_{p,Bi_{dT}}}{T}\right)\\ &= 16,560 - 23.79T - 1.2T\ln T + 8.225\times 10^{-3}T^2 - 10.55\times 10^5 T^{-1}\\ &= -RT\ln X_{Bi(liquidus)}.\end{aligned} \tag{3.72}$$

Or

$$\ln X_{Bi(liquidus)} = \frac{-1992}{T} + 2{\cdot}861 + 0.144\ln T - 9{\cdot}892\times 10^{-4}T + 1{\cdot}269\times\frac{10^5}{T^2}.$$

$$\int_{544}^{T}(1.2 - 16.45\times 10^{-3}T + 21.1\times 10^5 T^{-2})\mathrm{d}T = 1{\cdot}2T - 544\times 1{\cdot}2 - \frac{16.45\times 10^{-3}T^2}{2} + \frac{1.65\times 10^{-3}}{2}\times(544)^2 - \frac{21{\cdot}1\times 10^5}{T} + \frac{21.1\times 10^5}{544}.$$

This equation is drawn as broken line (i) in Fig. 3.13.

Similarly, if the small solid solubility Bi in Cd is ignored,

$$\Delta G^{\circ}_{m(Cd)} = -RT \ln X_{Cd(liquidus)}.$$

$$\Delta H^{\circ}_{m(Cd)} = 6400J \ at \ \ T_{m,Cd} = 594K \text{ and thus,}$$

$$\Delta S^{\circ}_{m(Cd)} = \frac{6400}{594}.$$

$$= 10.77 \frac{J}{K} \text{at 594 K.}$$

The constant pressure molar heat capacities are:

$$C_{\beta_1 Cd(S)} = 22.2 + 12.3 \times 10^{-3} T \ J/K,$$

and $C_{p,\ Cd(l)} = 29.7$ J/K.

Thus,

$$\Delta C_{\beta,Cd} = C_{\beta,Cd(l)} - C_{\beta,Cd(s)} = 7{\cdot}5 - 12.3 \times 10^{-3} T J/K \Delta G^{\circ}_{m(Cd)}$$

$$= \Delta H^{\circ}_{m(Cd),594} + \int_{594}^{T} \Delta C_{p,Cd} dT - T\left(\Delta S^{0}_{m(Cd),594} + \int_{594}^{T} \frac{\Delta C_{p,Cd}}{T} dT\right.$$

$$= 4155 + 37.32T - 7.5T \ln T + 6.15 \times 10^{-3} T^2 J = -RT \ln X_{Cd(liquidus)}. \quad (3.72)$$

Or

$$\ln X_{Cd(liquidus)} = \frac{-495}{T} - 4 \cdot 489 + 0 \cdot 90 \ln T - 7.397 \times 10^{-4} \ T.$$

In Fig. 3.13 it is drawn as the broken line, (ii).

Raoult's liquid composition: Intersect lines i and ii intersect. It is simultaneously saturated with Cd and Bi at 406K. If the liquids were ideal, this would be the eutectic temperature.

$$\int T^{-2} dT = -T^{-1}.(-1) \frac{1}{T^2}.$$

The actual liquidus lines: They lie above the calculated values. Actual eutectic temperature is 419 K.

From Eq. (3.71), $\Delta G^0_{m(A),419k} = 2482 \ J$ and from Eq. (3.72),

$$\Delta G^0_{m(Cd),419k} = 1898 \ J.$$

Thus, from Eq. (3.70), the actual eutectic melt,

$$a_{Bi} = \exp\left(\frac{-2482}{8 \cdot 3144 \times 419}\right) = 0.49,$$

and

$$a_{Cd} = \exp\left(\frac{-1898}{8.3144 \times 419}\right) = 0.58$$

The actual eutectic composition is $X_{Cd} = 0.55,\ X_{Bi} = 0 \cdot 45.$

Thus, the activity coefficients are $\gamma_{Bi} = \frac{0.49}{0 \cdot 45} = 1.09$ and $\gamma_{Cd} = \frac{0.58}{0 \cdot 55} = 1.05$.

That is, the liquidus temperatures rise for positive deviations from Raoultian ideality.

3.6.3 Effect of Pressure on Phase Transformation and Phase Equilibria

Heat of mixing of ideal gases is zero because the interatomic forces are absent between the gases of ideal gases. In contrast, chemical affinity between gases for each other when mixed is present.

When hydrogen and oxygen gases in the presence of a catalyst are mixed, a significant amount of heat is released. This is addressed in one of two ways.

(a) the mixture of H_2 and O_2 is non-ideal. Then, the fugacities of the components H_2 and O_2 are ($a_i = \frac{f_i}{f_i^o}$; f_i is the fugacity in the state concerned and f_i^o is the fugacity in the standard state) estimated as follows. For an ideal condensed solution and the vapor in equilibrium, $f_i = P_i$, where P_i is the partial pressure of "i". Or,
(b) H_2 and O_2 have reacted to an extent with each other and H_2O results as a product.

If the pressure of the system is low enough, the equilibrium state at a given T is the partial pressures exerted by H_2, O_2 and H_2O in the system. Both the treatments are thermodynamically equivalent. But the latter is more convenient and practical.

The equilibrium state is where the Gibbs free energy is minimum.

In reacting gases the total change in the Gibbs free energy has a contribution from a change in Gibbs free energy due to the chemical reactions and a contribution from the mixing of the product gases.

Knowing the change in this change in Gibbs free energy with composition (which ranges from the pure unmixed reactant gases to the pure unmixed product gases) the equilibrium state in any system of reactive gases can be found out.

Here the concept of the **Equilibrium constant of a Reaction** is defined.

Relation between equilibrium constant and the standard Gibbs free energy change for a reaction:

Examine at constant T and P the reaction, $A_{(g)} + B_{(g)} \leftrightarrow 2C_{(g)}$. At any instant during the reaction,

$$G' = n_A\overline{G}_A + n_B\overline{G}_B + n_C\overline{G}_C, \tag{3.73}$$

where n_A, n_B, n_C are the numbers of moles of A, B, C present in the system at that instant. $\overline{G}_A$, $\overline{G}_B$, $\overline{G}_C$ are partial molar Gibbs free energies of A, B, C in the gas mixture at the same instant.

Problem How to minimise G'? Find n_A, n_B, n_C values. This represents the equilibrium state at given T and P. Then, the reaction will stop.

Consider one mole of A and 1 mole of B reacting to produce 2 moles of C (total of gas).Then,

$n_A = n_B$ and $n_C = 2 - n_A - n_B = 2(1 - n_A)$.

Equation (3.73) can be rewritten as,

$$G' = n_A\overline{G}_A + n_B\overline{G}_B + 2(1 - n_A)\overline{G}_C.$$

But, $\overline{G}_i = G_i^0 + RT\ln P + RT\ln X_i$, and $X_A = \frac{n_A}{2}, X_B = \frac{n_A}{2}, X_C = \frac{2(1-n_A)}{2} = (1 - n_A)$.

Therefore,

$$\begin{aligned} G' = {} & n_A\left(G_A^0 + G_B^0 - 2G_C^0\right) + 2G_C^0 + 2RT\ln P \\ & + 2RT\left[n_A\ln\left(\frac{n_A}{2}\right) + (1 - n_A)\ln(1 - n_A)\right] \end{aligned}$$

or

$$\begin{aligned} G' - 2G_c^0 = {} & n_A\left(-\Delta G^0\right) + 2RT\ln P \\ & + 2RT\left[n_A\ln\left(\frac{n_A}{2}\right) + (1 - n_A)\ln(1 - n_A)\right], \end{aligned} \tag{3.74}$$

where $\Delta G^0 = 2G_C^0 - G_A^0 - G_B^0$. This is called the standard Gibbs free energy change for the chemical reaction at temperature T.

The standard Gibbs free energy change for a reaction at temperature T is the sum of the Gibbs free energies of the reactants in their standard states or it is equal to the difference between the sum of the Gibbs free energies of the reaction products in their standard states.

In this case, ΔG^0 = difference between the Gibbs free energy of 2 moles of C at P = 1 atmosphere and T, and 1 mole of A and 1 mole of B in the same conditions.

If the total pressure of the system is 1 atmosphere, then the equation becomes

$$G' - 2G_c^0 = n_A\left(-\Delta G^0\right) + 2RT\left[n_A\ln\left(\frac{n_A}{2}\right) + (1 - n_A)\cdot\ln(1 - n_A)\right]. \tag{3.75}$$

Difference between the Gibbs free energy of the 2-mole system, $n_A = n_B$ and the Gibbs free energy of the system and it consists of 2 moles of C = Left hand side of Eq. (3.75). 2 points.

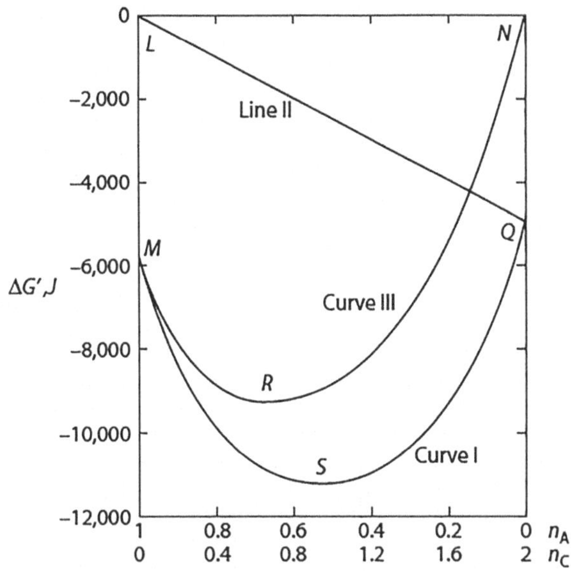

Fig. 3.14 The variations of the contributions to the decrease in the Gibbs free energy due to chemical reaction (line II), the contribution to the decrease in the Gibbs free energy due to gas mixing (curve III), with the extent of the reaction $A_{(g)} + B_{(g)} = 2C_{(g)}$ for which $\Delta G = -5000$ J at 500 K. Curve I is the sum of curves II and III [1]

1. The change in the Gibbs free energy due to the chemical reaction, i.e., due to the disappearance of the reactants is given by the right-hand side of Eq. (3.75).
2. The decrease in the Gibbs free energy caused by mixing of the gases is given by the second term on the right-hand side of Eq. (3.75).

For the reaction, $A_{(g)} + B_{(g)} = 2C_{(g)}$, at 500 K and 1 atmosphere pressure (Fig. 3.14),

ΔG^0_{500K} for the reaction is taken as -5000 J at 500 K.

Reference for Gibbs free energy is arbitrarily chosen as $G^o_A + G^o_B = 0$. Then, $2\,G^0_C = -5000\,J$

Here, the y-coordinate $\Delta G'$ is the difference between Gibbs free energy of the system comprising 1 mole of A and 1 mole of B before mixing of A and B.

Then, point L ($n_A = 1$, $n_B = 1$ before mixing) is located at $\Delta G' = 0$.

Then, Q ($n_C = 2$) is at $\Delta G' = -5000$ J.

M represents the decrease in the Gibbs free energy due to the mixing of 1 mole of A and 1 mole of B before any chemical reaction between the two occurs, i.e.,

$$\Delta G(L \rightarrow M) = \sum_i n_i RT \ln\left(\frac{p_i}{P_i}\right) = RT\left[n_A \ln\left(\frac{p_A}{P_A}\right) + n_B \ln\left(\frac{p_B}{P_B}\right)\right]$$

But $n_A = n_B = 1$ *and* $P_A = P_B = P_{mixture} = 1$. Here, the partial pressure is represented as p_A *or* p_B.

Or $p_A = p_B = 1/2$.

Therefore,

$$\Delta G(L \rightarrow M) = 8.3144 \times 500 \times 2\, ln\, 0.5 = -5763\, J$$

Or

$G' + 5000 = 1 \times 5000 - 5763$ J,and hence,

$$G' = -5763\, J$$

And so,

$$\Delta G' = G' - \left(G_A^0 + G_B^0\right) = -5763\, J.$$

The variation of G′ with n_A = sum of II (given by the first term on the right-hand side of Eq. (3.75) is given by curve I, the decrease in the Gibbs free energy (due to chemical reaction) and Curve III (given by the second term on the right-hand side of Eq. (3.75), the decrease in Gibbs free energy due to gas mixing).

The contribution from the chemical reaction in decreasing the Gibbs free energy linearly increases with n_C. But the contribution to the change in the total Gibbs free energy due to gas mixing goes through a maximum at the composition (R). R corresponds to where the gas mixture permits maximum randomization. Beyond R, following, further chemical reaction, the gas mixing contribution decreases (due to further increase in n_C and decrease of n_A and n_B), the randomness. Eventually, at composition S the sum of the two contributions (due to a decrease in the Gibbs free energy) is a maximum.

Beyond S, the increase in curve III is more than the decrease in line II, i.e., the total Gibbs free energy would increase. So, the Gibbs free energy has its minimum value at S. This is the equilibrium state.

The minimum in curve I is obtained from the criterion that at the minimum $\left(\frac{\Delta G'}{\partial n_A}\right)_{T,P} = 0$ and as

$$G' = n_A\overline{G}_A + n_A\overline{G}_B + 2(1 - n_A)\overline{G}_C$$
$$\frac{\partial G'}{\partial n_A} = \overline{G}_A + \overline{G}_B - 2\overline{G}_C = 0$$

That is, the criterion for reaction equilibrium is

$$\overline{G}_A + \overline{G}_B = 2\overline{G}_C \tag{3.76}$$

Equation (3.76) can be written as

$$G_A^0 + RT\ln p_A + G_B^0 + RT\ln p_B = 2G_C^0 + 2RT\ln p_C, \tag{3.77}$$

where, p_A, p_B, p_C are respectively the partial pressures of A, B and C, which occur at the reaction equilibrium.

Rearranging Eq. (3.77), $2G_C^0 - G_A^0 - G_B^0 = -RT \ln \frac{p_C^2}{p_A p_B}$

or

$$\Delta G^0 = -RT \ln \frac{p_C^2}{p_A p_B}. \tag{3.78}$$

The equilibrium partial pressure quotient of the reactants and products occurs as a logarithmic term in Eq. (3.78) is the Equilibrium Constant, K_p, for the Reaction, i.e.,

$$\left(\frac{p_C^2}{p_A p_B}\right)_{eq} = K_p \tag{3.79}$$

and hence

$$\Delta G^0 = -RT \ln K_p. \tag{3.80}$$

As ΔG^0 is a function only of T, it follows from Eq. (3.79) that K_p is a function only of T. For example, from Fig. 3.14 $\ln K_p = -\frac{\Delta G^0}{RT} = \frac{5000}{8.3144 \times 500} = 1.203.$

Therefore, $K_p = 3.329$

Now, $K_p = \frac{p_C^2}{p_A p_B} = \frac{X_C^2 p^2}{X_A^P X_B^P} = \frac{X_C^2}{X_A X_B} = \frac{(1-n_A)^2}{(n_A^2/4)} = 3.329.$

Hence, $n_A = 0.523$ (the other solution to the quadratic, $n_A = 11.4$ is not physically meaningful).

That is, reaction equilibrium occurs with $n_A = n_B = 0.523$ and $n_C = 0.954$ or when the chemical reaction A + B = 2C has proceeded to 47.7% completion. [$n_C = (2 - 2n_A) = (2 - 1.046) = 0.954$; $(1 - 0.523) \times 100 = 47.7\%$]. The minimum in curve I in Fig. 3.14 occurs at $n_A = 0.523$.

If ΔG^0 for the reaction was zero (an appropriate temperature is involved), there would be no chemical reaction contribution to $\Delta G'$. The $\Delta G'$ variation with n_A,will be revealed by curve III (Fig. 3.14), that is, the criterion for reaction equilibrium is the maximization of the system randomness. It is seen at composition R. From Eq. (3.80), if $\Delta G^0 = 0$, then $K_p = 1$ and thus

$$1 = \frac{4(1-n_A)^2}{n_A^2} + n_A = \frac{2}{3}.$$

Therefore, the minimum in curve III occurs at $n_A = n_B = n_C = \left(\frac{2}{3}\right)$. That is, maximum randomness occurs when all three species are present in an equal amount.

Important Curve I minimum in Fig. 3.14—the equilibrium state at P = 1 atmosphere and 500 K—is the only point on curve I is the only significant point in classical thermodynamics.

The point S is the only state on curve I, point S corresponds to the state—which lies on the equilibrium surface in P – T – Composition space—for the fixed values of P = 1 atmosphere and T = 500 K. A change in P and/ or T moves the equilibrium composition over the equilibrium surface in the P – T – Composition space.

3.6.4 The Effect of Temperature on the Equilibrium Constant

The minimum in curve I in the Fig. 3.14: It is deduced from the difference between the lengths of LM and NQ. LM represents the lowering of the Gibbs free energy due to mixing of the gases prior to the start of reaction and NQ = standard Gibbs free energy change, ΔG^0. Both the lengths are dependent on temperature, i.e.,

$$\text{LM} = 2\,\text{RT}\ (\ln\ 0.5);\ \text{and NQ} = \Delta G^0.$$

T, the minimum in the composition—T in curve I, which decides the vlue of K_p depends on the relative effects of temperature on the lengths of LM and NQ. When the amount of reactants is fixed, the length of LM linearly increases with T and the change in the length NQ with T is decided by the magnitude and sign of the change in the standard entropy. That is, the condition $\left(\frac{\partial \Delta G^0}{\partial T}\right)_\text{P} = -\Delta S^0$.

If ΔS^0 is negative, with increasing T the length LM increases while that of NQ decreases. Thus, the minimum (Curve I) moves to the left, i.e., K_p decreases with rising temperature.

Variation of K_p with T is obtained from the Gibbs–Helmholtz equation $\left[\frac{\partial\left(\frac{\Delta G^0}{\text{T}}\right)}{\partial T}\right]_\text{P} = \frac{\Delta H^0}{T^2}$.

[derived earlier as $\left[\frac{d\left(\frac{\text{G}}{\text{T}}\right)}{dT}\right]_\text{P} = -\frac{\text{H}}{T^2}$ from G = H – T].

As $\Delta G^0 = -\text{RT} \ln K_p$,

$$\frac{\partial}{\partial T}\left(\frac{-\Delta G^0}{RT}\right) = \frac{\partial \ln K_P}{\partial T} = \frac{\Delta H^0}{RT^2}, \tag{3.81}$$

or

$$\frac{\partial \ln K_P}{\partial\left(\frac{1}{T}\right)} = -\frac{\Delta H^0}{R}. \tag{3.82}$$

Equation (3.82) (known as the Van't Hoff equation) reveals that the effect of T on K_p depends on the sign and magnitude of ΔH^0.

The reaction is endothermic if ΔH^0 is positive. In addition, a rise in T increases the value of K_P.

The reaction is exothermic if ΔH^0 is negative . Then, K_P decreases with increasing T.

ΔH^0 dependence on T should be known for integrating Eq. (3.82), which, in turn, depends on ΔC_P.

The direction of K_P vs. T is referred to as Le Chatelier's principle. If heat is added at reaction equilibrium, then the equilibrium shifts in that direction of absorption of heat. Examine the gaseous reaction $Cl_2 = 2Cl$, which is endothermic. Therefore, ΔH^0 is positive.

And so, as $K_P = \frac{P_{Cl}^2}{P_{Cl_2}}$ rises with T, i.e., equilibrium moves towards that direction (requires heat absorption). For reaction $2Cl = Cl_2$, ΔH^0 is negative. $K_P = \frac{P_{Cl_2}}{P_{Cl}^2}$ is lower with a rising T. For both cases, a rise in T increases P_{Cl}, i.e., P_{Cl_2} is lowered.

From Eq. (3.82), if ΔH^0 does not depend on T, then ln $K_P \alpha \left(\frac{1}{T}\right)$.

3.6.5 The Effect of Pressure on the Equilibrium Constant

K_p, the equilibrium constant (Eq. (3.79), does not depend on P—because ΔG^0 is the difference between the Gibbs free energies of the pure products, each at unit pressure. By definition, it is independent of P.

The numbers of moles of species define the reaction equilibrium. Alternatively, it can be defined by the partial pressures of the species. The total pressure decides if the chemical reaction implies a change in the total number of moles. Once again the reaction $Cl_2 = 2Cl$ is scrutinized.

This reaction, when complete doubles of the number of moles available. The effect of a change in pressure results, once again, from the Le Chatèlier's principle.

At reaction equilibrium if the pressure is raised, then shift is towards that direction that lowers the pressure, i.e., that direction where the number of moles get lowered. Thus, if the pressure exerted on the Cl-Cl_2 system is raised the shift in the equilibrium will be towards the Cl_2 side.

That is, $K_P = \frac{P_{Cl}^2}{P_{Cl_2}}$. It is independent of pressure.

$$K_P = \frac{X_{Cl}^2 . P^2}{X_{Cl_2} . P} = \frac{X_{Cl}^2 . P}{X_{Cl_2}} = K_X \, P,$$

where K_X is the equilibrium constant expressed as mole fractions.

K_X is lowered to keep K_P constant if there is a rise in P. This lowering is accomplished when the reaction equilibrium moves to the Cl_2. side. Then, X_{Cl_2} rises and X_{Cl} is lowered.

In the reaction A + B = 2C, 2 moles of gas are present. Then, the equilibrium is not dependent on pressure and is given in mole fractions.

That is, $K_P = \frac{P_c^2}{P_A P_B} = \frac{X_c^2 \, P^2}{X_A P \, X_B P} = \frac{X_C^2}{X_A X_B} = K_X$.

If P ≠ 0 (Eq. (3.74)), then 2RT ln P (non-zero term) enhances or depresses Curve I (Fig. 3.14). The position of the minimum as a function of composition is not altered.

If this change in pressure affects the K_X value, the change of the number of molecules decides the equilibrium. For the general reaction, aA + bB = cC + dD

$$K_P = \frac{P_C^c\, P_D^d}{P_A^a\, P_B^b} = \frac{X_C^c\, X_D^d\, P^c P^d}{X_A^a X_B^b P^a P^b} = K_X.\text{P}(\text{c}+\text{d}-\text{a}-\text{b}).$$

That is, K_X is independent of total pressure P only if c + d – a – b = 0.

In the case of reaction A + B = 2C, Eq. (3.75) becomes $\Delta G^{'} = G^{'} - 2G_C^0 = n_A(-\Delta H^0)$.

$$\Delta G' = G' - 2G_C^0 = n_A(-\Delta H^0) + n_A(T\Delta S^0) + 2RT\left[n_A \ln\left(\frac{n_A}{2}\right) + (1-n_A)\ln(1-n_A)\right]$$

where $\Delta H^0 = -2500\ J$ and $\Delta S^0 = 5\ J/K$ (assumed values).

Or

$$\left[n_A(-\Delta H^0)\right] + T\left\{n_A \Delta S^0 + 2R\left[n_A \ln\left(\frac{n_A}{2}\right) + (1-n_A)\ln(1-n_A)\right]\right\}. \tag{3.83}$$

By minimizing $H^{'}$ at $n_A = 0$ and maximizing $T\Delta S^{'}$ at $n_A = 0.597$ the equilibrium is reached.

On the RHS the first term equals the enthalpy contribution to the change in Gibbs free energy.

$n_A\ \Delta S^0$ = entropy change due to a reaction.

$2R\left[n_A \ln\left(\frac{n_A}{2}\right) + (1-n_A)\ln(1-n_A)\right]$ is the entropy change because of gas mixing.

Then, the total entropy contribution becomes

Let $H^0 = -2500J,\ \ \Delta S^0 = 5\frac{J}{K}$.

If, as before, $G_A^0 + G_B^0 = 0,\ \ \Delta G' = G'$ (scale is displaced by $2G_c^0 = -5000$ J). Then, $\Delta G^{'}$ curve is identical to curve I of Fig. 3.14.

The minimum in the $\Delta G^{'}$ curve is a compromise between the minimum value of $H^{'}$ at $n_A = 0$ and maximum value of $\text{T}\Delta S^{'}$ at $n_A = 0.597$.

From Eq. (3.81), $[(\frac{ln K_P}{\partial T}) = \left(\frac{\Delta H^0}{RT}\right)]$—(since ΔH^0 is negative)—K_P decreases with increasing T.

$$\frac{\partial \ln K_P}{\partial T} = \frac{\Delta H^0}{RT^2};\ \ln K_P = \frac{-\Delta H^0}{RT} = \ln\left[\frac{(1-n_A)^2}{(n_A^2/4)}\right]$$
$$= \frac{2500}{8.314 \times 500}.\text{Solving for } n_A, n_A = 0.597.$$

3.7 Thermodynamics of Electrochemical Cells

All chemical reactions imply a change in the state of oxidation or "valence state" of some or all the participating atoms. "Valence state" (recall valency electrons) are not affected by the nature of the bonding atoms. For example, HF bonding is 50% ionic and 50% covalent. Still, H is +1 and F is −1.

Change in valence state means the removal or addition of electrons. Thermodynamic driving force of any reaction is related to the ease with which the required changes of valence of the participating atoms can occur. Example: $AO + B = BO + A$.

This reaction lowers the valence state of A from +2 to 0. The valence state of B rises from 0 to +2, i.e., two electrons are transferred from B to A.

Therefore, $A^{2+} + B = B^{2+} + A$. So, a free energy change implies electron transfer energetics. The equation of the reaction is the sum of $A^{2+} + 2^{'} = A$.

$B - 2^{-} = B^{2+}$. The electronic reaction is as given in Fig. 3.15.

Two electrons leave a B atom, travel from right to left along (a), reach A + AO mixture; convert an A^{2+} ion into A atom. At the same time, an O^{2-} ion leaves the A + AO mixture and via (b) reach the B + BO mixture. Thus the overall system charge neutrality is assured. So, the overall reaction is $A^{2+}O^{2-} + B = B^{2+}O^{2-} + A$, referred to an electrochemical reaction.

By keeping an external opposing voltage until no electric current flows, the electric voltage along (a), the driving force for electron transport, is known (at this point the electrochemical reaction will stop).

The relationship between the two kinds of forces permits ΔG for the reaction to be known.

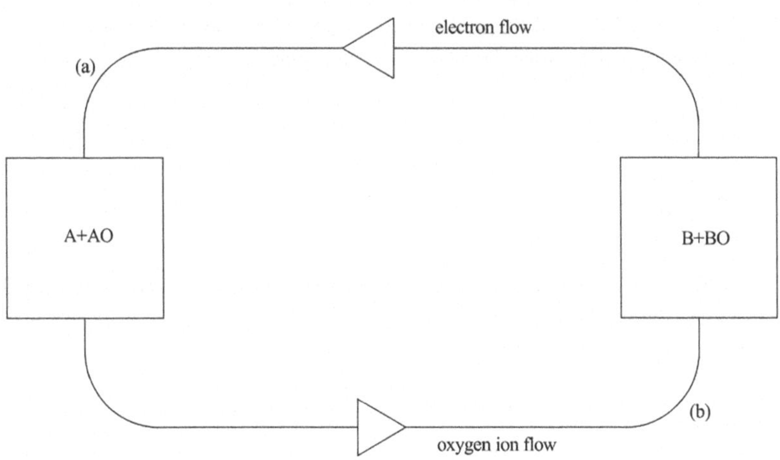

Fig. 3.15 An electrolytic reaction—A schematic representation [1]. (**a**) Only electrons can pass (electronic conductor). (**b**) Only oxygen ion can pass (ionic conductor)

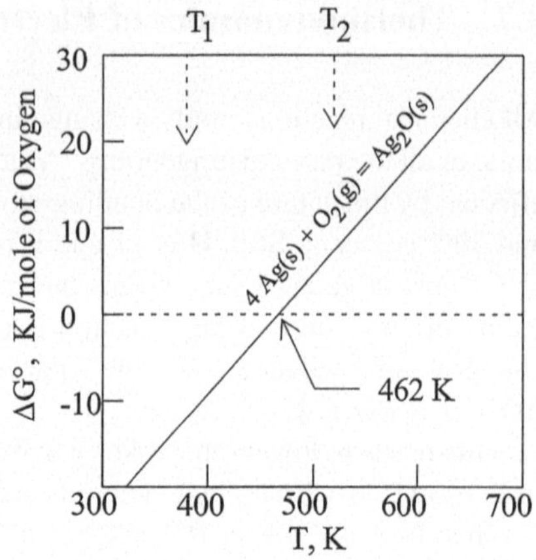

Fig. 3.16 The oxidation of silver is shown in an Ellingham diagram [1]

Normally, pure A, pure AO, pure B and pure BO are in thermodynamic equilibrium only at a unique temperature when the two Ellingham diagrams intersect. Here 4 phases can be at equilibrium at any temperature, when the chemical driving force equals the opposing electric driving force.

3.7.1 Ellingham Diagram

Ellingham experimentally determined the variation of ΔG^0 vs. T for the oxidation and sulfidation of a series of metals and found that a simple equation $\Delta G^0 = A + BT$ was valid (Fig. 3.16).

Standard enthalpy change independent of T is A. ΔH^0 and the constant B were equated with the negative of the temperature independent standard entropy change. (This is a linear approximation because the real equation involves T, T^2, T^{-1}….).

3.7.2 The Relation Between Chemical and Electrical Driving Forces

In a reversible process at constant T and P, the Gibbs free energy lowering is equal to w'_{max}, the work (apart from the work of expansion) done by the system. For a small increment,

$$-dG' = \delta w'_{max}.$$

The system performs electrical work, i.e., transfer across a vltage difference of an electric charge. The work performed = charge transported, q (coulombs) x electric potential difference, $\Delta\Phi$ (volts). Such work is measured in Joule.

A galvanic Cell converts a chemical reaction into electric work. Cell reaction describes the *overall chemical reaction*. The charge carried by 1 gram ion (Avagadro's number of ions) of unit positive charge is 96,487 Coulombs and is represented as F, Faraday's constant. Therefore, $\delta w' = ZF\Delta\phi dn$; dn indicates the given ions of valence Z transferred across a voltage difference $\Delta\Phi$ present between the cell electrodes.

For a reversible transport reaction the Electromotive force, EMF, ξ, of the cell, equals the electric potential difference between the electrodes.

Then

$$\delta w' = ZF\xi dn = -dG'. \tag{3.84}$$

For the transportation of 1 mole of ions, Eq. (3.84) becomes,

$$\Delta G = -ZF\xi. \tag{3.85}$$

This is the *Nernst equation effect of concentration (or activity) on EMF.*

Consider the reaction $Zn + CuSO_4 = Cu + ZnSO_4$. The direction depends on the states of the reactants and the products.

Take T = 298 K. Then, $Zn_{(S)} + CuSO_{4(\text{saturated aqueous solution})} = Cu_{(S)} + ZnSO_{4\,(\text{saturated aqueous solution})}$.

This is thermodynamically equivalent to $Zn_{(S)} + CuSO_{4(s)} = Cu_{(S)} + ZnSO_{4(s)}$.

At 298 K, $\Delta G = \Delta G^0_{298k} = -213{,}040$ J and thus Eq. (3.85) becomes,

$$\xi = -\frac{\Delta G^0}{ZF} = \frac{213{,}040}{2 \times 96{,}487} = 1 \cdot 104 \; volts,$$

where ξ, is the EMF of the cell. If the reactants and products are in their standard states it is the *Standard EMF of the cell.*

And so, Zn is dipped into a saturated aqueous solution of $ZnSO_4$ and Cu immersed into a saturated aqueous solution of $CuSO_4$. The externally applied voltage required to balance the chemical driving force of the cell reaction at 298 K is 1.104 volts.

For a general reaction, $aA + bB = cC + dD$, if the reactants and products are not in their standard states,

$$\Delta G = \Delta G^0 + RT \ln \frac{a_C^c a_D^d}{a_A^a a_B^b}. \tag{3.86}$$

From Eq. (3.85), the EMF of the cell is,

$$\xi = \xi_0 - \frac{RT}{ZF} \ln \frac{a_C^c a_D^d}{a_A^a a_B^b}.$$

Thus, with pure Zn and Cu in a Daniell Cell (one that converts chemical energy into electrical energy), the EMF is given as, $\xi = \xi_0 - \frac{RT}{ZF} \ln \frac{a_{ZnSO_4}}{a_{CuSO4}}$, which, at 298 K, gives, $\xi = 1.104 - 0.0296 \log \frac{a_{Znso_4}}{a_{CuSO4}}$.

For the EMF of the cell = 0, the activity quotient will be

$$10^{1.104/0.0296} = 1.98 \times 10^{37}.$$

If the $ZnSO_4$ solution is saturated, i.e., $a_{ZnSO_4} = 1$, the activity of $CuSO_4$ in the catholyte, with respect to the saturated solution as the standard case, will be 5×10^{-38}. In the external circuit between the electrodes equilibrium requires no biasing emf.

The diffusion of ions from one compartment to the next (an irreversible process) is caused by the concentration gradient that exists across the parent diaphragm. The concentration gradient causes a rise in potential, that is known as the liquid junction potential. The latter should be minimized by the use of a salt bridge (a junction connecting the anodic and cathodic compartments) or a galvanic cell (electrolytic solution) between the anolyte and the catholyte.

3.8 Thermodynamics of Point Defects in Solids

In the linear irreversible thermodynamics framework, the driving forces for fluxes and reactions are the chemical (electrochemical) potential gradients of the independent components of a non-equilibrium (reacting) system. Structure element (SE) = Reaction particles in the crystal. They do exist only inside the crystal lattice and so are not independent.

3.8.1 Components of a Crystal in a Thermodynamic Sense

The fluxes of more than one SE in the crystal are made from the flux of one independent chemical component. Local reactions between SE's could also be present. Thermodynamic derivations must be carefully proposed considering the crystal structure constraints symmetry.

Balance of chemical species, balance of sites and charges should be correctly allowed. Correct use of These balances should be correctly used to get the building elements.

Thermodynamics of point defects (=irregular SE's): The role of the point defects as carriers of elementary steps in mass transport should be carefully described.

3.8.2 *Thermodynamics of Crystals*

3.8.2.1 Point defect thermodynamics

A Phenomenological Approach

Let a crystal be in equilibrium with n chemical components (k = 1,2,...,n). At any given P and T, a Gibb's function, G, can be proposed as a homogenous function.

$$G = \sum_{k=1}^{n} \mu_k \cdot n_k,$$

where μ_k = Chemical potential, and n_k = number of moles of component k (already known).

The first and the second laws of thermodynamics suggest that

$$dG = \sum_{K=1}^{n} \mu_k, dn_k - SdT + VdP.$$

Therefore, at a given P and T,

$$\sum_{K=1}^{n} n_k.d\mu_k = 0; \quad \sum_{k=1}^{n} N_k \cdot d\mu_k = 0, \tag{3.87}$$

which is the Gibbs - Duhem equation.

It reduces the number of independent chemical potentials μ_k *from n to* $(n-1)$.

The chemical potential of component k is equal to the partial molar Gibbs energy (Eq. 3.86).

$$\mu_k = \left(\frac{\partial G}{\partial n_k}\right)_{n_j \neq n_k}. \tag{3.88}$$

From Eq. (3.85) for a nearly stoichiometric n component compound p.

$$\sum_{k=1}^{n} N_k \cdot \mu_k = G_p^0, \quad \sum_{k=1}^{n} N_k \ln a_k = \frac{\Delta G_p^0}{RT}.$$

Or

$$\Delta G_P^0 = G_P^0 - \sum_{k=1}^{n} N_k \cdot \mu_k^0 \tag{3.89}$$

where a_k = activity of k.

The small fractions of point defects, the fractions, N_k, are known as the number of lattice sites for component k in an ideal compound, also equal to the number of unit cells in the compound lattice.

Question Without breaking any structural constraints, let a single component k be placed in the crystal from an external reservoir. Then, the question is how the numbers of sublattice sites can be corelated?

Represent sublattice atoms (ions) of component k by x (x = 1,2,…,k). They follow the exchange reaction between the crystal sublattice and the external reservoir (= buffer beta), i.e.,

$$k(\beta) \Leftrightarrow k(x). \tag{3.90}$$

Since k(β) can be exchanged with sublattice x or λ, e.g., $k(x) \Leftrightarrow k(\lambda)$.

For equilibrium condition,

$$\mu_{k(\beta)} = \mu_{k(x)} = \mu_{k(\lambda)} (= \mu_k). \tag{3.91}$$

Therefore, from Eq. (3.88),

$$dG = \sum_{k=1}^{n} \mu_{k(x)} dn_k. \tag{3.92}$$

And since $dn_k = \sum_{x=1}^{K} dn_{k(x)}$

$$dG = \sum_{n=1}^{K} \sum_{=1}^{n} \mu_{k(x)} \cdot dn_{k(x)}. \tag{3.93}$$

Therefore,

$$\mu_{k(x)} = \left(\frac{\partial G}{\partial n_{k(x)}} \right)_{n_{l(\lambda)} \neq n_{k(x)}}. \tag{3.94}$$

This differentiation differs from the differentiation of Eq. (3.88) because it is more specific.

There are only two ways of adding a particle *k* on the sublattice x.

1. Sublattice x has vacancies, or
2. By adding particle k at a potential site while vacancies are formed simultaneously at other sublattices simultaneously so that the proportions of lattice sites in the various sublattices are retained.

The number of unit cells of the lattice is raised by 1 in the latter case.

Let Z_α be the number of sites in sublattice in α (α = x, λ,...). The ratio then is

$$\left(\frac{Z_x}{Z_\lambda}\right) = m_{x,k} \tag{3.95}$$

$m_{x,\lambda}$ is usually a rational number.

For k sub-lattices, (k−1) equations are present (e.g., Eq. (3.95). They form the structural constraints. The structural elements are symbolized by S_x^q. S can represent particles of components k or a vacancy V. In comparison with a perfect crystal q is the effective electrical charge.

Effective charges:

X'=neutral; $'$ $''$ $'''$ is singly, doubly and triply negatively charged SE; * ** *** is singly, doubly and triply positively charged SE.

Therefore, the exchange reaction in Eq. (3.90) between the buffer and the crystal becomes,

$$K(\beta) + v_x^X = k^X \tag{3.95}$$

or

$$K(\beta) = \left[k_x^X - v_x^X\right]. \tag{3.96}$$

Therefore, Eq. (3.91) becomes for equilibrium,

$$\mu_k = \mu_{k(x)} = \mu\left[K_x^X - V_x^X\right]. \tag{3.97}$$

$[k_x^X - V_x^X] = \text{k(x)}$ is called the "building element".

Virtual chemical potentials to individual SE's: Eq. (3.97) becomes,

$$\mu_k = \mu_{k(x)} = \mu_{k_x^X} - \mu_{V_k^X}. \tag{3.98}$$

Equation (3.93) becomes

$$\mathrm{d}G = \Sigma_{k=1}^{K} \Sigma_{k=1}^{n} \left(\mu_{k_x} X - \mu_{V_x} X\right).dn_{k(x)}. \tag{3.99}$$

Therefore, measurable chemical potential of the building element k(x) is,

$$\mu_{k(x)} = \mu_{k_x^x} - \mu_{V_x^x} = \left(\frac{\partial G}{\partial n_{k(x)}}\right) \text{ at constant } n_{k_\lambda}. \tag{3.100}$$

$$\lambda \neq x; \lambda = 1, 2, 3, \ldots., K.$$

Therefore, there is no experiment to determine the individual SE's chemical potential.

Point defect thermodynamics:

From Phase Rule, crystals containing only one component (A) become invariant by the predetermination of 2 thermodynamic variables, e.g., P & T.

When the vacancy mole fractions, N_V, are relatively small,

$$\mathrm{M_V} = \mathrm{M_V(P, T)} + \mathrm{RT\ ln\ N_V}; N_V = \left[\frac{e^{-(\mu_V^0)(A) - \mu_V^0(vac.)}}{RT}\right]. \tag{3.101}$$

Standard value of the vacancy chemical potential of a crystal only depends on T very faintly.

Therefore, $\mathrm{N_V}$ is roughly α $e^{\left(\frac{n}{T}\right)}$.

Interstitials also have this kind of Arrhenius dependence.

Solid electrolytes:

These are solids with predominantly ionic transference at least on a certain range of their component activities. Electronic conductivity is from electrons and electron holes.

All ionic constituents contribute to ionic conductivity.

A crystal is a solid electrolyte if,

$$c_e \cdot u_e + c_h \cdot u_h < < \sum |z_i| \cdot c_i \cdot u_i \left(= \sum |z_{p_i}| \left(c_{p_i} \cdot u_{p_i}\right)\right. . \tag{3.102}$$

c_i = concentration of (ionic) component i, u_i = electrochemical mobility of i (= V_i/$\underline{E}$); V_i is the velocity of particle; $\underline{E}$ is the electrical field vector; Z_i is the valence number of particles, i; p_i's are those point defects that render the corresponding ions mobile.

Ionic defects move much slower than electronic defects. Therefore, predominance of electrolytic conduction is present when $C_{el} << C_p$, i.e., ionic point defects must be the majority defects.

Ionic crystals are necessarily compounds.

Problems

1. In a binary solution of A and B, the molar fractions of A and B are 0.6 and 0.4, respectively. The activity of component A, is given by the expression $a_A = x_A$. Calculate the activity of component B, assuming ideal behavior

Solution:

For an ideal solution:

$a_A = x_A$

$a_B = x_B$

Thus, the activity of component B is **0.4**.

The Gibbs Free Energy of Formation of a Solution

2. In a binary solution at a constant temperature of 298 K, the Gibbs free energy of mixing, ΔGmix is –500 J/mol. Calculate the Gibbs free energy of formation of the solution, ΔGfif the pure components have standard Gibbs free energies of formation of 0 J/mol

Solution:

For ideal solutions, the Gibbs free energy of mixing is given by:

$\Delta G_{mix} = \Delta G_f \text{ (solution)} - (x_A \, \Delta G_f \text{ (pure A)}) + (x_B \, \Delta G_f \text{ (pure B)})$

Since the standard Gibbs free energies of formation for pure components A and B are 0 J/mol

$\Delta G_{mix} = \Delta G_f \text{ (solution)}$

Thus, ΔG_f (solution) = −500 J/mol

3. A solution of two components has a molar fraction of xA = 0.5 and xB = 0.5. If the number of available microstates for the mixture is 1000, calculate the entropy change of the solution using a statistical model

Solution:

$\Delta Smix = -R \, (x_A \ln x_A + x_B \ln x_B)$

Substiture the values:

$\Delta Smix = -8.314 \, (0.5 \ln 0.5 + 0.5 \ln 0.5)$

$= 8.314 \times 1.386$

$= 11.5$ J/K

Thus the entropy change of the solution is 11.5 J/K

4. In a two-phase system of water and ice at 0 °C, how many degrees of freedom exist for the system?

Solution:

The Gibbs phase rule is given by:

$F = C - P + 2$

Where:

- F = degrees of freedom
- C = number of components (water)
- P = number of phases (liquid and solid)

Substitute the values:

$F = 1 - 2 + 2 = 1$

Thus, the system has **1 degree of freedom**

5. For a binary solution at a fixed temperature, the Gibbs free energy of mixing is given by:

$\Delta G_{mix} = RT \, (x_A \ln x_A + x_B \ln x_B)$

If $x_A = 0.4$ and $x_B = 0.6$, calculate the Gibbs free energy of mixing at 300 K

Solution:

Substitute the values:

$$\begin{aligned}\Delta G_{mix} &= 8.314 \times 300 \, (0.4 \ln 0.4 + 0.6 \ln 0.6) \\ &= 2494.2 \times (-0.366) \\ &= -914.3 \text{ J/mol}\end{aligned}$$

Thus, the Gibbs free energy of mixing is **−914.3 J/mol**

6. The standard electrode potential for the reaction Cu^{2+} + 2e- Cu is E° = 0.34 V. Calculate the Gibbs free energy change for the reaction at 298 K

Solution:

$$\begin{aligned}\Delta G^\circ &= -nFE^\circ \\ \Delta G^\circ &= -(2)(96485)(0.34) \\ &= -65668.4 \text{ J/mol}\end{aligned}$$

Thus, the Gibbs free energy change is **−65668.4 J/mol**

(continued)

7. From the Third Law of Thermodynamics, we expect that the configurational entropy of a system in equilibrium goes to zero as the temperature goes to zero

Solution

First we write $G = G(T, \eta)$,

$$dG = \left(\frac{(\partial G}{\partial T}\right)_\eta dT + \left(\frac{(\partial G}{\partial \eta}\right)_T d\eta = 0$$

$$\frac{d\eta}{dT} = \frac{\left(\frac{\partial G}{\partial T}\right)_\eta}{\left(\frac{\partial G}{\partial \eta}\right)_T} = \frac{S}{\left(\frac{\partial G}{\partial \eta}\right)_T}$$

Since both terms go to zero in the quotient (at equilibrium), we need to apply L'Hopitals's rule:

$$\text{Lim of} \left(\frac{d\eta}{dT}\right) \text{ as } T \rightarrow 0 = \frac{\left(\frac{\partial s}{\partial \eta}\right)_\eta}{\left(\frac{\partial^2 G}{\partial \eta^2}\right)_T} = \frac{0}{} = 0$$

8. Copper and gold form the complete range of solid solution at temperatures between 410 °C and 889 °C, and, at 600 °C, the excess molar Gibbs free energy of formation of the solid solutions is given by

$G^{XS} = -28,280\ X_{Au}X_{Cu}J$

Calculate the partial pressures of Au and Cu exerted by the solid solution of $X_{Cu} = 0.6$ at 600 °C

The saturated vapor pressure of solid copper is given by

$\ln p^{\circ}_{Cu}\ (\text{atm}) = \frac{40,920}{T} - 0.86 \ln T + 21.67$

and the saturated vapor pressure of solid gold is given by

$\ln p^{\circ}_{Cu}\ (\text{atm}) = -\frac{45,650}{T} - 0.3066 \ln T + 10.81$

Solution

The solid solutions are regular with $\alpha = -28,280\ \text{J}\ 0.4^2$

$\ln \gamma_{Cu} = \frac{\alpha}{RT} X^2_{Au} = \frac{28,280 \times 0.4^2}{8.3144 \times 873} = -0.624$

Thus,

$\gamma_{Au} = 0.536$ and $a_{Cu} = \gamma_{Cu}X_{Cu} = 0.536 \times 0.6 = 0.322$

Similarly,

$\ln \gamma_{Cu} = \frac{28,280 \times 0.6^2}{8.3144 \times 873} = -1.403$

Thus,

$\gamma_{Au} = 0.246$ and $a_{Au} = 0.246 \times 0.4 = 0.098$

Therefore at 873 K,

$p^{\circ}_{Cu} = 3.35 \times 10 - 14$ atm

$p^{\circ}_{Au} = 1.52 \times 10 - 16$ atm

From Eq. (9.12), $ai = \frac{p_i}{p_i{}^0}$ and thus the partial pressure exerted by the alloy are

$p_{Cu} = 0.322 \times 3.35 \times 10^{-14} = 1.08 \times 10^{-14}$ atm

$p_{Au} = 0.098 \times 1.52 \times 10^{-16} = 1.50 \times 10^{-16}$ atm

9. One mole of solid Cr_2O_3 at 2500 K is dissolved in a large volume of a liquid Raoultian solution of Al_2O_3 and Cr_2O_3, in which $X_{Cr_2C_3} = 0.2$ and which is also at 2500 K. Calculate the changes in enthalpy and entropy caused by the addition.

The normal melting temperature of Cr_2O_3 is 2538 K, and it can be assumed that $\Delta S_{m,Al_2O_3} = \Delta S_{m,Cr_2O_3}$

Solution

When a **solid is added to a melt**, the overall process is:

Solid Cr_2O_3 → Liquid Cr_2O_3 (X = 0.2)

This happens in two steps:

1. **Fusion (melting)** of 1 mol. Cr_2O_3 at its melting point Tm = 2538 K,
2. **Mixing** into a solution of X = 0.2 at T = 2500

But, since the process occurs **at 2500 K**, and **melting point is 2538 K**, **undercooling** must be accounted for.

So, the enthalpy and entropy change is the **Gibbs energy change** for transferring solid Cr_2O_3 at 2500 K into the melt at the same temperature and composition

From standard thermodynamics of ideal solutions:

$\Delta G = \Delta G^0_{fusion} + RT \ln XCr_2O_3$

(continued)

	$\Delta G_{fusion} = \Delta H_m \left(1 - \frac{T}{T_m}\right)$ We will compute ΔG using: Let: ΔH_m: molar enthalpy of melting of Cr_2O_3 • $T = 2500\,K$ • $Tm = 2538\,K$ Also, since: $\Delta S_m = \frac{\Delta H_m}{T_m}$ We now write: $\Delta G = \Delta H_m \left(1 - \frac{T}{T_m}\right) + RTlnX$ We can then use this to compute entropy and enthalpy changes: $\Delta G = \Delta H_m + 0$ (mixing is ideal, enthalpy of mixing = 0) However, since it occurs at $\mathbf{T < T_m}$, we must use: $\Delta H = \Delta H_m \left(\frac{T}{T_m}\right)$ Because the actual heat absorbed during melting at $T < T_m$ is reduced. $\Delta H = \Delta S_m \times T$ So: Entropy change of the system has two components: $\Delta S = \underbrace{\Delta S_{fusion}}_{\Delta S_m \frac{T}{T_m}} + \underbrace{\Delta S_{mix}}_{RlnX}$ $= \Delta S_m \frac{T}{T_m} - RlnX.$ Let's proceed with: $\Delta S_m = 60$ J/mol.K Enthalpy change: $\Delta H = \Delta S_m \cdot T = 60 \times 2500$ $= 150,000\ J/mol$ Entropy Change: $\Delta S = 60 \times \frac{2500}{2538} - 8.314\ \ln\ (0.2)$ $\Delta S = 59.09 + 13.38 = 72.47\ J/mol.K$
10.	In a binary solution of A and B, the molar fractions of A and B are 0.6 and 0.4, respectively. The activity of component A, aA, is given by the expression aA = xA. Calculate the activity of component B, assuming ideal behavior **Solution:** For an ideal solution: aA = xA aB = xB Thus, the activity of component B is **0.4**. **The Gibbs Free Energy of Formation of a Solution**

(continued)

11. Calculate the heat required to form a liquid solution at 1356 K, starting with 1mole of Cu and 1 mole of Ag at 298 K. At 1356 K, the molar heat of mixing of liquid Cu and liquid Ag is given by $\Delta H^{m} = -20{,}590X_{cu}X_{Ag}$

Solution

Heat each metal from 298 K to its melting point

Metal	Melting point (K)	Specific heat (J/mol.K)
Cu	1356 K	24.5
Ag	1235 K	25.4

We'll assume approximate values of Cp (valid since small variation across this range)

Heat of fusion to melt each metal

Metal	Heat of fusion (ΔH, KJ/mol
Cu	13.1
Ag	11.3

Cu is already at 1356 K after melting

Ag melts at 1235 K, so must be heated further from 1235 K → 1356 K

The given molar enthalpy of mixing:

$\Delta H^{m} = -20590X_{cu}X_{Ag}$

For 1 mole Cu and 1 mole Ag,

$$X_{cu} = X_{Ag} = \frac{1}{2}$$

$\Delta H^{m} = -20590X_{cu}X_{Ag}$

$= -5147.5$ J/mol of solution

Heating solids to melting points

Cu (298 → 1356 K):

$q_1 = 1mol.24.5\frac{J}{mol.K} \times (1356 - 298) = 24.5 \times 1058 = 25{,}921\ J$

Ag (298 → 1235 K)

$q_2 = 1mol \times 25.4\frac{J}{mol.K} \times (1235 - 298) = 25.4 \times 937 = 23{,}810J$

Melting

$\text{Cu} : q_3 = 13.1\,\text{kJ} = 13{,}100\text{J}$

$\text{Ag} : q_4 = 11.3\,\text{kJ} = 11{,}300\text{J}$

Heating Ag (liquid) from 1235 → 1356 K

Assume

$Cp \approx 30$ J/mol. k for liquid Ag

$q_5 = 1 \times 30 \times (1356 - 1235) = 30 \times 121 = 3{,}630J$

Enthalpy of Mixing

$q_6 = \Delta H^{m} = -5147.5J$

Total Heat Input

$Q = q_1 + q_2 + q_3 + q_4 + q_5 + q_6$

$=72{,}613.5$ J

12. Use the Gibbs–Duhem equation to show that, if the activity coefficients of the components of a binary solution can be expressed as

$\ln\gamma_A = \alpha_1 X_B + \frac{1}{2}\alpha_2 X_B^2 + \frac{1}{3}\alpha_3 X_B^3 + \ldots.$

(continued)

And

$ln\ \gamma_B = \beta_1 X_A + \frac{1}{2}\beta_2 X_A^2 + \frac{1}{3}\alpha_3 X_A^3 + \ldots$

Over the entire range of composition, then $\alpha_1 = \beta_1 = 0$, *and that, if the variation can be represneted by the quadrtic terms alone, then* $\alpha_2 = \beta_2$

Solution

We are given the expressions for the **activity coefficients** of components A and B in a **binary solution** and are asked to use the **Gibbs–Duhem equation** to show:

1. That $\alpha_1 = \beta_1 = 0$
2. If the expansion is truncated at the quadratic term, then $\alpha_2 = \beta_2$

Given

$\ln \gamma_A = \alpha_1 X_B + \frac{1}{2}\alpha_2 X_B^2 + \frac{1}{3}\alpha_3 X_B^3 + \ldots.$ (1)

$ln\ \gamma_B = \beta_1 X_A + \frac{1}{2}\beta_2 X_A^2 + \frac{1}{3}\alpha_3 X_A^3 + \ldots$ (2)

And since it's a binary solution:

$X_A + X_B = 1 => dX_A = -dX_B$

The Gibbs-Duhem equation for a binary solution at constant T and P:

$X_A d(\ln\gamma_A) + X_B d(\ln\gamma_B) = 0$ (3)

We now substitute the expressions for $\ln\gamma_A$ and $\ln\gamma_B$ differentiate them, and substitute into Eq. (3)

Differentiate $\ln\gamma_A$ with respect to X_B

$\frac{d(ln\gamma_A)}{dX_B} = \alpha_1 + \alpha_2 X_B + \alpha_3 X_B^2 + \ldots$ (4)

Differtiate ($\ln\gamma_B$) with respect to X_B,using X_A=1-X_B

$ln\gamma_B = \beta_1(1 - X_B) + \frac{1}{2}\beta_2(1 - X_B)^2 + \frac{1}{3}\beta_3(1 - X_B)^3 + \ldots$

Now differentiate ($\ln\gamma_B$) with respect to X_B

$\frac{dln\gamma_B}{dX_B} = -\beta_1 - \beta_2(1 - X_B) - \beta_3(1 - X_B)^2 - \ldots.$ (5)

Now substitute eqs. (4) and (5) into (3)

$X_A\left(\frac{dln\gamma_A}{dX_B}\right) + X_B\left(\frac{dln\gamma_B}{dX_B}\right) = 0$

Substitute $X_A = 1 - X_B$

$(1 - X_B)(\alpha_1 + \alpha_2 X_B + \alpha_3 X_B^2 + \ldots) + X_B\left(-\beta_1 - \beta_2(1 - X_B) - \beta_3(1 - X_B)^2 - \ldots\right)$ (6)

Now simplify this equation and group terms by powers of X_B

$(1 - X_B)(\alpha_1 + \alpha_2 X_B + \alpha_3 X_B^2 + \ldots) = \alpha_1(1 - X_B) + \alpha_2 X_B(1 - X_B) + \alpha_3 X_B^2(1 - X_B) + \ldots.$

Expand the second term

$X_B\left(-\beta_1 - \beta_2(1 - X_B) - \beta_3(1 - X_B)^2 - \cdots.. = -\beta_1 X_B - \beta_2 X_B(1 - X_B) - \beta_3 X_B(1 - 2X_B + X_B^2) - \cdots\right.$

Now, collect all terms and group powers of X_B : *constant, liner, quadratic, etc.*

After collecting the coefficient of X_B^0(i.e., constant term) is: α_1

The coefficient of X_B^1 is: $-\alpha_1 + \alpha_2 - \beta_1$

To satisfy the identity for all values of X_B, each coefficient must be zero:

$\alpha_1 = 0\ (constant\ term)$

$-\alpha_1 + \alpha_2 - \beta_1 = 0$

$\alpha_2 = \beta_1$

If expansion is truncated at quadratic terms, then:

$$ln\,\gamma_A = \frac{1}{2}\alpha_2 X_B^2$$

$$ln\,\gamma_B = \frac{1}{2}\beta_2 X_A^2$$

From Gibbs-Duhem equation, differentiating and comparing:

$X_A.\alpha_2 X_B = -X_B.\beta_2 X_A$

$\alpha_2 = \beta_2$

Reference

1. Gaskell, D. R. (2009). *Introduction to the thermodynamics of materials*. Taylor and Francis.

Chapter 4
Some Thermodynamic Equations Needed for Kinetics of Materials

Contents

4.1 Heterogeneous Reaction Kinetics

4.1.1 Transport across Phase Boundaries

Atomic dimensions define the smallest width of an interface. Therefore, their kinetics is confined to a narrow space. So, relative to the behavior in the bulk changes across interfaces in (thermodynamic) potential are large. So the transport rate is far in excess.

Potential change across an interface can be determined experimentally easily. It is a major task to keep the potential probes very close to interface. Although a comparable problem exists in electrode kinetics, it is easier to handle as fluids are involved.

All the particle fluxes should be zero if bulk α is in equilibrium. The fluxes involved are:

K. A. Padmanabhan et al., *Thermodynamics and Kinetics of Materials*, Engineering Materials, https://doi.org/10.1007/978-3-032-14527-7_4

$$j_i^\alpha = c_i^\alpha . b_i^\alpha \left[\sum_n \nabla \beta_{i,n}^\alpha + \sum_{k \neq i} \beta_{i,k}^\alpha . j_k^\alpha \right] \tag{4.1}$$

where $c_i^\alpha . b_i^\alpha$ is the transport coefficient of i in α and b is the mobility.

The summation of the action of n potential gradients appears as the first term on the RHS. The friction between the i particles appears as the second term. j_k; $\beta_{i,k}$ are the friction (cross) coefficients.

At equilibrium individually the fluxes j_k disappear. This necessitates that,

$$j_i^\alpha = c_i^\alpha . b_i^\alpha \Sigma \Delta p_{i,n}^\alpha = 0,$$

∇p_i^α is the potential of component i in α. That is,

$$\sum_n \nabla p_{i,n}^\alpha = 0, \tag{4.2}$$

The gradients, $\nabla p_{i,n}^\alpha$, individually disappear when the potentials are independent.

This stipulates that the extended equilibrium phases are homogeneous.

Assume that there are two phases, α + β. Let $\Delta\xi^b$, be the thickness of the interface phase.

Then, for equilibrium,

$$\frac{1}{\Delta\xi^b} \cdot \sum \Delta p_{i,n}^b = 0; i.e., \sum_n \Delta p_{i,n}^b = 0. \tag{4.3}$$

Δ stands for thejump across boundary, b. Of the several chemical species, at least one $\Delta p_{i,n}^b$ term is $\Delta\mu_i^b \left(= \mu_i^\alpha - \mu_i^\beta \right)$. Electric, elastic, etc. potential differences could also exist. For interface equilibrium involving charge paticles i, the interface is in equilibrium when,

$$\Delta\mu_i^b + \Delta\varphi_i^b = 0, \tag{4.4}$$

where $\varphi_i = z_i. F. \varphi$.

$\Delta\mu_i^b$ is ~1 eV (due to chemical equilibrium differences). The necessary electric field can be built with a very small amount of charge considering the number of "*i*" atomic paticles involved (even in the 2D) system. And so, Eq. (4.4) is equivalent to $\Delta\eta_i^b = 0$; η_i = electrochemical potential.

In a heterogeneous system, $\Delta\mu_i^b$ exists across the interface. According to Eq. (4.4) at equilibrium an electrical potential drop $\Delta\varphi^b$ is always present across a phase boundary.

Let (n) charged species i be in equilibrium across a phase boundary. For a complete solution, the set of Eq. (4.4) should be simultaneously solved for

$i = 1,2,\ldots,n$. This will ensure that the electroneutral components are constant across the interface. The charge distribution in space will be treated independently.

4.2 Four Basic Kinetic Situations

4.2.1 *Homogeneous Reactions: Point Defect Relaxation—An Example*

In binary ionic crystal, AgBr, Frnkel defects are formed.

Due to thermal activation, regular lattce sites are left by, Ag_{Ag}, the silver ions. They then enter interstitial (i) sites (Ag^{*}) and leave vacancies in the silver lattice sublattice.

Frenkel 2D defect pairs in AgBr are presented in Fig. 4.1 below.

$$Ag_{Ag} + V_i = Ag_i^{\cdot} + V'_{Ag};$$

a vacant cation site is shown.

The defect concentration is fixed at equilibrium, which is influenced by T or P, which takes one to a new equilibrium.

Defect relaxation causes equilibration and the chemical reaction involving atomic structural elements is,

$$Ag_{Ag} + V_i = Ag_i^{\cdot} + V'_{Ag}. \tag{4.5}$$

Here, matter, sites and charges are balanced.

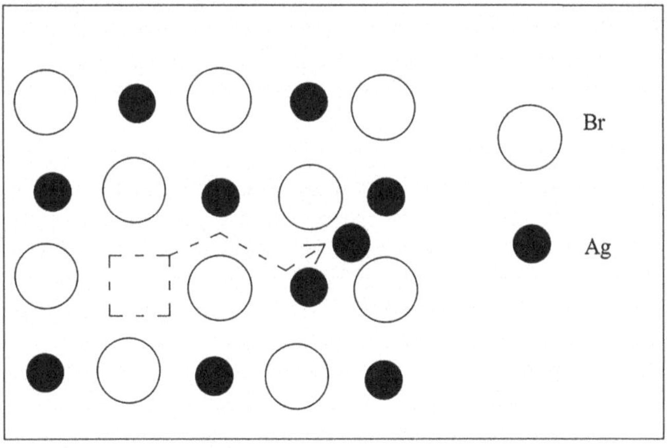

Fig. 4.1 Frenkel defect pairs in AgBr [1]

The time derivative of concentration considers the site exchange of frequency, ν, between a vacancy and an ion on a different sublattice and is given by,

$$c_i = \overline{\nu}.c_{Ag}.N_{V_i} - \overline{\nu}.c_i.N_V \tag{4.6}$$

Expressed in mole fraction,

$$\dot{N}_i = +\overline{\nu}.(N_{Ag}.N_{V_i}) - \overline{\nu}(N_i.N_V) \tag{4.7}$$

On the RHS each product represents the average fraction of silver ions that get exchanged with a neighbor vacancy. The Site and charge balances are given by,

$$N_{Ag} + N_V = 1; N_{V_i} + N_i = 1; N_i = N_V. \tag{4.8}$$

Since, N_V, N_i << 1, from Eq. (4.3)

$$\dot{N}_i = \overline{\nu}.\left(1 - \frac{\overleftarrow{\nu}}{\overrightarrow{\nu}}.N_i.N_V\right). \tag{4.9}$$

At equilibrium, $\dot{N}_i = 0$ *and* $N_i(eq.) = N_V(eq.) = \frac{1}{(N_0^2)}$, and therefore,

The ratio $\left(\overleftarrow{\nu}/\overrightarrow{\nu}\right) = \frac{1}{(N^0)^2}$, and

$$\dot{N}_i = \overrightarrow{\nu}.\left(1 - \frac{N_i}{N^0}.\frac{N_V}{N^0}\right). \tag{4.10}$$

The defect fraction to the equilibrium value, called the reference state, is obtained by keeping $N_i = N_V = (N + \delta)$.

Then, Eq. (4.10) becomes,

$$\delta = \frac{2 \cdot \overrightarrow{\nu}}{N^0}\left(\delta + \frac{\delta^2}{N^0}\right). \tag{4.11}$$

For sufficiently long times (i.e., $\delta \to 0$), integration of Eq. (4.11) leads to

$$\delta(t) = \delta(0)e^{-(2\overrightarrow{\nu}t)/N^0}. \tag{4.12}$$

Therefore,

$$\tau_R = \frac{N^0}{\left(2\overrightarrow{\nu}\right)} = \left(\frac{1}{\left(2N^0\overleftarrow{\nu}\right)}\right), \tag{4.13}$$

gives Frenkel relaxation time.

4.2.2 Point Defects in a Binary Compound in Steady State Flux

Gibbs Phase Rule: Three variables, e.g., *P,T,* μ_k, (*k* is a component index)define local equilibrium.

So, on the opposite faces of a compound crystal, at constant *P* and *T*, if μ_k gets fixed, on opposite surfaces of a compound crystal, e.g. AX (a metal oxide), at two different levels, all (local) equilibrium functions of state reach different values.

As point defect concentrations are functions of state, on either side of the crystal surface different pint concentrations exist. Till a steady state is reached in the crystal lattice, mobile point defects move down their concentration gradient.

Drifting cation vacancies is accompanied in the opposite direction by a cation counter-flux.

Thermal motion, which is random, is ignored. Immobility is attributed to anions.

In a linear theory, the flux (if there are no other restrictions) of, e.g., vacancies of A in the compound AX is given by,

$$j_V = C_V \cdot v_V = C_V \cdot \left(b_V \cdot \underline{K_V}\right), \tag{4.14}$$

From irreversible thermodynamics,

$\underline{K_V} = -\nabla\mu_V (= -RT\nabla\, ln\, N_V)$, when vacancies do not interact and their concentrations are small. Insert $\nabla\mu_V$ in Eq. (4.14) and denote $b_V RT = D_V$. Then, Fick's First Law of diffusion results. D_V is the vacancy diffusion coefficient and b and D are related by the Nernst-Einstein relation. For non-interacting, vacancies of dilute concentrations, D_V is a constant.

From Eq. (4.14),

$$j_V = -D_V.\nabla C_V = -D_V.\frac{C''_V - C'_V}{\Delta\xi} \tag{4.15}$$

where $\Delta\xi$ is the sample thickness.

Equation (4.15) is the steady state equation. Then, ΔC_V is constant and independent of $\Delta\xi$.

Since $\mu_A(\xi')$ and $\mu_A(\xi'')$ are fixed at the corresponding surfaces, μ_V can be presented using μ_A $\left(or\ \mu_{X_2}\right)$. This results from the A vacancy formation reaction in equilibrium,

$$A_A = A + \left(V'_A + h^{\cdot}\right) \; or \; \left(A_A - V'_A\right) = \left(A + h^{\cdot}\right). \tag{4.16}$$

where A_A is the (regular) structure element and A is the chemical component A. $h^{\cdot}$ stands for an electron hole, formed to maintain electroneutrality ($N_{V'} = N_{h^{\cdot}}$).

From the site balance, we know that $N_{A_A} \cong 1$. From equilibrium condition (Eq. (4.16)) the chemical potential gradient of component A is (in view of $\nabla\mu_h = \nabla\mu_V$ due to $N_{V'} = N_{h^{\cdot}} \ll 1$)

$$\nabla\mu_A = -2 \cdot \nabla\mu_V = -2 \cdot RT \cdot \nabla \ln N_V. \tag{4.17}$$

Integrating Eq. (4.17) across the crystal, one obtains,

$$C'_V = C''_V \cdot e^{-\left[\frac{\left(\mu'_A - \mu''_A\right)}{2RT}\right]}. \tag{4.18}$$

where the primes symbolize the two opposite surfaces. Substituting Eq. (4.18) into Eq. (4.15) yields,

$$j_V = -\frac{D_V \cdot C''_V}{\Delta\xi}\left(1 - e^{-\left[\left(\mu'_A - \mu''_A\right)/2RT\right]}\right), \tag{4.19}$$

or

$$j_V \cong \frac{D_V \cdot C''_V}{\Delta\xi}; \mu'_A - \mu''_A > > RT. \tag{4.20}$$

The difference of the component potentials ($\mu'_A - \mu''_A$) is found in 2 ways.

1. Use component A reservoirs with different activities.
2. Use component $X\left(\frac{1}{2}X_2\right)$ reservoirs of different activities (partial pressures).

As $A + \frac{1}{2}X_2 = AX$,

$$\mu_A + \frac{1}{2}\mu_{X_2} = \mu^0_{AX} \; or \; d\mu_A + \frac{1}{2}d\mu_{X_2} = 0.$$

The cation vacancy (and cation) flux deal with only the cation sublattice. The anions are assumed to be immobile and so are considered in the transport reaction. In case 1, the AX crystal is stationary and $\Delta\mu_A$ of the A reservoir drives the fluxes.

Vacancy flux in the direction of the oxidizing surface with the higher μ_{x_2} shifts the entire AX crystal (case 2). This is known when Eq. (4.16) is applied to a surface equilibrium. That is,

$$\frac{1}{2}X_2 = [X_X + V_A'] + h^{\cdot}. \tag{4.21}$$

From equilibrium condition of Eq. (4.21), one derives,

$$\frac{1}{2}\nabla\mu_{X_2} = 2 \cdot RT.\nabla\, ln\, N_V \tag{4.22}$$

which is in accordance with Eq. (4.17).

Equation (4.21) reveals that on the oxidizing side, the defect combination, $[X_A + V_A']$ is added to the crystal (see the elements with the parenthesis in Eq. (4.21)). An opposite reaction is present in the reducing side.

$[X_A + V_A']$ represents a "lattice molecule". It is deducted from the crystal surface. For every vacancy one lattice molecule traverses across the crystal from ξ' to ξ''.

The different component activities at ξ' and ξ'' impose a defect in the AX crystal. This introduces an elastic inhomogeneity in the crystal.

Elastic stresses influence. The chemical potential μ_V affects the elastic stresses and thus provides a driving force for the flux, i.e., flow caused by a gradient caused by a stress, which is ignored.

4.2.3 The Kinetics of an Interface Reaction

Phases α and β, that are separated by interfaces, experience in the form of (1) stationary interfaces facing fluxes of components across them, or (2) a chemical reaction moves the interface between α and β at the phase boundary.

Catalytic reactions are not considered here.

Kinetic description of interfaces: The equilibrium state is the reference state. Across an interface components k net fluxes are zero. Component k in equilibrium experiences zero driving forces (as the mobilities at these interfaces are infinite).

For crystals at constant T and P but containing electrically charged structure elements, this implies that $\Delta\eta_i = 0$ (i is the reversible (charged) carrier of type i).

Then the equilibrium condition becomes,

$$-Z_i \cdot F \cdot \Delta\varnothing^0 = \Delta\mu_i \tag{4.23}$$

i.e., at equilibrium an abrupt change is present in electrical potential across an interface.

For each charged component i in equilibrium ($\Delta\eta_i = 0; i = 0, 1, 2, 3, \ldots$) electrical neutrality ensures that the electroneutral components are also in equilibrium. In a linear theory, when equilibrium across an interface does not exist ($\Delta\eta_i$ is not equal to zero), a flux of charged species i is present. That is,

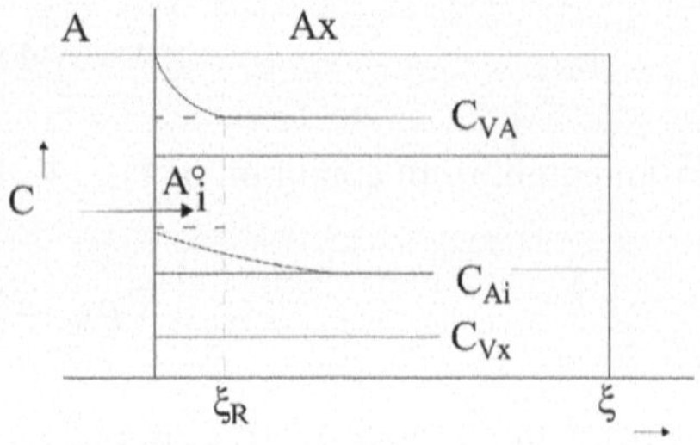

Fig. 4.2 Variations in concentration of Point Defects and Reactants Across a Moving Interface (ξ)

$$j_i = -l_i \cdot \Delta\eta_i. \tag{4.24}$$

Here l_i is the interface "conductivity" of species i. In the kinetic theory of interfaces, the meaning of $l_i(\alpha/\beta)$ is not easily comprehended.

Figure 4.2 presents a simple model.

Flow of A cations can be present across the boundary ($\xi = 0$) into a semi-conducting compound AX. Relaxation of point defect between $0 < \xi < \xi_R$ causes $V'_A + A_i^0 = A_A$; ξ_R is equal to the width of relaxation zone.

Let the Schottky disordered AX crystal be in contact with (anode) metal A. When equivalent fractions of cation and anion vacancies exist, "Schottky disorder" is present.

Assume

$$D_i < < D_V \left(i = A_i^{\cdot}, V = V'_A\right).$$

If an ion (electric) flux is injected into the interstitial sublattice under load, i.e., $j_V(\xi = 0) = 0$, the defects flux has a diffusive and a field term, viz.,

$$j_V = -D_V \cdot \left(\nabla C_V - \frac{F \cdot C_V}{RT} \cdot \nabla_\phi\right). \tag{4.25}$$

Therefore, at $\xi = 0$,

$$C_V = C_V^0 e^{(F/RT)(\phi - \phi^0)} = C_V^0 \cdot e^{\left(\frac{F}{RT}\right) \cdot \Delta U}, \tag{4.26}$$

where ΔU is the charge in the interfacial voltage drop relative to its equilibrium value. And $C_V^0 = C_V(eq)$ at $\xi = \infty$.

For interstitials i, in analogy with Eq. (4.25),

$$j_i = D_i \cdot \left(\nabla C_i + \frac{F \cdot C_i}{RT} \nabla \phi\right) \tag{4.27}$$

$$= -D_i\left(\nabla C_i + \frac{C_i}{C_V}\nabla C_V\right). \tag{4.28}$$

The second part of Eq. (4.27) follows from Eq. (4.26).

From the requirement of electrical neutrality and the definitions of a (linearized) defect recombination zone of width ξ_R, Eqs. (4.27) and (4.28) lead to,

$$j_i(0) = \frac{-2D_i \cdot \Delta C_V}{\xi_R} = -\frac{2 \cdot D_i C_V^0}{\xi_R} \cdot \left(e^{F/RT \cdot \Delta U} - 1\right) \tag{4.29}$$

and $Z_i \cdot F \cdot j_i(0)$ is the steady state electrical current across the interface, driven by the applied voltage ΔU.

If we set $\xi_R = \sqrt{2\tau_R \cdot D_i}$, where τ_R = relaxation time of Schottky defects for attaining equilibrium, Eq. (4.6) yields for $DU < < \left(\frac{RT}{F}\right)$,

$$j_i(0) = -\frac{2D_i \cdot C_V^0}{\sqrt{2 \cdot \tau_R \cdot D_i}} \cdot \frac{F \cdot \Delta U}{RT} = j_V^0(\xi > \xi_R). \tag{4.30}$$

This general methodology, leads to the kinetic equation for a simple A/AX interface model. The electrodes in electrochemistry behave similar to these A/AX interface.

4.2.4 *Kinetics of Compound Formation:* A + B = AB

Local equilibrium at the A/AB and AB/B interfaces of the A/AB/B reaction couple is considered here (Fig. 4.3).

While drawing the above two diagrams, the following assumptions have been made:

1. Reaction geometry is linear, and
2. Interfaces between the reactants and the product AB are planar.

Increase in layer thickness,

$$\Delta\xi = V_{AB} \cdot (|j_A| + |j_B|), \tag{4.31}$$

Here, V_{AB} is the molar volume.

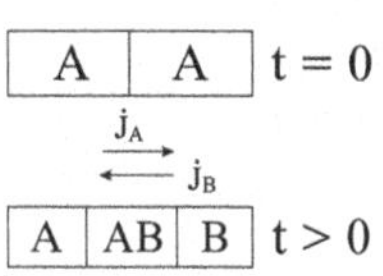

Fig. 4.3 Schematic of Interdiffusion Leading to AB Phase Formation Between A and B

Each flux of A and B can be replaced by the product of a transport coefficient (L_i) and a driving force (X_i). For example,

$$j_A = -L_A \cdot \nabla \mu_A = -\overline{L}_A \cdot \frac{\Delta \mu_A}{\Delta \xi}. \tag{4.32}$$

The RHS results from integration. When local equilibrium is present, the average value, $\overline{L}_A$, of the transport coefficient, across the reaction layer, is obtained from the thermodynamic parameters at the interfaces A/AB and AB/B. Therefore, it does not depend on the reaction layer thickness $\Delta\xi$.

Insert Eq. (4.32) in Eq. (4.31). A parabolic rate law results as,

$$\Delta G(t) = \sqrt{2 \cdot V_{AB} \cdot (L_A \cdot \Delta \mu_A + L_B \cdot \Delta \mu_B)}.\sqrt{t}. \tag{4.32a}$$

Since $\Delta \mu_A = \Delta \mu_B = \Delta G^0_{AB}$, one finally has,

$$\Delta \xi(t) = \sqrt{2 \cdot V_{AB} \cdot \Delta G^{\circ}_{AB} \cdot (L_A + L_B)} \cdot \sqrt{t}. \tag{4.33}$$

The increase in $\Delta\xi$ will occur at interface A/AB if $L_A/L_B << 1$ and it will occur at AB/B if $L_A \gg L_B$.

The parabolic rate laws in heterogeneous solid-state reactions result from two conditions: (a) a linear geometry should be present, and (b) phase boundaries exist, along with local equilibrium.

In ionic crystals, a flux coupling due to the conditions of electro-neutrality is present. That is,

$$j_A + \alpha j_B = 0. \tag{4.34}$$

From Eq. (4.32) and Eq. (4.34), again one concludes that the reaction kinetics is parabolic. However, the rate constants for Eq. (4.33) for this case is different.

The above 4 are the basic kinetic concepts. They are the starting point of more complicated kinetic problems.

4.3 Kirkendall Effect

Differential transport is an inhomogeneous single sublattice crystal occupied with components A and B as well as vacancies is considered here.

Kirkendall shift X_k of weld plane (dashed) is as shown because A diffuses more quickly in B than B does in A. Pores are created on the side of the more rapidly diffusing component, while on the other side the material 'swells'.

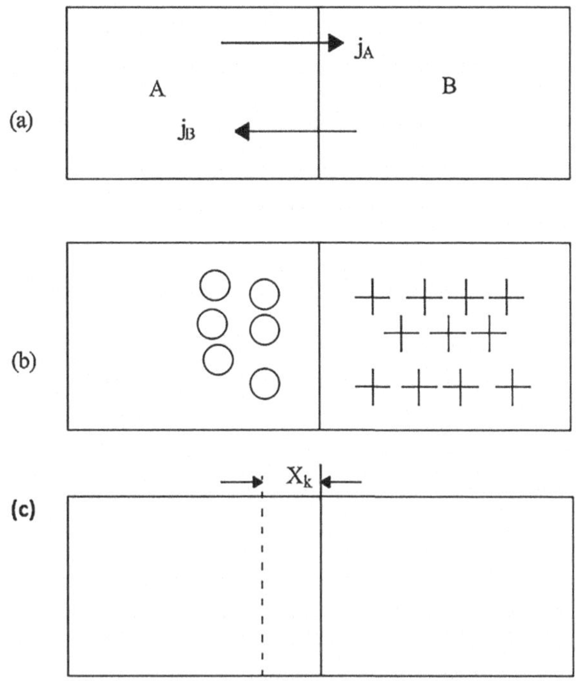

Fig. 4.4 **(a)** Diffusion fluxes of species A and B across the interface. **(b)** Schematic representation of inert markers (circles) initially placed at the interface; lattice sites indicated by plus signs. **(c)** Displacement of the Kirkendall marker plane by a distance X_k, due to unequal diffusion rates of A and B

The two starting materials are welded together so that small oxide inclusions or small wires of an inert high melting point metal mark the position of the weld plane permanently.

If the two components diffuse at different rates, see Fig. 4.4. Then, after an anneal, extra material will be found on the side of the slower component measured from the weld plane.

The material, therefore, "swells" in all 3 dimensions.

On the other side, more than the equilibrium concentration of vacancies are created as a result of loss of the more rapidly diffusing component. These extra vacancies coagulate to form pores and the material shrinks.

If the whole specimen is held externally, the weld plane markings move in the opposite direction.

Kirkendall effect is the name given to this phenomenon.

The system of reference is referred to, after L. Darken, as Darken's notation.

The $\tilde{x}$ coordinates are measured from the left-hand end of the diffusion couple and the x-coordinates from the weld plane.

The flux of B through the weld plane in the x- coordinate system is,

$$j_B(x=0,t) = -\tilde{D}_B \frac{\partial C_B}{\partial x}\bigg|_{x=0} = \tilde{D}_B \frac{\partial C_B}{\partial \tilde{x}}\bigg|_{\tilde{x}=L(t)}. \tag{4.35}$$

In the $\tilde{x}$ system an additional transport term is involved.

$$j_B(\tilde{x}=L,t) = -\tilde{D}_B \frac{\partial C_B}{\partial \tilde{x}}\bigg|_{\tilde{x}=L} + vC_B|_{\tilde{x}=L}. \tag{4.36}$$

v is the velocity of the weld plane at the point $\tilde{x}=L(t)$ with respect to the left-hand end of the specimen.

Similar equations can be written for component A.

The continuity equation for the total density of atoms is,

$$\frac{\partial C}{\partial t} = \frac{\partial C_B}{\partial t} + \frac{\partial C_A}{\partial t} = -\left(\frac{\partial j_B}{\partial \tilde{x}} + \frac{\partial j_A}{\partial \tilde{x}}\right) = \frac{\partial}{\partial \tilde{x}}\left\{\tilde{D}_A \frac{\partial C_A}{\partial \tilde{x}} + \tilde{D}_B \frac{\partial C_B}{\partial \tilde{x}} - Cv\right\}. \tag{4.37}$$

If the total density remains constant, $\left(\frac{\partial c}{\partial t}\right) = 0$, and the expression inside the curly brackets is a constant, it must be zero since at left hand end of the specimen, $\tilde{x}=0$ and the concentration gradients and v vanish. Thus, the speed of the weld plane marking (at $\tilde{x}=L$) is given by,

$$v = \frac{1}{c}\left(\tilde{D}_A \frac{\partial C_A}{\partial \tilde{x}} + \tilde{D}_B \frac{\partial C_B}{\partial \tilde{x}}\right) = \left(\tilde{D}_A - \tilde{D}_B\right)\frac{\partial v_A}{\partial \tilde{x}}. \tag{4.38}$$

Taking into account that C = constant, $\frac{\partial C_A}{\partial \tilde{x}} = -\frac{\partial C_B}{\partial \tilde{x}}$ *and* $v_i = \frac{C_i}{C}$.

Substituting Eq. (4.38) in Eq. (4.36) for B, an equation is obtained of the form of Fick's second law.

$$\frac{\partial C_B}{\partial t} = \frac{\partial}{\partial \tilde{x}}\left(\tilde{D}_B \frac{\partial C_B}{\partial \tilde{x}} - \tilde{D}_A \frac{C_B}{C}\frac{\partial C_A}{\partial \tilde{x}} - \tilde{D}_B \frac{C_B}{C}\frac{\partial C_B}{\partial \tilde{x}}\right) = \frac{\partial}{\partial \tilde{x}} \times \left[\left(\frac{\tilde{D}_B C_A + \tilde{D}_A C_B}{C}\right)\left(\frac{\partial C_B}{\partial \tilde{x}}\right)\right]. \tag{4.39}$$

$$\tilde{D} = v_A \tilde{D}_B + v_B \tilde{D}_A. \text{(known as Mantano evaluation)} \tag{4.40}$$

Equations (4.38) and (4.40) are known as the Darken equations.

Using them it is possible to calculate $\tilde{D}_A, \tilde{D}_B$ separately after having measured $\tilde{D}$ *and* v.

For example, in the case of a Cu—22 (at. %) alloy at 785 °C,

$$\frac{\tilde{D_{Zn}}}{\tilde{D_{Cu}}} = 2.3 + \frac{\tilde{D_{Zn}}(\nu_{Zn} = 0.22)}{\tilde{D_{Zn}}(\nu_{Zn} \to 0)} = 17.$$

From Gibbs–Duhem equation,

$$\nu_A \frac{d\, ln\, \gamma_A}{d\nu_A} = \nu_B \frac{d\, ln\, \gamma_B}{d\nu_B}. \tag{4.41}$$

Thus, Darken's equations can be tested by measuring $\tilde{D}(\nu_B), v(\nu_B), D_A, D_B$ and $\gamma_B(\nu_B)$ for the desired alloy composition. This has been experimentally verified for Au-Ni alloys.

For Kirkendall effect, vacancy concentration is not in equilibrium. μ_V is not constant (equal to zero) everywhere. Therefore, deviations are present, as Darken's equations are derived for equilibrium conditions. This is referred to as the Thermodynamic factor.

The effect of thermodynamic factor is very pronounced in systems with a tendency to decompose. In the case of a spinodal, it is defined within the two-phase region by the condition $\frac{\partial^2 F}{\partial \nu_B^2} = 0$. But,

$$\nu_A \frac{\partial^2 F}{\partial \nu_B^2} = \frac{\partial \mu_B}{\partial \nu_B} = \frac{k_1}{\nu_B}\left(1 + \frac{d \ln \gamma_B}{d \ln \nu_B}\right), \tag{4.42}$$

is effectively a thermodynamic factor.

Below the spinodal, $\frac{\partial^2 F}{\partial \nu_B^2} < 0$. That is, the individual diffusion coefficient $\tilde{D_B}\left(= D_B\left(1 + \frac{d \ln \gamma_B}{d \ln \nu_B}\right)\right)$ reverses its sign and leads up to "up-hill" diffusion and creates concentration differences (instead of removing them, as in the normal sense of Fick's first law).

4.4 Reaction Kinetics

Kinetics cannot be understood without thermodynamics, which decides the feasibility.

Kinetics decides the rate of a reaction. But the driving force is decided by thermodynamics. For example, Free Energy Change, Equilibrium Constant, or Partial Pressure. Any study of the rate of a real process must take into account both thermodynamic properties and the kinetic factors.

Factors affecting reaction rates (mass of reactants transformed per unit time) depend on:

1. Concentration, more accurately the thermodynamic activities of reactants.
2. Temperature, *T*, of reactions—an increase in *T* increases the reaction rate.

3. Physical state of the reactants: whether it is a gaseous phase or in a solution (homogeneous reactions) or in separate phases (heterogeneous reactions) or in separate phases (heterogeneous reaction as in a sponge iron in a rotary kiln); whether in solid blocks (low surface area). The higher the surface area, faster the reaction rate.
4. Presence of a catalyst or an inhibitor. (The topc of catalysis—a subject in itself—is not considered in this book.)
5. Means of excitation, e.g., light, UV, radioactive radiation etc.

4.4.1 Rate Law and Order of Reaction

Simple reaction where one constituent breaks up into a number of product constituents. For example, $A = B + C + D + \ldots$.

If there is no appreciable reverse reaction, then the rate of reaction is represented by—$\left\{\frac{d[A]}{dt}\right\}$, where $[d[A]]$ represents the concentration of the constituent. Negative sign is present since the concentration decreases.

The Rate law: The rate of reaction is proportional to the concentration of the reactant(s). Hence,

$$\frac{d[A]}{dt} = k[A], \tag{4.43}$$

Or

$$\frac{\mathrm{d}[A]}{[A]} = kdt.$$

Let $[A] = [A]_0$ at t = 0.

On integrating,

$$ln[A] = -Kt + C.$$

$$C = ln\,[A]_0\ or\ \frac{[A]}{[A]_0} = e^{-kt}. \tag{4.44}$$

If in time t, x amount of A has reacted, then, $[A] = [A]_0 - x$ and $[A]_0 - x = [A]_0 e^{-kt}$ Or

$$x = [A]_0\left(1 - e^{-kt}\right).$$

$$k = \frac{1}{t}\, ln\, \frac{[A]_0}{[A]} = -\left(\frac{2\cdot 303}{t}\right) log_{10}\left(\frac{[A]_0 - x}{[A]_0}\right) = -\left(\frac{2.303}{t}\right) log_{10}\left(1 - \frac{x}{[A]_0}\right). \tag{4.45}$$

A plot of $-\,log\left(1 - \frac{x}{[A]_0}\right)$ vs. t would be a straight line of slope (k/2.303), from which k can be found. This is called a "First Order Reaction".

By a similar logic, for a bimolecular reaction of the type

$$A + B = C + D + \ldots$$

and starting with equal concentrations of A and B, (i.e. $[A]_0 = [B]_0$) we get, $\frac{1}{\left(1 - \frac{x}{[A]_0}\right)} = 1 + akt$ and this applies to a second order reaction.

In general, the order of a reaction is derived from a study of the variation of the reaction rates with time, concentration and temperature.

4.4.1.1 Decomposition of an Oxide

Consider $MO = M + \frac{1}{2}O_2$

Degree of reduction: Ratio of oxygen removed from the oxide to the total removable oxygen and it is represented by α.

$$\alpha = 1 - \frac{[(MO)]}{[MO]_0}.$$

A plot of $log_{10}(1 - \alpha)$ vs t at any given T should be a straight line with a slope of (k/2.303), from which the value of the velocity constant of a reaction at that temperature can be calculated. This way the reaction is found to be of the first order.

If the plot is not a straight line, the reaction is not of the first order, even though it is a unimolecular reaction. Very often, reactions take place in several steps and are much more complex than what they appear at first glance.

Temperature Dependence of Velocity Constant

The Arrhenius equation:

Arrhenius in 1887 proposed that $k = ce^{-(E/RT)}$; where E is the Activation Energy, that is, the energy barrier to a chemical reaction.

Therefore,

$$ln\, k = c' - \frac{E}{RT}\ or\ \log k = c'' - \left(\frac{E}{4.576}\right)\frac{1}{T}, \tag{4.45}$$

if R and E are expressed in calories. (In SI units, J.mol.$^{-1}$ or kJ.mol.$^{-1}$ is preferred. 1 calory = 4.184 Joules).

Therefore, for evaluating E, plot log k vs. $1/T$ and measure the slope. When this relation is not linear, the reaction is not of first order.

For second order, plot $\frac{1}{(1-\alpha)}$ vs. $\left(\frac{1}{T}\right)$.

To determine the order of a reaction, a reduced time plot is used. Here, α is plotted against dimensionless time by taking the ratio of the actual time to the time required to attain $\alpha = 0.5$.

In different regions different mechanisms may be important.

Variation of Reaction Rate with T

First, Second, Third or higher order reactions or fractional order reactions are possible.

Zero order of a reaction is present when there is no effect of concentration on the reaction rate.

A thumb rule for metallurgical reactions: If we increase T by 10 °C, chemical reaction rates are doubled.

On analysing the variation in the velocity constant at 1000 °C, it is found that the above reasoning is valid only if E is 220 kcal.mol^{-1}.

If $E \sim$ 55–60 kcal.mol^{-1} the increase in the chemical reaction rate on increasing T by 10 °C, is only 20%.

4.4.1.2 Role of Diffusion

In a gas-solid reaction, the reaction product, if solid, tends to form a barrier to the interaction process. The gaseous species must diffuse through the product layer in order that further reaction is continued.

Alternatively, the solid product may dissolve in a product layer to interact with the gas phase.

If these diffusion rates are sufficiently fast, the reactions may still proceed, with the chemical reaction rate being the controlling step, and the activation energy of the chemical reaction would determine the variation of reaction rate with temperature (chemical reaction rate-controlled process). If the diffusion rates are too slow, the activation energy of chemical reaction would not have any role in the progress of reaction and one would have a situation where the progress of reaction is "diffusion controlled".

In the intervening regions where both chemical reaction rate and rates of diffusion have influence on the overall progress of a reaction, it is said that the reaction progress is in the "mixed control" domain.

It is important to find ways to circumvent diffusion steps in reactions involving condensed phases (either solids or liquids) as diffusion barriers often make industrial exploitation of a process unviable.

Problems

1.	A heterogeneous reaction is occurring at the surface of a solid materials. The rate is expressed by the equation $r = k.C^{m}_{solid}$, where, r is the rate of reaction, k is the rate constant, C_{solid} is the concentration of reaction at the surface and n is the reaction order. If the concentration of the solid reactant at the surface is Csolid = 0.4 mol/L and the reaction order n = 2, find the rate of reaction if the rate constant k = 5 $mol^{-1}Ls^{-1}$. **Solution:** Using the rate equation $r = k.C^{m}_{solid}$ Substitute the values: R = 5 × $(0.4)^2$ = 0.8 mol/L/s
2.	In a homogeneous reaction involving point defects relaxation, the rate constant for the relaxation process is given by $k = Aexp\left(-\frac{Q}{RT}\right)$, where, A is the pre exponential factor; Q is the activation energy; R is the universal gas constant and T is the temperature in Kelvin. Given that A = 1 × $10^{13}s^{-1}$, Q = 150 kJ/mol, and T 1200 K. Calculate the rate constant k. **Solution:** Substitute the values in to the equation, $k = (1 \times 10^{13})\, exp\left(-\frac{150000}{8.314 \times 1200}\right)$ $K = 3.06 \times 10^{6} s^{-1}$
3.	In a binary compound, the steady –state flux of point defects is given by: $J = D.\left(\frac{C_{initial} - C_{final}}{L}\right)$ Where, J is the flux of point defects, D is the diffusion coefficient, $C_{initial}$ is the initial concentration of defects; C_{final} is the final concentration of defects and L is the length over which the diffusion occurs. Given that $D = 1 \times 10^{-9} m^2/s$; $C_{initial} = 2 \times 10^{23} m^{-3}$, $C_{final} = 1 \times 10^{23} m^{-3}$ and L = 0.01 m. calculate the steady state flux. **Solution:** Substitute the values into the equation $J = (1 \times 10^{-9}).\left(\frac{2 \times 10^{23} - 1 \times 10^{23}}{0.01}\right)$ $J = 10^{16} m^{-2} s^{-1}$
4.	The Kirkendall effect describes the movement of interdiffusing components in a material. The rate of diffusion of component A is given by $J_A = D_A.\left(\frac{C_A}{L}\right)$ Where, J_A is the flux of component A; D_A is the diffusion coefficient of component A, C_A is the concentration gradient of A and L is the characteristics length. Given $D_A = 1 \times 10^{-9} m^2/s$, $C_A = 2 \times 10^{23} m^{-3}$ and L = 0.01 m, calculate the flux of component A. **Solution:** Substitute the values into the equation:

(continued)

$$J_A = (1 \times 10^{-9}).\left(\frac{2 \times 10^{23}}{0.01}\right)$$
$$= 2 \times 10^{16} m^{-2} s^{-1}$$

Reference

1. Yan, Q., Li, C., Lin, C. P., Zhao, Y. L., & Zhang, M. H. (2018). Visible light response AgBr/Ag_3PO_4 hybrid for removal of anionic dye and tetracycline hydrochloride in water. *Journal of Materials Science: Materials in Electronics., 29*. https://doi.org/10.1007/s10854-017-8174-x

Chapter 5
Kinetics of Diffusion

Contents

5.1 Atomic and Molecular Motion by Diffusion

Diffusion phenomenon can be viewed either on macroscopic or microscopic scales.

At the macroscopic level, fluxes and particle concentration profiles are arrived at as functions of driving forces and phenomenological coefficients. Atomic physics is

K. A. Padmanabhan et al., *Thermodynamics and Kinetics of Materials*, Engineering Materials, https://doi.org/10.1007/978-3-032-14527-7_5

used to understand diffusion on a microscopic scale, where atomic mechanisms and particle jump frequencies replace phenomenological coefficients.

It is well known that when a system is not in equilibrium, it experiences net forces that cause fluxes. Then, a coupling between these forces and fluxes, which leads to differential equations that predict the evolution of concentration fields, is achieved for important systems. Solving these equations for different initial and boundary conditions for important systems forms the next goal. Then, the techniques are extended to multi-component systems.

Using the principles of statistical mechanics, the displacements of migrating particles are analyzed at the level of microscopic diffusion. The next step is to develop atomistic models for diffusion and diffusivities in bulk non-crystalline and crystalline materials, which also involve line and planar defects.

In irreversible thermodynamics, entropy and entropy production, and the coupling between forces and fluxes are examined. This results in conjugate forces and fluxes.

The entropy flux, $\overrightarrow{J_S}$, is,

$$\overrightarrow{J_S} = \frac{1}{T}\left(\vec{J}_u - \sum_j \psi_j \overrightarrow{J_j}\right); \overrightarrow{J_u} = \overrightarrow{J_Q} + \sum_j \psi_j \overrightarrow{J_j}, \tag{5.1}$$

which leads to

$$\dot{\sigma} = \overrightarrow{J_Q} \cdot \nabla \frac{1}{T} - \sum_j \frac{\overrightarrow{J_j}}{T} \nabla \psi_j. \tag{5.2}$$

Here $\overrightarrow{J_u}$ is the flux of internal energy density, $\overrightarrow{J_Q}$ is the heat flux, $\overrightarrow{J_j}$ is the flux of component j, ψ_j is a tensor, and $\dot{\sigma}$ which is expressed per unit volume corresponds to the entropy production rate. That is, Eq. (5.2) expresses the rate of unit volume entropy production. Multiplying by T, one gets,

$$T\dot{\sigma} = -\frac{\vec{J}}{T} \nabla \mathrm{T} - \sum_j \overrightarrow{J_j} \nabla \psi_j. \tag{5.3}$$

Every term on the RHS is concerned with a flux or a gradient and are scalar products. Moreover, the unit of all terms on the RHS is density of energy dissipation, J. $m^{-3}s^{-1}$, i.e., a product of flux and gradient of potential. In the case of entropy production rate, "conjugate forces" is the term used for the paired forces and fluxes. These are presented below.

These forces and fluxes are for unconstrained extensive quantities. But, in many systems, constraints exist that connect the changes in their extensive quantities. To simplify the calculations, the material is considered to have similar properties in all

Table 5.1 Fluxes, conjugate forces, and force-flux relations, where there are no constraints on components, (*i*) [1]

Quantity (extensive property)	Net rate of flow ("flux")	Concerned forces ("conjugate force")	Force and flux relations[a] (empirical)
Thermal	$\overrightarrow{J_Q}$	$-\frac{1}{T}\nabla T$	Fourier's $\overrightarrow{J_Q} = -K\nabla T$
As applied to components, *i*	$\overrightarrow{J_i}$	$-\nabla\mu_i = -\nabla\varphi_i$	Modified Fick's $\overrightarrow{J_i} = -M_i c_i \nabla\mu_i$
Electricity	$\overrightarrow{J_q}$	$-\nabla\varphi$	Ohm's $\overrightarrow{J_q} = -\rho\nabla\varphi$

[a] K = Proportionality constant between the flux and the temperature difference, M_i = constant connecting flux and chemical potential gradient for component, *i*, ρ = Proportionality constant between current and voltage (gradient)

directions ("isotropic"), and the fluxes are parallel to the forces. (Not valid for materials with direction-dependent properties ("anisotropic")).

Commonly used empirical force-flux laws that apply for some conditions are given in Table 5.1, for example, the relation proposed by Fourier for the transfer of heat, diffusion in bulk at constant T in a modified form suggested by Fick, the relation for density of current at constant T in electricity due to Ohm. M_i, is the velocity of component *i* due to force of magnitude 1.

5.2 Spontaneous or "Irreversible" Thermodynamics: A Basic Postulate

Production of entropy locally is positive near equilibrium.

$$\dot{\sigma} = \frac{\partial S}{\partial t} + \nabla\overrightarrow{J_S} \geq 0. \tag{5.4}$$

Based on Table 5.1, the entropy production for some cases can be known. For just heat flow, from Eq. (5.3) and Fourier's heat-flux law,

$$\overrightarrow{J_Q} = -K\nabla T \tag{5.5}$$

which leads to

$$T\dot{\sigma} = \frac{K|\nabla T|^2}{T}. \tag{5.6}$$

The "mod sign" points out that thermal conductivity is positive.

If the only operating process is diffusion,

$$T\dot{\sigma} = \sum_{i=1}^{N_c} M_i c_i |\nabla \mu_i|^2. \tag{5.7}$$

The "mod sign" indicates that mobility is positive.

5.3 Linear Irreversible Thermodynamics

A temperature gradient will produce a heat flux and an electric potential gradient—the thermoelectric effect. That is, if heat flow can be present, an electrical potential gradient can lead to a flux of heat. Different types of fluxes and forces can be coupled together. Moving atoms/particles have mass, electrons conduct electricity, and flow of phonons leads to heat conduction. Exchange interactions between particles can give rise to momentum. A generalization of the empirical force-flux equations can result from formulating the coupling effects.

5.4 Fluxes and Forces: General Coupling

If F_i represents the sum of all the driving forces, J_Q, the heat flux, F_Q, the conjugate force (which could also be a function of the other forces), and the forces that drive the process are small and are close to equilibrium, each flux is presented linearly around the point of equilibrium ("Taylor series expansion"), F_Q=F_q=F_1= .... =F_N.

If the relation between the forces and the fluxes is linear, then,

$$J_\alpha = \sum_\beta L_{\alpha\beta} F_\beta \ (\alpha, \beta = \mathrm{Q}, \mathrm{q}, 1, 2, \ldots.., N_c). \tag{5.8}$$

Here,

$$L_{\alpha\beta} = \frac{\partial Q_\alpha}{\partial F_\beta} \tag{5.9}$$

$L_{\alpha\beta}$ is the direct coefficient, which is estimated at F_β (for all β, $F_\beta = 0$).

In Eqs. (5.8) and (5.9), the direct coefficients are the diagonal terms, $L_{\alpha\alpha}$. The terms that are not along the diagonal cause the coupling effects, also known as the cross effects.

Equations (5.6) and (5.8) together lead to a relation for the production of entropy close to equilibrium.

$$T\dot{\sigma} = \sum_{\beta}\sum_{\alpha} L_{\alpha\beta} F_{\alpha} F_{\beta}. \tag{5.10}$$

If an insulator (a pure substance) is an insulator and has only a gradient that is constant and has no other fields that is thermal in character, then from Eq. (5.8) and Table 5.1,

$$J_Q = L_{\alpha\alpha}\left(-\frac{1}{T}\nabla T\right). \tag{5.11}$$

From the first equation in Table 5.1, the thermal conductivity K is given be,

$$K = \frac{L_{QQ}}{T}. \tag{5.12}$$

Here L_{QQ} is the direct coefficient. The material is assumed to be isotropic. Therefore, forces and fluxes are zero order tensors (scalars). Terms like $\overrightarrow{J_Q} = (\partial \overrightarrow{J_Q} / \partial \overrightarrow{F_Q}) \overrightarrow{F_Q}$ are tensors of order two multiplying first order tensors (vectors), and the equations that result will be linear.

The general force-flux equations for electronically conducting materials are:

$$J_Q = L_{QQ} + L_{Qq} F_q. \tag{5.13}$$

$$J_q = L_{qQ} F_Q + L_{qq} F_q. \tag{5.14}$$

When the gradient in temperature is constant and there are no electrical contacts at the ends of the specimen (steady state) the current has a charge density, 0. Therefore, $J_q = 0$ and the force present is,

$$F_q = \frac{L_{qQ}}{L_{qq}} F_Q \tag{5.15}$$

will arise. F_q means an electrical potential gradient, $\nabla\varphi$ along the bar. From Eqs. (5.15) and (5.13),

$$\overrightarrow{J_Q} = \left[L_{QQ} - \frac{L_{Qq}L_{qQ}}{L_{qq}}\right] F_Q = -\left[\frac{L_{QQ}}{T} - \frac{L_{Qq}L_{qQ}}{TL_{qq}}\right] \nabla T = -\mathrm{K}\nabla \mathrm{T}. \tag{5.16}$$

Then, Fourier's law holds good. The thermal conductivity K is proportional to the direct coefficient L_{QQ} (Eq. 5.14). The electrical flow of charge and the coefficients, both direct and coupled, are related.

5.5 Force-Flux Relations for Constrained Extensive Quantities

Often, changes in an extensive quantity are coupled to changes in others. An important case: substitutional atoms that do not have sinks or sources in a crystal. To cite an example, dislocations (the components occupy just the sites corresponding to those occupied by atoms in each crystal lattice). When a component moves away from a site, it must be replaced. This gives rise to the so-called "network constraint". Atom-vacancy exchange mechanism wherein every time a vacancy leaves its lattice point, another atom occupies that space or vice versa is a case in point. In other words, mutual dependence exists between the component fluxes. Due to the non-existence of lattice points, in amorphous materials the N components can occupy or leave anywhere without any constraint. The dN_i is independent of interstitial solutes in crystalline materials. In other words, in a dilute solution, atoms of interstitial carbon may be inserted or eliminated without restriction.

For a network,

$$\sum_{i=1}^{i=N_c} dN_i. \tag{5.17}$$

Solving this equation and substituting in $Tds = du + dw - \sum_{i=1}^{i=N_c} \mu_i dN_i$, results in

$$\mathrm{T\,ds} = \mathrm{du} + \mathrm{dw} - \sum_{i=1}^{i=N_C-1} \left(\mu_i - \mu_{N_c}\right) \mathrm{d}c_i. \tag{5.18}$$

Repeating the process used for Eq. (5.18) one can get from Eq. (5.3) for the *ith* component diffusion in a network inside a crystal that is constrained, a conjugate force,

$$\vec{F_i} = -\nabla\left(\mu_i - \mu_{N_c}\right). \tag{5.19}$$

This component force J_i, is based on the difference between the partial molar Gibbs energy of component *i, and* N_C, the gradient in substitutional diffusion through a vacancy exchange mechanism. During migration, state of a site changes from atom of type *i* to vacancy occupation, and causes the difference.

At constant temperature, when an electric field is absent, in a three-component material corresponding to components $i = 1, 2,$ *and 3 (A, B, C)*, respectively, if the N_c^{th} component is component C, Eq. (5.18), together with Eq. (5.19) suggest that,

Table 5.2 Components, *i*, of network-constrained systems: fluxes and conjugate forces [1]

Quantity	Flux	Force of a conjugate nature
Thermal terms	$\overrightarrow{J_Q}$	$-\frac{1}{T}\nabla$ T
Along direction *i*	$\overrightarrow{J_i}$	$-\nabla\left(\mu_i-\mu_{N_c}\right)=-\nabla\varphi_i$
Charge	$\overrightarrow{J_q}$	$-\nabla\varphi$

$$\begin{aligned}\overrightarrow{J_A} &= -L_{AA}\nabla(\mu_A-\mu_C)-L_{AB}\nabla(\mu_B-\mu_C)\\ \overrightarrow{J_B} &= -L_{BA}\nabla(\mu_A-\mu_C)-L_{BB}\nabla(\mu_B-\mu_C)\\ \overrightarrow{J_C} &= -L_{CA}\nabla(\mu_A-\mu_C)-L_{CB}\nabla(\mu_B-\mu_C).\end{aligned} \tag{5.20}$$

If B is the Nth component,

$$\begin{aligned}\overrightarrow{J'_A} &= -L_{AA}\nabla(\mu_A-\mu_B)-L_{AC}\nabla(\mu_C-\mu_B)\\ \overrightarrow{J'_B} &= -L_{BA}\nabla(\mu_A-\mu_B)-L_{BC}\nabla(\mu_C-\mu_B)\\ \overrightarrow{J'_C} &= -L_{CA}\nabla(\mu_A-\mu_B)-L_{CC}\nabla(\mu_C-\mu_B).\end{aligned} \tag{5.21}$$

As $\overrightarrow{J_i}$ and $\overrightarrow{J'_i}$ are the same and the terms arising from the gradient are non-zero, Equations (5.20) and (5.21) imply that,

$$\sum\nolimits_{j=1}^{N_c} L_{ij}=0. \tag{5.22}$$

If the lattice network and the fluxes pertain to the same coordinate system, the constraint from the network stipulates that,

$$\sum_{i=1}^{N_c}\overrightarrow{J_i}=0. \tag{5.23}$$

Therefore, a further condition gets imposed that

$$\sum\nolimits_{1=1}^{N_c} L_{ij}=0. \tag{5.24}$$

That is, any column or row entries of the matrix L_{ij} when added together leads to the result, zero.

The fluxes and forces of a conjugate type that result when a network constraint is the only constraint are presented below Table 5.2.

Further possible constraints: Suppose 1 is a ionic species that is distributed in a non-uniform manner and has no constraint from any network. Then, the local electric field ensures that every ion is surrounded by an electric field due to an electrostatic force, which results from its chemical potential gradient. The force terms that give rise to the potential is the total of the chemical potential and the charged ion electro-potential (electro-chemical potential). Therefore,

$$\overrightarrow{F_1} = -\nabla(\mu_1 + q_1\varphi). \tag{5.25}$$

Additional forces to the chemical potential force, e.g., if a magnetic field were present, can be a magnetic moment.

5.6 Diffusion Potential

The force driving diffusion of a component depends on the stored energy and the added component. The sum of all potentials and the chemical potential results in the diffusing component's diffusion potential φ, which equals its total conjugate force. Therefore, the flux of component 1 that has a conjugate force is,

$$\overrightarrow{F_1} = -\nabla\varphi_1. \tag{5.26}$$

For electrochemical potential (Eq. (5.25)),

$$\varphi_1 = \mu_1 + q_1\text{φ}. \tag{5.27}$$

5.7 Symmetry Principle of Onsager

The forces and the fluxes are connected by 3 axiomatic statements.

1. The entropy rate change and the local entropy production rate is derived by assuming thermodynamic variations under equilibrium and equilibrium on a localized scale is assumed.
2. Non-negative production of entropy.
3. There is linear dependence between flux and all the driving forces.

These postulates are beyond the scope of the first and second laws of Fick.

These propositions subscribe to statistical theory of reversible fluctuations and support Onsager's symmetry principle, which states that coupling coefficients are symmetric when the fluxes and forces are conjugates. That is,

$$L_{\alpha\beta} = L_{\beta\alpha}. \tag{5.28}$$

This simplifies the coupled force-flux equations, which are experimentally verified. Eq. (5.28) ensures that all the eigenvalues of Eq. (5.8) are real numbers. Also, Eq. (5.10) which is quadratic, and Eq. (5.4) suggest that the kinetic matrix, $L_{\alpha\beta}$, is "positive definite" and has all non-negative eigenvalues. Therefore,

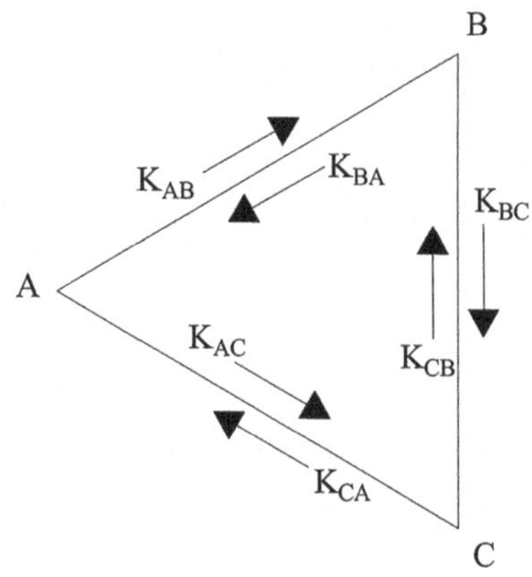

Fig. 5.1 Molecules of type *A, B, and C* being converted (schematic)

$$\frac{\partial J_\alpha}{\partial F_\beta} = \frac{\partial J_\beta}{\partial F_\alpha}. \tag{5.29}$$

That is, the flux change in some quantity, caused by a change in the force driving for another, equals the change in the flux of the other (second) quantity achieved by altering the force that drives the first, *c.f.*, Maxwell relations.

In systems when microscopic reversibility is close to equilibrium: From statistical-mechanics basis, Onsager's symmetry principle is derived, i.e., a correlation (time average) between a force that drives type α and fluctuations of β is achieved which is like flipping between α and β.

Suppose *A, B, and C* are three isomers. Without a change of composition, it is possible to convert each isomer into any one of the other two. Consider these molecules in a closed system and maintained at constant T, constant P. Then, the conversion rate of one to another is proportional to the number of the concerned species. The proportionality constant is the rate constant, K (Fig. 5.1).

A, B, and C change in numbers (the rates),

$$\begin{aligned}
\frac{dN_A}{dt} &= -(K_{AC} + K_{AB})N_A + K_{BA}N_B + K_{CA}N_C. \\
\frac{dN_B}{dt} &= K_{AB}N_A - (K_{BC} + K_{BA})N_B + K_{CB}N_C. \\
\frac{dN_C}{dt} &= K_{AC}N_A + K_{BC}N_B - (K_{CA} + K_{CB})\ N_C.
\end{aligned} \tag{5.30}$$

At equilibrium the time derivatives in Eq. (5.30) vanish. When a system is closed ($N_A + N_B + N_C = N^{tot}$) the solution for the equilibrium is

$$N_A^{eq} = \frac{K_\alpha N^{tot}}{K_\alpha + K_\beta + K_\gamma}; N_B^{eq} = \frac{K_\beta N^{tot}}{K_\alpha + K_\beta + K_\gamma}; N_C^{eq} = \frac{K_\gamma N^{tot}}{K_\alpha + K_\beta + K_\gamma}. \tag{5.31}$$

Here,

$$\begin{aligned} K_\alpha &= K_{BA}K_{CA} + K_{BA}K_{CB} + K_{CA}K_{BC}. \\ K_\beta &= K_{CB}K_{AB} + K_{CB}K_{AC} + K_{AB}K_{CA}. \\ K_\gamma &= K_{AC}K_{BC} + K_{AC}K_{BA} + K_{BC}K_{AB}. \end{aligned} \tag{5.32}$$

Near the equilibrium, let $Y_A = (N_A - N_A^{eq})$. Introducing similar relations for *A, B and C* in Eq. (5.30),

$$\frac{dY_A}{dt} = -(K_{AC} + K_{AB})Y_A + K_{BA}Y_B + K_{CA}Y_C. \tag{5.33}$$

For *B and C* also similar equations can be written.

When one assumes the law due to Henry, a constant activity coefficient and expressing the chemical potential in an expansion near equilibrium (small values of $(\frac{Y_A}{N_A^{eq}})$) (the chemical potential is, $\mu_i = \mu_i^0$ + kT $\ln(\gamma_i X_i)$), one gets,

$$\Delta\mu_A^{eq} = (\mu_A - \mu_A^{eq}) = \text{kT}\ \ln\left(1 + \frac{Y_A}{N_A^{eq}}\right) \approx \left(\frac{kTY_A}{N_A^{eq}}\right). \tag{5.34}$$

Using Eqs. (5.34) and (5.33) and similar operations for *B and C* one obtains,

$$\begin{aligned} \frac{dY_A}{dt} &= -\left(\frac{(K_{AC} + K_{AB})A_A^{eq}}{kT}\right)\Delta\mu_A^{eq} + \left(\frac{K_{BA}N_B^{eq}}{kT}\right)\Delta\mu_B^{eq} + \left(\frac{K_{CA}N_C^{eq}}{kT}\right)\Delta\mu_C^{eq}. \\ \frac{dY_B}{dt} &= \left(\frac{K_{AB}N_A^{eq}}{kT}\right)\Delta\mu_A^{eq} - \left(\frac{(K_{BC} + K_{BA})N_B^{eq}}{kT}\right)\Delta\mu_B^{eq} + \left(\frac{K_{CB}N_C^{eq}}{kT}\right)\Delta\mu_C^{eq}. \\ \frac{dY_C}{dt} &= +\left(\frac{K_{AC}N_A^{eq}}{kT}\right)\Delta\mu_A^{eq} + \left(\frac{K_{BC}N_B^{eq}}{kT}\right)\Delta\mu_B^{eq} - \left(\frac{(K_{CA} + K_{CB})N_C^{eq}}{kT}\right)\Delta\mu_C^{eq}. \end{aligned} \tag{5.35}$$

These are relationships that are linear between the potential differences, $(\mu_i - \mu_i^{eq})$, which navigate the Y_i's to the proximity of equilibrium. The relevant rates, $(\frac{dY_i}{dt})$ appear as (using Onsager coefficients),

$$
\begin{aligned}
\frac{dY_A}{dt} &= L_{AA}F_A + L_{AB}F_B + L_{AC}F_C. \\
\frac{dY_B}{dt} &= L_{BA}F_A + L_{BB}F_B + L_{BC}F_C. \\
\frac{dY_C}{dt} &= L_{CA}F_A + L_{CB}F_B + L_{CC}F_C.
\end{aligned} \tag{5.36}
$$

In a complex system, in the presence of microscopic reversibility every elementary process, including rates of forward and backward processes, establishes equilibrium in the entire system. As both forward and backward fluxes are equal (Fig. 5.1),

$$
\begin{aligned}
\frac{K_{BA}}{K_{AB}} &= \frac{N_A^{eq}}{N_B^{eq}} = \frac{K_\alpha}{K_\beta} = \frac{K_{BA}K_{CA} + K_{BA}K_{CB} + K_{CA}K_{BC}}{K_{CB}K_{AB} + K_{CB}K_{AC} + K_{AB}K_{CA}}. \\
\frac{K_{CB}}{K_{BC}} &= \frac{N_B^{eq}}{N_C^{eq}} = \frac{K_\beta}{K_\gamma} = \frac{K_{CB}K_{AB} + K_{CB}K_{AC} + K_{AB}K_{CA}}{K_{AC}K_{BC} + K_{AC}K_{BA} + K_{BC}K_{AB}}. \\
\frac{K_{AC}}{K_{CA}} &= \frac{N_C^{eq}}{N_A^{eq}} = \frac{K_\gamma}{K_\alpha} = \frac{K_{AC}K_{BC} + K_{AC}K_{BA} + K_{BC}K_{AB}}{K_{BA}K_{CA} + K_{BA}K_{CB} + K_{CA}K_{BC}}.
\end{aligned} \tag{5.37}
$$

A comparison of Eq. (5.36) with Eqs. (5.35) and (5.37) reveals that $L_{ij} = L_{ji}$. That is, the symmetry of the Onsager coefficients reflects microscopic reversibility. The reaction rates and the forces are often combined, Eqs. (5.4) and (5.35), to show that reaction rates and the forces are conjugates.

5.8 Some Important Examples

Forces of different kinds can drive processes that can give rise in chemical components fluxes.

- Charged species flow because of an electrostatic field.
- A local volume dilation can be caused by a solute atom, which then to reduce the compression can flow to lower compression regions.
- Flow of chemical components is from higher to lower regions of chemical potential.
- Fick's first law results from a response of a chemical potential gradient to a flux, i.e., $\overrightarrow{J_i}$, and its concentration gradient, ∇c_i, or $\overrightarrow{J_i} = -D\ \nabla c_i$, when D is diffusivity in the bulk.

Different forces can drive a chemical species. The processes at the microscopic level that account for the value of D could be due to different mechanisms. Therefore, the forces that drive the processes can be collectively attributed to the generalized diffusion potential, defined as φ, and discussed earlier.

The flux of components in a solution could be the result of several processes and cannot diffuse independently. Therefore, different coefficients defined in specific frames of reference will be needed.

5.9 Diffusion Along Gradient in Concentration

When at constant, uniform temperature, in a single-phase that has a gradient of concentration, uninfluenced by other fields, grain boundaries etc., are present, the component's potential due to diffusion is identical to its potential that is chemical in nature (partial molar Gibbs energy). This potential gradient gives rise to diffusion. Then, as already discussed, a flux due to diffusion that is proportional to the gradient of the potential due to diffusion comes up. But a gradient in concentration is easier to obtain from an experiment than a potential gradient arising from diffusion. Therefore, in a thermodynamic model, the gradient of a chemical potential is more easily determined as a gradient in concentration. The result is a flux due to diffusion which is proportional to the gradient in concentration. The term that couples the flux and the gradient in concentration is D, known as a "diffusion coefficient" or "diffusivity". Therefore, the first law due to Fick is presented as

$$\vec{J} = -D\nabla c. \tag{5.38}$$

Again, this relationship must be expressed with a reference frame.

5.10 Self-diffusion: No Chemical Effects During Diffusion

During self-diffusion in all three states of matter (pure gas, liquid, or solid), the components present a homogeneous medium and radioactive tracer isotopes are used. The concentration of the tracer in the concentration profile that evolves is measured to determine self-diffusivity.

If atom-vacancy exchange mechanism is responsible for self-diffusion, the flux law due to Fick applies. In modeling terms, such a network is constrained by 3 components consisting of inert atoms, radioactive atoms and vacancies. Atomic planes are used to quantify the fluxes by counting the atoms crossing per unit time per unit area. During self-diffusion the crystal is rigidly fixed. Therefore, these planes form a C- reference frame (crystal frame) for flux determination. For a fixed temperature, vacancies are the N_cth component, using Eqs. (5.8) and (5.19) for one-dimensional isotropic diffusion one gets,

$$J_1^c = L_{11}F_1 + L_{1*1}F_{*1} = -L_{11}\frac{\partial(\mu_1-\mu_V)}{\partial x} - L_{1*1}\frac{\partial(\mu_{*1}-\mu_V)}{\partial x}$$
$$J_{*1}^c = L_{11}F_1 + L_{1*1}F_1 = -L_{1*1}\frac{\partial(\mu_1-\mu_V)}{\partial x} - L_{1*1}\frac{\partial(\mu_{1*1}-\mu_V)}{\partial x}. \quad (5.39)$$
$$\text{and } J_1^c + J_{*1}^c + J_V^c = 0.$$

The radioactive species is indicated by a star. The superscript c indicates that C-frame is used to measure the flux of *i.* These equations apply for vacancy-exchange mechanisms. Vacancies are assumed to be in thermodynamic equilibrium. The equations written are for an atom exchanging position with a vacancy. Therefore, $\mu_V = 0$. $c_1 + c_{*1} + c_V =$ A; A is a constant. Vacancy to atom ratio is $<10^{-4}$. The chemical potentials of species 1 and *1 is expressed in terms of Raoultian behavior ($\gamma_1 = \gamma_{*1} = 1$). Therefore, $\partial\mu_i/\partial x = (kT\partial \ln c_i)/\partial x$. The assumed uniform vacancy equilibrium ensures that $c_1 + C_{*1} = (A - c_V)$ is uniform as well. Equation (5.39) becomes,

$$J_1^c = -kT\left[\frac{L_{11}}{c_1} - \frac{L_{1*1}}{c_{*1}}\right]\frac{\partial c_1}{\partial x}.$$
$$J_{*1}^c = -kT\left[\frac{L_{*1*1}}{c_{*1}} - \frac{L_{*11}}{c_1}\right]\frac{\partial c_{*1}}{\partial x}. \quad (5.40)$$

As $J_V^c = 0$ as well as $J_1^c + J_{*1}^c = 0$,

$$J_{*1}^c = -kT\left[\frac{L_{11}}{c_1} - \frac{L_{1*1}}{c_{*1}}\right]\frac{\partial c_{*1}}{\partial x} = -*D\frac{\partial c_{*1}}{\partial x}. \quad (5.41)$$

Equation (5.41) shows that Fick's law is obeyed by self-diffusion of the radioactive tracers. *D is the radioactive tracer diffusivity. Unless otherwise specified, diffusion is treated as isotropic.

The vacancy chemical potential is defined by $\mu_V = \left(\frac{\partial G}{\partial N_V}\right)_{T,P}$, i.e., a vacancy is created at a source, e.g., dislocation or an interface, because of a large free-energy change. (Dislocations are sources of vacancies when the climb is in one direction or they act as vacancy sinks if they climb in the other.) It is assumed that the source and sink behavior is complementary. If the change in free energy = 0, i.e., μ_V, vacancy formation will be at a minimum free energy state. Then, in the system μ_V is uniformly zero, i.e., density of available vacancy sources is enough.

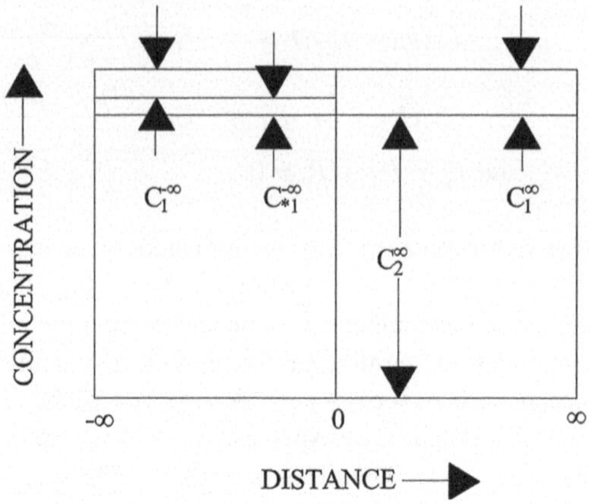

Fig. 5.2 Measuring diffusion using a diffusion couple in the presence of a c_{*1} gradient in a chemically homogeneous material [1]

5.11 Homogeneous Binary Solution and Self-Diffusion of a Component

Let atoms of types 1 and 2 comprise a binary solution that has a gradient in concentration of a radioactive isotope to measure self-diffusion. The tracer has a concentration gradient, c_{*1}. But both $(c_1 + c_{*1})$ and c_2 are uniform—see Fig. 5.2.

The law of Fick for the self-diffusion of a radioactive tracer, therefore, becomes,

$$J_{*1}^{c} = -\mathrm{k\,T}\left[\frac{L_{11}}{c_1} - \frac{L_{1*1}}{c_{*1}}\right]\frac{\partial c_{*1}}{\partial x} = - * \mathrm{D}\frac{\partial c_{*1}}{\partial x}. \tag{5.42}$$

5.12 Diffusion of Substitutional Particles in a Chemical Concentration Gradient—Kirkendall Effect

Consider an inhomogeneous solution in which a substitutional solute of type i diffuses. Then, both the solute particles and host particles experience inter-diffusion on the substitutional sites. If one species experiences faster diffusion, the domain loses net mass and shrinks. In contrast, the domain containing the species that diffuses more slowly acquires greater mass and increases in size. This represents the flow of mass, known as the Kirkendall effect, reported first in metals—see Fig. 5.3.

A vacancy mechanism causes interdiffusion along x. The Kirkendall effect leads to the flow of crystal. Compared with the edges of the sample, its local planes exhibit relative speed $v\,(x)$. Right through the diffusing crystal, there is no unique C-frame. The distribution of the solute ($i = 1$), which is faster diffusing is shown in Fig. 5.3a. The fluxes should satisfy the equation in a (local) C-frame,

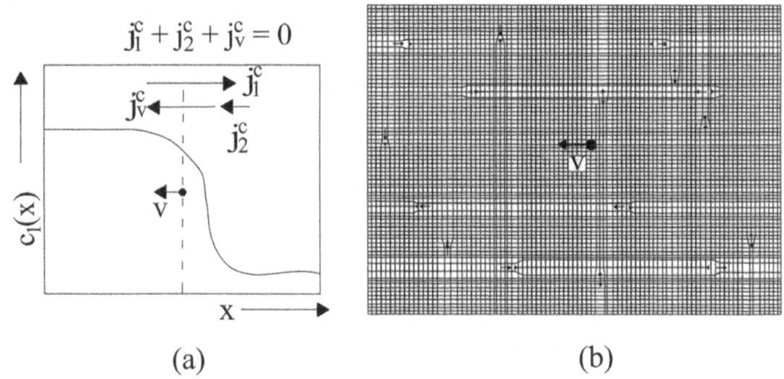

Fig. 5.3 Effect due to Kirkendall seen in a two-phase alloy (schematic). Atom-vacancy exchange mechanism is assumed to exist. Dislocation motion at some time after the diffusion couple's initial condition is illustrated. (**a**) Concentration as a function of distance in the zone of diffusion is plotted. A marker that is inert is embedded and it moves at a speed, v. The fluxes connected to the marker are indicated in the local C-frame. (**b**) Local arrangement of atom planes near the marker (across the zone of diffusion) is seen. Arrows indicate climb of dislocations of edge orientation that are present. Vacancies move from right to left [1]

$$\overrightarrow{J_1^c} + \overrightarrow{J_2^c} + \overrightarrow{J_V^c} = \text{O}. \tag{5.43}$$

When the Kirkendall effect is present, vacancies are required to establish a difference in the fluxes of vacancies of the two substitutional species, i.e., on one side of the markers continuous net generation of vacancies will be present and on the other side annihilation of vacancies. Vacancy creation and elimination can occur by dislocation climb - see Fig. 5.3 for edge dislocations. When atoms from the extra planes fill the incoming vacancies, vacancy-deletion is seen. This causes the extra planes to shrink. Vacancy creation and extra planes formation are present on the RHS by a reverse process—see Fig. 5.3b. This contraction-expansion gives rise to a mass flow seen as the movement of embedded inert markers.

Climb removes a crystal plane. Then, the plane contracts in the normal direction. In directions that are aligned to the plane of the interface, in contrast, specimen ends constrain (free) expansion, as here significant diffusion is absent. But interface cohesion between expanding and contracting regions is seen. As a result, in the direction normal to the diffusion direction, restricted dimensional changes are present. Compatibility stresses are generated in the plane[1]. Development of compatibility stresses outside the plane is absent, i.e., in the direction of diffusion," diffusion-couple" freely expands.

In a narrow diffusion zone in a specimen (compared with the specimen) in the diffusion direction, pure shear compatibility stresses are present. When the stresses

[1]The component that diffuses faster does so from the vapor phase into a thin sheet. Then, the thickness of the sheet is much thinner than the diffusion zone. This reduces the constraints significantly on expansion in a direction parallel to the interface on which diffusion takes place. Normal to the direction of diffusion large expansions of specimen are seen.

go beyond the yield stress, the cross-section of the diffusion specimen will remain constant, which will be allowed by plastic flow.

When there are species 1, species 2, and the vacancies at constant P and T with sites that are a part of a network except at dislocations or interfaces, the fluxes will be (from the Eqs. (5.44) and (5.19)),

$$\mu_i = \mu_i^0 + kT \ln(\gamma X_{ii}), \tag{5.44}$$

$$J_1^c = L_{11}F_1 + L_{12}F_2 = -L_{11}\frac{\partial \varphi_1}{\partial x} - L_{12}\frac{\partial \varphi_2}{\partial x} = -L_{11}\frac{\partial(\mu_1 - \mu_V)}{\partial x} - L_{12}\frac{\partial(\mu_2 - \mu_V)}{\partial x}.$$

Likewise,

$$J_2^c = L_{21}F_1 + L_{22}F_2 = -L_{21}\frac{\partial(\mu_1 - \mu_V)}{\partial x}. \tag{5.45}$$

Whence, $J_V^c = -\left(J_1^c + J_2^c\right)$ (as the sum of all three is equal to zero). Due to local equilibrium,

the Gibbs-Duhem equation becomes

$$c_1\frac{\partial \mu_1}{\partial x} + c_2\frac{\partial \mu_2}{\partial x} + c_V\frac{\partial \mu_V}{\partial x} = 0. \tag{5.46}$$

In a direction opposed to that of the species 1 that diffuses the fastest—Fig. 5.3—a net flux of vacancies develops. Vacancy concentrations under nonequilibrium conditions would not develop if dislocation climb were present. Usually, a sufficient density of dislocations is available to maintain near-equilibrium vacancy concentration. Then, to a good approximation, $\mu_V = 0$. Thus, $\nabla\mu_V = 0$ everywhere in the diffusion zone. Using Eqs. (5.45) and (5.46), with $\nabla\mu_V = 0$, one can obtain,

$$\begin{aligned} J_1^C &= -\left[L_{11} - \left(\frac{c_1}{c_2}\right)L_{12}\right]\frac{\partial \mu_1}{\partial x}. \\ J_2^c &= -\left[L_{22} - \left(\frac{c_2}{c_1}\right)L_{21}\right]\frac{\partial \mu_2}{\partial x}. \end{aligned} \tag{5.47}$$

Then it follows from Eq. (5.47) that the chemical potential gradients can be related to concentration gradients as

$$\mu_i = \mu_i^0 + \text{kT ln } (\gamma_i(\Omega)c_i). \tag{5.47}$$

This is obtained by combining Eq. (5.44) with the equation,

$$(\Omega) = \frac{V^{tot}}{N^{tot}} = \frac{V^{tot}/N_i}{N^{tot}/N_i} = \frac{1/c_i}{1/X_i} = \frac{X_i}{c_i}. \tag{5.48}$$

The net result is

$$\frac{\partial \mu_1}{\partial x} = \mathrm{kT}\left[1 + \frac{\partial\ ln\ \gamma_1}{\partial\ ln\ c_1} + \frac{\partial\ ln\ (\Omega)}{\partial\ ln\ c_1}\right]\frac{1}{c_1}\frac{\partial c_1}{\partial x}. \tag{5.49}$$

A local change of composition leads to a local volume expansion[2].

Diffusion is sustained by this expansion via the average site volume (Ω) derivative. The non-ideal nature of the solution appears in the above equation as the derivative of γ_1. A flux that is linearly and directly related to the gradient in concentration is obtained from Eqs. (5.49) and (5.47). In other words,

$$J_1^c = -kT\left[\frac{L_{11}}{c_1} - \frac{L_{12}}{c_2}\right]\left[1 + \frac{\partial\ ln\ \gamma_1}{\partial\ ln\ c_1} + \frac{\partial\ ln\ \gamma_1}{\partial\ ln\ c_1}\right]\frac{\partial c_1}{\partial x} = -D_1\frac{\partial c_1}{\partial x}. \tag{5.50}$$

Let the intrinsic diffusivity be D_1 (in the local C-frame). In a inhomogeneous material (chemically speaking) this is the form of the law due to Fick. A similar approach yields for component c_2,

$$J_2^c = -D_2\left(\frac{\partial c_2}{\partial x}\right).$$

In a homogeneous solution of concentration c_1, $^*D_1$ in Eq. (5.42) of species 1 is its self-diffusivity (like that of the same species in a inhomogeneous solution which has an intrinsic diffusivity that appears as D_1 in Eq. (5.50)). Often, in addition to the approximation of a concentration-independent average site volume (Ω), the off-diagonal terms, $\left(\frac{L_{12}}{c_2}\right)$ and $\left(\frac{L_{1*1}}{c_{*1}}\right)$ that have been coupled in Eqs. (5.42) and (5.50) are neglected in comparison with the direct term $\left(\frac{L_{11}}{c_1}\right)$. In this approximation,

$$D_1 \approx \left[1 + \frac{\partial\ ln\ \gamma_1}{\partial\ ln\ c_1}\right] * D_1. \tag{5.51}$$

The difference between D_1 and $^*D_{*1}$ due to the activity coefficient of component 1 is dependent on concentration. A chemical potential drives mass diffusion. For a nonideal solution diffusivity and gradient of concentration are made proportional. No such effect exists for "self-diffusion".

[2]This expansion is referred to as Vegard law, $\varepsilon_{ij} - \varepsilon_{ij}^0 = \varrho_{ij}\ (c - c_0)$; ε_{ij} is the strain due to the concentration change. ϱ_{ij} are the Vegard coefficients.

When the total volume of specimen is constant, diffusion due to Kirkendall effect needs just a single diffusivity (inter-diffusivity) when the total specimen volume, measured in a fixed volume, is constant (a single reference frame).

Chemical interdiffusion on a substitutional lattice, therefore, can be presented in two perspectives—constant volume framework, which is described by a single diffusivity (the inter-diffusivity) and a local concentration-dependent framework which uses two intrinsic diffusivities. They are related as,

$$\tilde{D} = c_1\Omega_1 D_2 + c_2\Omega_2 D_1. \tag{5.52}$$

D_1 and D_2 are the diffusivities of components 1 and 2. It is assumed frequently that $\Omega_1 = \Omega_2$. (Each diffusivity is usually a function of the local composition.)

Possible consequences of the Kirkendall effect: (1) If the membrane is allowed to move to the left, say, the pressure on both sides (left and right) will become the same. (2) For a fixed membrane, cavitation on the left side may be present to compensate for the lower pressure.

5.13 Mass Diffusion: The Effect a Potential Gradient in Electricity

An electrostatic potential gradient can give rise to a force to drive the mass diffusion of a species, e.g., charged ions can experience transport by diffusion induced by a gradient in potential, as in batteries that employ solid electrolytes (interstitial atoms in metals causes the electron-current-induced diffusion).

5.13.1 Ionic Conductors

Low concentration positively charged ions diffuse interstitially in an ionic material: In the absence of any field, let these ions have an inter-diffusivity of $\tilde{D}$. In the presence of $\vec{E} = -\nabla\varphi$ (an electric field, an electrochemical potential, $\varphi_1 = \mu_1 q_1 \varphi$, will result from the potential due to diffusion. According to the equation $J_\alpha = \sum_\beta L_{\alpha\beta} F_\beta$; ($\alpha$, β = Q, q, 1, 2, …, N_c), the flux of charged interstitials is

$$\vec{J_1} = L_{11}\vec{F_1} = -L_{11}\nabla\varphi_1 = -L_{11}\nabla(\mu_1 + q_1\varphi). \tag{5.53}$$

From $\mu_1 = \mu_1^0$ + kT ln $(K_1 c_1)$ and $D_1 = M_1 kT = L_{11}\frac{kT}{c_1}$ one can obtain the equation

$$\vec{J_1} = -D_1\nabla c_1 - \frac{D_1 c_1 q_1}{kT}\nabla\varphi. \tag{5.54}$$

When a significant concentration gradient is absent, the corresponding flux in charge is,

$$\overrightarrow{J_q} = q_1 \overrightarrow{J_1} = -\frac{D_1 c_1 q_1^2}{kT} \nabla \phi. \tag{5.55}$$

Analogous to the law due to Ohm, one obtains $\overrightarrow{J_q} = -\rho \nabla \varphi$, where ρ is electrical conductivity. Or

$$\mathrm{P} = \frac{D_1 c_1 q_1^2}{kT}. \tag{5.56}$$

In other words, ρ and diffusivity are directly related.

5.13.2 Electromigration

In metals a potential due to electricity induces electromigration (diffusion) brought about by an interaction between a species that diffuses and a flux of electrons that conducts. When a dilute solution of interstitial atoms in a metal experiences an electric field two fluxes appear, a conduction electrons flux, J_q, plus flux due to interstitials, J_1. When temperature is constant, $F_q = -\nabla \varphi \equiv \vec{E}$, and

$$\overrightarrow{J_1} = -L_{11} \nabla \mu_1 - L_{1q} \vec{E}. \tag{5.57}$$

The potential of a chemical nature due to interstitials is $\mu_1 = \mu_1^0 + kT \ln (K_1 c_1)$.

When the force that is induced by current is averaged by considering all the jumps of interstitials that diffuse, $\vec{F}^e$, which is an average force, proportional to $\vec{E}$ results, i.e.,

$$\left(\vec{F}^e \right) = \beta \vec{E}; \beta = \text{constant}. \tag{5.58}$$

If the dilute species obeys the law of Henry, the interstitial flux experiences a gradient in concentration and a force that is electrostatic drives the system. That is,

$$\overrightarrow{J_1} = -D_1 \left(\nabla c_1 - \frac{c_1 \beta}{kT} \vec{E} \right). \tag{5.59}$$

Uphill diffusion continues till the gradient due to concentration (α ∇c_1) and the term arising from electromigration cancel each other. All terms other than β can be experimentally measured while β is deduced from the rest of the quantities. Electromigration purifies metals by collecting interstitials at a specimen end. In IC (integrated circuit) devices electromigration carries the current and here the conductors are very narrow. The electric field generates a vacancy mechanism that causes substitutional atoms to migrate as directed by the electric field. When severe, electromigration can spoil an electronic device by transporting thin mass of metal conductors away from thin regions. When high current density is present in thin regions Joule heating can destroy a device in a rapid fashion.

5.14 Effect of a Thermal Gradient on Mass Diffusion

Concentration as well as temperature affect interstitial chemical potential. A dilute component 1 in a dilute interstitial can experience flow of heat as also diffusion of mass. That is,

$$\vec{J_1} = -L_{11} \nabla \mu_1 - \frac{L_{1Q}}{T} \nabla \mathrm{T}. \tag{5.60}$$

$$\mathrm{d}\mu_1 \ (c_1, T) = \left(\frac{\partial \mu_1}{\partial c_1}\right)_T \mathrm{d}c_1 + \left(\frac{\partial \mu_1}{\partial T}\right)_{c_1} \mathrm{dT}. \tag{5.61}$$

The partial atomic entropy is given by,

$$\overline{S_1} = \left(\frac{\partial \mu_1}{\partial T}\right)_{c_1}. \tag{5.62}$$

Therefore,

$$\mu_1 = \left(\frac{\partial \mu_1}{\partial c_1}\right)_T \nabla c_1 - \overline{S_1} \nabla \mathrm{T}. \tag{5.63}$$

Equations (5.64) and (5.65)

$$\mu_1 = \mu_1^0 + kT \ \ln \ (K_1 c_1) \tag{5.64}$$

and

$$D_1 = M_1 kT = L_{11} \left(\frac{kT}{c_1}\right) \tag{5.65}$$

, along with Eqs. (5.60) and (5.63) give

$$\vec{J_1} = -D_1 \nabla c_1 - \frac{D_1 c_1 Q_1^{trans}}{kT^2} \nabla \mathrm{T}. \tag{5.66}$$

Q_1^{trans} has the dimensions of energy and is referred to as the heat of transport,

$$Q_1^{trans} = \frac{L_{1Q}}{L_{11}} - T\overline{S_1}. \quad (5.67)$$

It follows from Eq. (5.66) that gradients in composition, temperature or both can induce mass diffusion. The heat of transport, Q_1^{trans}, decides the extent of coupling between temperature gradient and mass diffusion. Q_1^{trans} originates from the difference in the three energy states before a species jumps into an adjacent site, as well as during and after the jump. Like in an electromigration experiment, measurements are made to determine β. A temperature gradient causes a quasi-steady state, where $\overrightarrow{J_1} = 0$. Then, in that state the gradient in concentration is experimentally measured and the heat transferred is then calculated using, $Q_1^{trans} = -\frac{kT^2 \nabla c_1}{(c_1 \nabla T)}$. In case of carbon (an interstitial atom) in α-iron, the diffusion of the interstitial atoms is towards specimen's hot end, i.e., heat transport is negative.

5.15 Mass Diffusion Motivated by Capillarity

In the direct vicinity of components local interface curvature decides their diffusion potentials. If in interfaces curvatures are observed, the differences in the potential drives diffusion in a direction that reduces that favors the overall energy reduction in the system.

Figure 5.4 is an example of an undulating surface where a vacancy exchange mechanism causes self-diffusion in pure crystalline material. Just below the convex surface the boundary experiences a higher potential for atomic diffusion than where the boundary is concave thus establishing from the convex to the concave parts a higher diffusion current that passes through the bulk (see Fig. 5.4) reducing the total interfacial energy by smoothing the surface. Destruction or creation of vacancies at dislocations within the bulk is ignored as a justifiable approximation. The surface smoothing rate is then determined from (diffusion) equation and the atom flux using the boundary conditions at the surface and solving the diffusion equation.

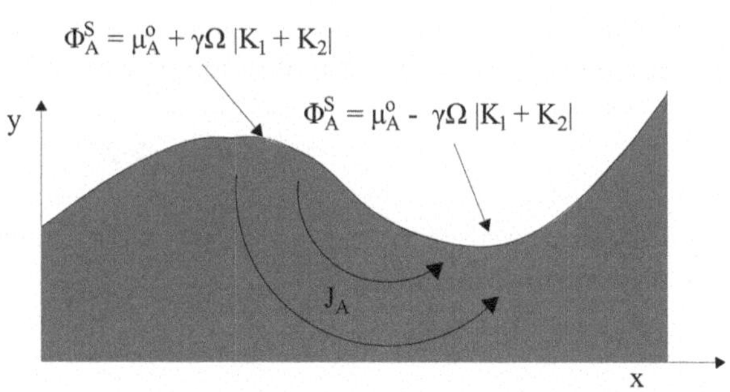

Fig. 5.4 Positively and negatively curved regions in a wavy surface produce curvature-dependent diffusion potential gradients. In turn, this causes surface smoothing by diffusional transport [1]

5.16 The Flux and Diffusion Equations

Host atoms and vacancies constitute two network-constrained components. When temperature is constant and an electric field does not exist, the flux equation reduces to,

$$\begin{aligned}\overrightarrow{J_A} &= L_{AA}\overrightarrow{F_A} = -L_{AA}\nabla\varphi_A = -L_{AA}\ \nabla(\mu_A - \mu_V).\\ \overrightarrow{J_V} &= -\overrightarrow{J_A}.\end{aligned} \tag{5.68}$$

The coefficients L_{AA} correspond to flat surfaces in a large crystal. In a system consisting of N_A atoms and N_V vacancies (a dilute solution),

$$\underline{G} = N_A\mu_A^0 + N_V G_V^f + kT\left[N_A\ ln\left(\frac{N_A}{N_A+N_V}\right) + N_V\ ln\left(\frac{N_V}{N_A+N_V}\right)\right]. \tag{5.69}$$

Here μ_A^0 is free energy per only atom A containing crystal without vacancies and having zero curvature surfaces (i.e., flat). Then, $G_V^f = H_V^f - TS_V^f$, where S_V^f is the G_V^f excludes the mixing free energy. The mixing free energy arising from entropy caused by a random vacancy distribution is represented by the last term.

The problem of smoothing the surface shown in Fig. 5.4 is often simplified by setting $\left(\frac{\partial c_A}{\partial t}\right) = 0$ because, to a good approximation, it is a field which describes diffusion in a state that is quasi-steady. The problem, therefore, amounts to finding a solution (within crystal) for (5.70), subject to the boundary conditions on φ_A (Note: The equation has the form of Laplace equation).

$$\nabla^2\varphi_A \tag{5.70}$$

5.17 Boundary Conditions

The boundary conditions on the diffusion potential $\varphi_A = \mu_A - \mu_V$ are found using earlier results. $\mu_V = 0$ because vacancies are in equilibrium at the surface. The surface work term $\Delta W = \gamma k\Omega_V$ (γ is the isotropic surface tension, Ω_V is the crystal expansion because of the vacancy addition, k is the local curvature) plus the chemical term, μ_A^0 represents the diffusion potential of atoms. Or,

$$\varphi_A^S = \mu_A^0 + \gamma k\Omega_V. \tag{5.71}$$

5.18 Mass Diffusion in Presence of Stress

Mobility is affected by stress, diffusion potential, and boundary conditions for diffusion. Stress, whose selected effects are studied in isolation, enables one to understand diffusion under stress.

5.18.1 Mobilities: Stress Effects

Small interstitial atoms diffuse isothermally among interstices between large host atoms in unstressed crystals. That is,

$$\overrightarrow{J_1} = -L_{11}\nabla\mu_1. \tag{5.72}$$

It is assumed that isotropic diffusion is present. Mobility, M_1 is a scalar.

In a uniform stress field, due to its energy being position-independent, a diffusing interstitial experiences no force. Flux is proportional to the gradient of chemical potential, i.e., $\overrightarrow{J_1} = -M_1c_1\nabla\mu_1$, if no other fields are present. But as stress affects the atomic migration rates in different directions in all types of crystals, the mobility M_1 will be a tensor. In other words, the host lattice will be distorted when one interstitial moves from its site to another interstitial site. The work to overcome the resistance to distortion involves a jump against a stress field. The stress fields of a fixed nature experience different distortions in different directions. Therefore, during these jumps, the stress field should be overcome by work, W. Arrhenius relationships determine the rate of a particular jump without stress, that is, the rate α $\exp\left(\frac{-G^m}{kT}\right)$. Here, to jump the stress barrier, free energy, G^m, is necessary. In presence of stress, W also becomes a part of the energy barrier. Or, α $\exp\left[\frac{-(G^m+W)}{(kT)}\right]$. For practically important cases, W/(kT) being small,

$\exp\left[\frac{W}{(kT)}\right] \approx 1 - \left(\frac{W}{kT}\right)$. Hence, the exponential term becomes $\exp\left[\frac{G^m}{(kT)}\right]\left[1 - \frac{W}{(kT)}\right]$. The interstitials making different types of jumps decide the interstitial overall mobility. A jump and W are linearly related. The final expression for mobility becomes,

$$M_{ij} = M_{ij}^0 + \sum\nolimits_{kl} M_{ijkl}\sigma_{kl}. \tag{5.73}$$

Similar arguments are valid for the substitutional atoms migration in a stress field.

Stress, for example, can form an atmosphere of a solute and atom around a dislocation.

When stress field is nonuniform, a force acts on an interstitial atom in a direction that lowers the interaction energy. If stress does not affect mobility, the force arises from the non-uniform stress field. In the process, the interstitial solute atoms diffuse faster in an inhomogeneous stress field of an edge dislocation (stress-induced diffusion), as shown in Fig. 5.5.

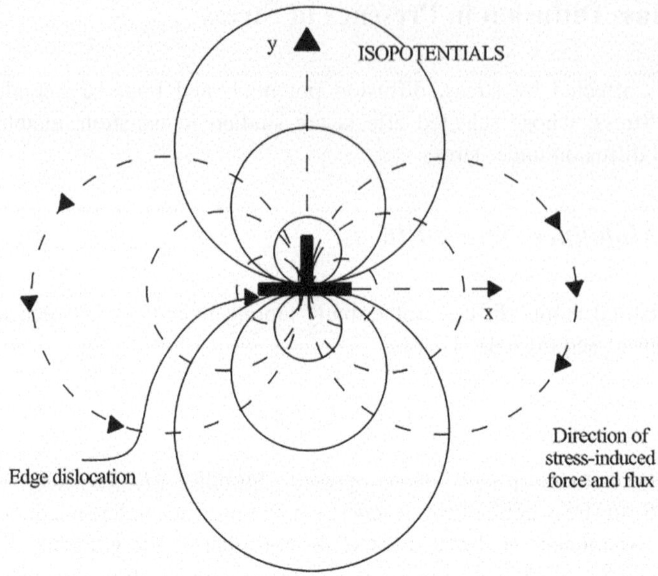

Fig. 5.5 Cylinders of iso-potential for the portion of potential due to diffusion of an interstitial atom in the stress field of an edge dislocation (hydrostatic) in an isotropic solid are indicated by solid lines. The interstitial atom experiences a force in a direction indicated by dashed lines and tangential arrows [1]

As shown in Fig. 5.6 an interstitial is oversized compared with the lattice atoms. This pushes outward the neighborhood atoms (positive center of dilation). This results in volume expansion. The force exerted on an interstitial atom is found by entropy production in a small cell embedded in the material.

According to Cottrell, the rate of solute atoms diffusing to dislocations and then pinning them in place is α (time $^{(2/3)}$). This explained quantitatively how solute pinning held back dislocations during strain aging.

5.19 Diffusional Creep: Stress Effects on Diffusion and the Role of the Boundary Conditions

During diffusion creep diffusion currents between sources and sinks transport atoms and change the specimen shape dictated by the stress imposed—see Fig. 5.7.

If a sample with a "bamboo" grain structure is acted upon by a stress $\overrightarrow{F_{app}}$ (tensile), the diffusion potential in the transverse grain boundaries gets reduced. In contrast, no normal component of stress is present on the surface of the specimen. So, as an approximation, applied stress is assumed not to affect the diffusion potential. If $\overrightarrow{F_{app}}$ is large, the surface diffusion potential is far more than at transverse

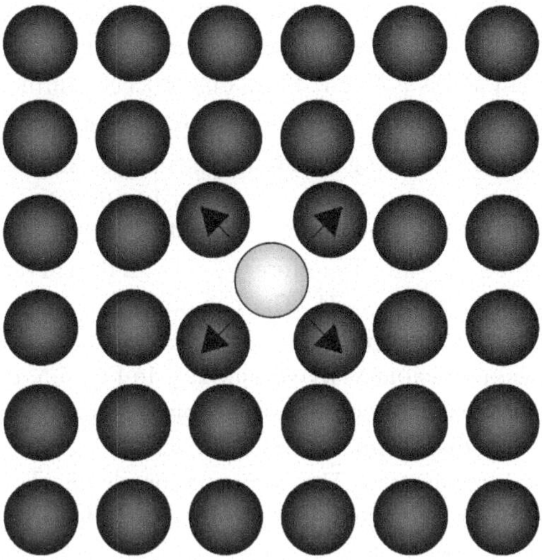

Fig. 5.6 Dilation due to an interstitial atom in a crystal. The outward displacements of the interstitials of the nearest neighbors are indicated by arrows [1]

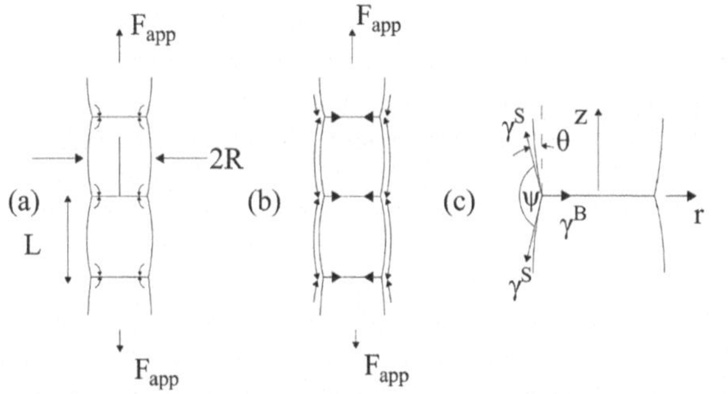

Fig. 5.7 A sample under a tensile, uniaxial stress, σ_{zz}, due to a force, $\overrightarrow{F_{app}}$. (**a**) displays fluxes of diffusion in the bulk, (**b**) shows diffusion connected with grain boundary as well as surface events which give rise to fluxes. They lead to elongation due to diffusion. (**c**) The grain boundary-surface junction (an enlarged view) [1]

boundaries. Surface atoms become sources of atoms and diffuse to the transverse boundaries, which are the sinks. This way, the external stress elongates the specimen. Sources and sinks action of dislocations due to the diffusion potential changes with the inclination of every dislocation. Diffusion potential gradients give rise to vacancy fluxes, leading to an edge dislocation climb. And so, a wire elongates in the direction of the tensile stress.

Elongation rate then reduces to a problem in diffusion determined by the boundary value. That is, the inclination of the applied stress to the sample along with its magnitude decides the boundary conditions at the sources and sinks. The method of obtaining elongation rate of the polycrystalline wire when the metal is of cubic shape and bulk diffusion is present assumes a vacancy exchange mechanism.

Diffusion and flux Equations: For inducing diffusional creep, small stresses are applied. Therefore, vacancy concentration variation throughout the specimen is small and is ignored. Then, one proceeds further by using $\varphi_A = \mu_A - \mu_V$, $\overrightarrow{J_A} = -\frac{D_V X_V}{(\Omega)kT}\nabla\varphi_A$, and $\nabla^2\varphi_A$ (equations derived earlier for the case where diffusion leads to surface smoothing assuming quasi-steady state diffusion and ignoring the generation and destruction of vacancies at dislocations).

Boundary conditions are as follows. The surface of a cylindrical wire is a vacancy source and sink. Therefore, $\mu_V = 0$. At the curved surface the diffusion potential, φ_B^S, is given by Eq. 5.71. The condition $\mu_V = 0$ is valid at the grain boundaries. The boundary traction $\sigma_{nn} = \widehat{n}^T.\sigma.\widehat{n}$. By an atom insertion, the tractions can be altered. The grain expands by volume Ω_A. In Fig. 5.5, the boundary orientation is such that its normal and the stress-direction are parallel, i.e., $\sigma_{nn} = \sigma_{zz}$.

The displacement does work, $\sigma_{nn}\Omega_A = \sigma_{zz}\Omega_A$. The potential energy is reduced by the same amount. This term and the chemical term, μ_A^0, are added together. So, the diffusion potential along the grain boundary is,

$$\varphi_A^B = \mu_A^0 - \sigma_{nn}\varphi_A - \mu_V = \mu_A^0\sigma_{nn}\Omega_A. \tag{5.74}$$

With a stress enhancement, φ_A^B decreases. σ_{nn} and external force are directly related. For very large σ_{nn} $\varphi_A^B < \varphi_A^S$. Atoms will diffuse at a quasi-steady state from the surface to the boundaries. In other words, the bamboo wire displays a quasi-steady state diffusional transport of a viscous material. Complete solutions are available for the elongation rates caused by fluxes arising from grain boundary- and surface-diffusion (Fig. 5.5a, b).

Problems

1.	The flux of a diffusing species is given by: $J = -D\frac{dC}{dx}$ Where, J is the diffusion flux$^{mol}/_{m^2s}$, D is the diffusion coefficient $(^{m^2}/_s)$, $\frac{dC}{dx}$ is the concentration gradient$(^{mol}/_{m^3.m})$. If the diffusion coefficient $D = 1\times10^{-9}\,{}^{m^2}/_s$ and the concentration gradient $\frac{dC}{dx} = 5\times10^6\,{}^{mol}/_{m}$, Calculate the diffusion flux **Solution:** Substitute the given values into the flux equation $J = -(1\times10^{-9}).(5\times10^6) = -5\times10^{-3}\,{}^{mol}/_{m^2s}$
2.	The diffusion potential ϕ_{diff} is related to the concentration gradient of a species as follows: $\phi_{diff} = -\frac{RT}{zF}\,ln\left(\frac{C_2}{C_1}\right)$ Where, R is the gas constant, T is the Temperature(K), z is the change number of the species, F is the Faraday constant and C_1 and C_2 are the concentrations of the species at point 1 and 2. If T = 300 K, z = 1, C_1 = 0.5 mol/L, C_2 = 0.2 mol/L, calculate the diffusion potential **Solution:** Substitute the values in the equation $\Phi_{diff} = \frac{8.314\times300}{1\times96485}\,ln\left(\frac{0.2}{0.5}\right) = 0.0225\text{V}$

Reference

1. Balluffi, R. W., Allen, S. M., & Craig Carter, W. (2005). *Kinetics of materials.* A John Wiley & Sons.

Chapter 6
Diffusion Equation

Contents

6.1 Fick's Second Law

The diffusion equation describes the evolution of the concentration field due to a flux. The solution to this differential equation, with appropriate boundary and initial conditions, gives the time- and spatial-dependence of concentration. Various forms of the diffusion equation for the flux, consistent with Fick's law, are obeyed in different situations, e.g., the diffusivity is constant, a function of concentration, a function of time, or a function of direction. If the diffusive flux is $\vec{J}$, the diffusion equation of the general form is,

K. A. Padmanabhan et al., *Thermodynamics and Kinetics of Materials*, Engineering Materials, https://doi.org/10.1007/978-3-032-14527-7_6

$$\frac{\partial c}{\partial t} = \dot{n} - \nabla \vec{J}. \quad \left(\vec{J} \text{ is referred to a V-frame.} \right) \tag{6.1}$$

$\dot{n}$ is an added source or sink term corresponding to the rate of creation per unit volume of the diffusing material, by a chemical reaction or fast-particle irradiation. Often, $\dot{n}$ in Eq. (6.1) is zero. When Fick's law applies and $\dot{n} = 0$, Eq. (6.1) takes the general form,

$$\frac{\partial c}{\partial t} = -\nabla.\ \vec{J} = \nabla.\ (D\nabla c). \tag{6.2}$$

This is a consequence of the conservation of the diffusing species and is sometimes referred to as Fick's second law. The fluxes determine the accumulation within a volume at its boundary. In one dimension,

$$\frac{\partial N}{\partial t} = \int_{x_1}^{x_2} \frac{\partial c}{\partial t} A dx = \int_{x_1}^{x_2} \frac{\partial}{\partial x} A dx = (J(x_1, t) - (J(x_2, t))A. \tag{6.3}$$

Here A is the area through which diffusion occurs and N is the number of particles. In three dimensions,

$$\frac{\partial N}{\partial t} = \int_V \frac{\partial c}{\partial t}\, dV = \int_V -\nabla \vec{J}\, dV = -\int_{\partial V} \vec{J}\left(\vec{r}, t\right).\vec{n}\, dA. \tag{6.4}$$

Here, in the final integral, $\vec{J}\left(\vec{r}, t\right)$ is a time-dependent value of flux at the oriented surface ∂V that bounds V. Figure 6.1 shows how $c\ (x, t)$ changes locally (geometric interpretation). The above equations applicable for the entire concentration field present mathematically "the conservation constraint".

Figure 6.1 points out that concentration field evolution according to Fick's law, i.e., $\left(\frac{\partial c}{\partial t}\right)$ is proportional to the curvature of the concentration field. Typical boundary conditions are,

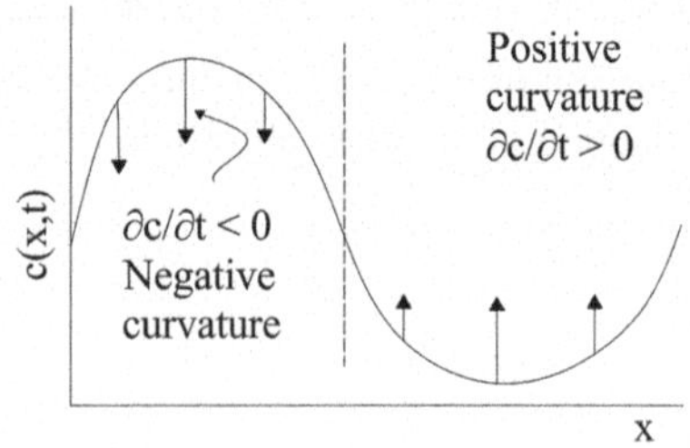

Fig. 6.1 Evolution of concentration field according to Fick's law. $\partial c/\partial t$ is proportional to the curvature of the concentration field

$$c\ \left(\vec{r} = \vec{r_B}\right) = f(t) = c_B(t) \quad \text{or} \quad \vec{J}\left(\vec{r} = \vec{r_B}\right)\ \hat{n}_B = g(t) = J_B(t). \tag{6.5}$$

Here $\hat{n}_B$ is the normal to the boundary. The initial conditions have the form,

$$c(x, y, z, t = t_0) = c\left(\vec{r}, t = t_0\right) = h(x, y, z) = h\left(\vec{r}\right) = c_0(x, y, z). \tag{6.6}$$

The simplest case is when in Eq. (6.2) D is constant. D could also be a function of concentration, particularly in concentrated solutions in which interactions among the solute atoms could be significant. Also, when the temperature of the diffusing body changes with time D could be a function of time. In anisotropic materials D may also depend upon the direction of diffusion.

6.1.1 Linearization of the Diffusion Equation

The diffusivity, when expanded about its average value, $\tilde{D_0}$ appears as follows.

$$\tilde{D}(c) = \tilde{D_0}(c) + \frac{\tilde{D_1}}{\langle c\rangle}\ \Delta c + \ldots \tag{6.7}$$

$$\text{Where } \Delta c = c - \langle c\rangle, \text{and } \tilde{D_1} = \left.\frac{\partial\tilde{D}}{\partial c}\right|_{c=\langle c\rangle} \langle c\rangle. \tag{6.8}$$

Then the diffusion equation becomes

$$\frac{\partial c}{\partial t} = \tilde{D_0}\nabla^2 c + \frac{\tilde{D_1}}{\langle c\rangle}\Delta c\nabla^2 c + \frac{1}{\langle c\rangle}\nabla c.\ \tilde{D_1}\ .\nabla c + \ldots. \tag{6.9}$$

For small Δc and small $|\nabla c|$ the first order approximation is

$$\frac{\partial c}{\partial t} = \tilde{D_0}\nabla^2 c. \tag{6.10}$$

This equation for diffusion assumes constant diffusivity.

6.2 Fick's Law and the Heat Equation

Evolution of a temperature field during heat flow also has the form

$$\frac{\partial h}{\partial t} = -\nabla \cdot \overrightarrow{J_Q}; c_P \frac{\partial T}{\partial t} = -\nabla \cdot (-K \nabla T); \frac{\partial T}{\partial t} = \nabla \cdot \left(\frac{K}{c_P} \nabla T \right) = \nabla \cdot (k \nabla T). \quad (6.11)$$

Here h is the enthalpy density, c_P the heat capacity per unit volume, $\frac{K}{c_P}$, the thermal diffusivity, K. By assumption, no enthalpy is stored by a phase change and c_P is constant.

Therefore, all results obtained from the form of Fick's second law applies to heat as well as concentration evolution. It should be noted that the equations for thermal and mass diffusion are different. The mass diffusion equation, $\frac{\partial c}{\partial t} = \nabla. \nabla \cdot D \nabla c$, represents a partial-differential equation for the density of an extensive quantity. In the thermal case, $\frac{\partial T}{\partial t} = \nabla \cdot k \nabla$T pertains to an intensive quantity.

6.2.1 The Case Where the Diffusivity Constant

Different methods are available to solve the diffusion equation for several initial and boundary conditions when D is constant and therefore has the simple form

$$\frac{\partial c}{\partial t} = \mathrm{D} \nabla^2 c \quad (6.12)$$

This is a second-order linear partial differential equation. The solution has uniqueness and existence theorems as well as maximum and minimum values for many sets of initial and boundary conditions. The heat by "diffusion"—thermal conduction _ and mass diffusion differential equations have the same form. Please note that concentration comes in the place of temperature and mass diffusivity is substituted by thermal diffusivity, K. So, many heat-flow boundary-value problem solutions are also used as solutions to the respective mass diffusion problems. The solutions to Eq. (6.10) are given below for simple boundary conditions. If the localized sources can be visualized as diffusing out into an infinite medium, the solution will be a spreading Gaussian distribution. That is,

$$c\left(\vec{r}, t\right) = \frac{n_d}{2^d (\pi D t)^{(\mathrm{d}/2)}} \cdot e^{-\left(\vec{r} . \vec{r}\right)(4Dt)}. \quad (6.13)$$

Here d represents the dimensionality of the space involved in the diffusion problem. n_d is the source strength (i.e., the total number of particles in the source). When a distribution of sources represents the initial condition, superposition by integration of the individual solutions leads to the final solution. When planar orthonormal surfaces define the boundaries, trigonometric series appear as the solutions. For diffusion in a cylinder, the solution is s sum over Bessel function. For diffusion problems with spherical symmetry, the solution takes the form of trigonometric functions. All such problems are solved by the variable separable method.

Table 6.1 Selected laplace transform pairs (after Balluffi et al. [1])

$\widehat{c}(x,p) = \int_0^\infty e^{-pt} c(x,t)dt$ $q \equiv \sqrt{p/D}$	
$\frac{1}{p}$	1
$\frac{1}{p^{\nu+1}}$ $\nu > -1$	$\frac{t^\nu}{\Gamma(\nu+1)}$
$\frac{1}{p+\alpha}$	$e^{-\alpha t}$
$\frac{\omega}{p^2+\omega^2}$	$\sin\omega t$
$\frac{p}{p^2+\omega^2}$	$\cos\omega t$
e^{-qx}	$\frac{xe^{-x^2}/(4Dt)}{\sqrt{4\pi Dt^3}}$
$\frac{e^{-qx}}{q}$	$\sqrt{\frac{D}{\pi t}}e^{-x^2}/(4Dt)$
$\frac{e^{-qx}}{p}$	$\text{erfc}\left(\frac{x}{\sqrt{4Dt}}\right)$
$\frac{e^{-qx}}{pq}$	$\sqrt{\frac{4Dt}{\pi}}e^{-x^2/(4Dt)} - \text{erfc}\left(\frac{x}{\sqrt{4Dt}}\right)$

A third method is solution by Laplace transforms. For complicated problems, e.g., those that involve boundary conditions that involve time, Laplace transforms are used. Back-transformation to the desired solution constitutes the difficult part here. Luckily, Laplace transform tables and tables of integrals are available for many problems (Table 6.1).

6.3 Steady State

Diffusion in a steady state with a composition profile that is time-independent constitutes a simple case. This is often found in finite samples at very long times when the boundary conditions are constant. Then $\partial c/\partial t = 0$. All local accumulation vanishes (no divergence), and the diffusion equation becomes the Laplace equation,

$$\nabla^2 c = 0. \tag{6.14}$$

The solutions for such equations are known as harmonic functions.

6.3.1 One Dimension

What is considered is diffusion through an infinite flat plate of thickness L, with $0 < x < L$. Boundary conditions

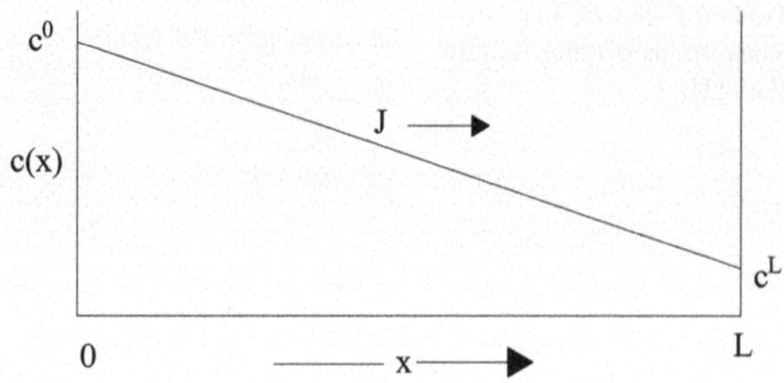

Fig. 6.2 Concentration, $c(x)$ vs. x. Steady state diffusion through a plate [1]

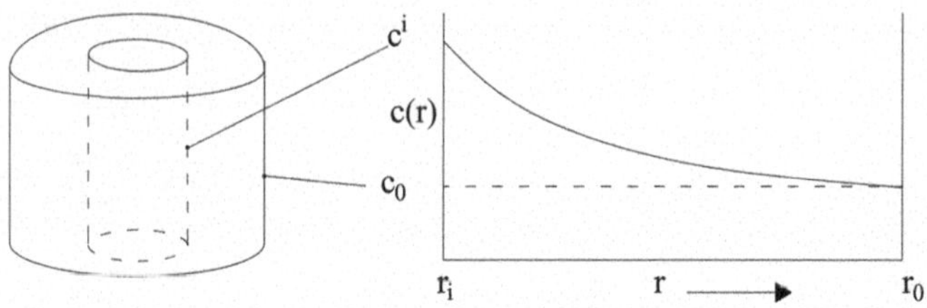

Fig. 6.3 Steady state diffusion in cylindrical shell. Concentration, $c(r)$ vs. r [1]

$$c(0,t) = c^0; c(L,t) = c^L. \tag{6.15}$$

Integrating the one-dimensional Laplacian, $\left(\frac{d^2c}{dx^2}\right) = 0$, one gets, $c(x) = a_1x + a_2$. Here a_1 and a_2 are integration constants. Use the boundary conditions, given in Eq. (6.15) a_1 *and* a_2 are obtained. Then, the one-dimensional steady-state solution is

$$c(x) = c^0 - \left(c^0 - c^L\right)\frac{x}{L}. \tag{6.16}$$

This means that the concentration varies linearly across the plate—see Fig. 6.2.

The flux is proportional to the slope and is constant, i.e.,

$$J = -D\frac{dc}{dx} = D\left(\frac{c^0 - c^L}{L}\right). \tag{6.17}$$

6.3.2 *Cylindrical Shell*

Consider steady-state diffusion through a cylindrical shell with inner radius r^{in} and outer radius r^{out} as in Fig. 6.3

The boundary conditions are,

$$c(r^{\text{in}}, \theta, z, t) = c^{\text{in}}; c(r^{\text{out}}, \theta, z, t) = c^{\text{out}}. \tag{6.18}$$

The Laplacian operator operating on $c(r, \theta, z)$ in cylindrical coordinates is,

$$\nabla^2 c = \frac{1}{r}\frac{\partial}{\partial r}\left(r\frac{\partial c}{\partial r}\right) + \frac{\partial^2 c}{\partial \theta^2} + \frac{\partial c^2}{\partial z^2}. \tag{6.19}$$

The boundary conditions are independent of θ and z. So, the solution must satisfy,

$$\frac{d}{dr}\left(\frac{1}{r}\frac{dc}{dr}\right) = 0. \tag{6.20}$$

Integrating twice one obtains

$$c(r) = a_1 \ln r + a_2. \tag{6.21}$$

Applying the boundary conditions, one gets

$$c(r) = c^{\text{in}} - \left[\frac{c^{\text{in}} - c^{\text{out}}}{\ln(r^{\text{out}} - r^{\text{in}})} \ln\left(\frac{r}{r^{\text{in}}}\right)\right]. \tag{6.22}$$

$$\text{The flux } J = -D\left(\frac{dc}{dr}\right) = D\left[\frac{c^{\text{in}} - c^{\text{out}}}{\ln(r^{\text{out}}/r^{\text{in}})}\right]\frac{1}{r}; \text{i.e., it depends inversely on } r. \tag{6.23}$$

Total current of particles entering the inner surface per unit length of the cylinder, $I = [J\,((r^{\text{in}})2\pi r^{\text{in}}]$, is the same as the total current leaving the outer surface, which is a condition for the steady state.

6.3.3 Spherical Shell

The Laplacian operator in spherical coordinates on c (r, θ, φ) is

$$\nabla^2 c = \frac{1}{r^2}\left[\frac{\partial}{\partial r}\left(r^2\frac{\partial c}{\partial r}\right) + \frac{1}{\sin\theta}\frac{\partial}{\partial \theta}\left(\sin\theta\frac{\partial c}{\partial \theta}\right) + \frac{1}{\sin^2\theta}\frac{\partial^2 c}{\partial \theta^2}\right]. \tag{6.24}$$

The steady state solution with boundary conditions dependent only on r is obtained by integrating the above equation twice and determining the two constants using the boundary conditions.

6.3.4 Variable Diffusivity

If steady state conditions prevail, the equation will be of the form $\nabla\,(D\,\nabla\,c) = 0$. By integrating, for one dimension the solution will be

$$c(x) = c(x_1) + a_1 \int_{x_1}^{x} \frac{d\xi}{D(\xi)}. \tag{6.25}$$

6.4 Non-steady State Diffusion

Here the diffusion profile is time dependent. Solutions to non-steady state diffusion problems are complicated. They are not considered in this book.

6.4.1 Solutions Involving the Error Function

The boundary and initial conditions are:

$$c(x,0) = \begin{Bmatrix} c_0 \; x > 0 \\ 0 \; x < 0 \end{Bmatrix}$$

$$\frac{dc}{dx} c(x = \mp\infty\,, t) = 0. \tag{6.26}$$

A uniform distribution of point, line, or planar sources simulates the initial distribution and is placed along $x > 0$ (see, Fig. 6.4) to get a solution. The amount of diffusant (strength) contributed by each source, is (c_0 dx).

The distance between the general point x and the source is $(\xi - x)$. At a general position x the contribution from a source at some other position ξ is

$$c(x,t) = \frac{c_0}{\sqrt{(4\pi Dt)}} e^{-(\xi - x)^2/(4Dt)}. \tag{6.27}$$

The solution satisfying Eq. (6.16) is the integral over all sources, viz.,

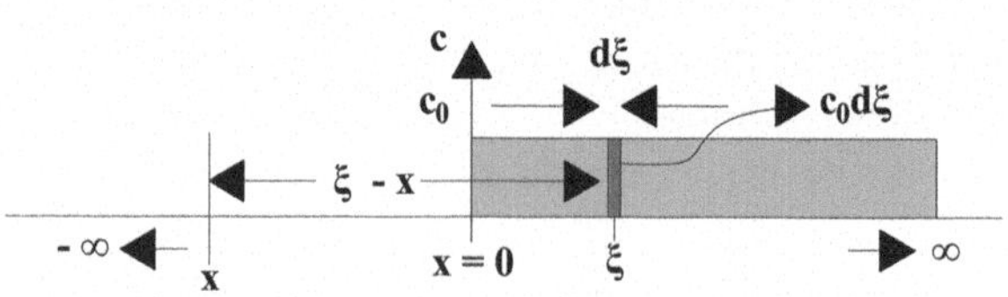

Fig. 6.4 Diagram used to determine the contribution at the general point x of a local source of strength ($c_0 d\xi$) located at ξ [1]

$$c(x,t) = \frac{c_0}{\sqrt{(4\pi Dt)}} \int_0^{\infty} e^{-(\xi - x)^2/(4Dt)}\, \mathrm{d}\xi. \tag{6.28}$$

Substitute $u = \frac{(\xi - x)}{\sqrt{(4Dt)}}$. Error function, (z) is defined as $\frac{2}{\sqrt{\pi}} e^{-\xi^2} d\xi$. Then, from the properties of even integrands,

$$c(x,t) = \frac{c_0}{\sqrt{\pi}} \int_{-\infty}^{x/2(Dt)} e^{-u^2} du = \frac{c_0}{2} + \frac{c_0}{2} \operatorname{erf}\left(\frac{x}{2\sqrt{(Dt)}}\right). \tag{6.29}$$

Summations over point-, line-, or planar-source solutions are obtained by the general method of Green functions (not examined here).

6.4.2 Fourier Series

If a function $u(x)$ exists on the interval $-L < x < L$, $u(x)$ can be expressed as

$$u(x) = \frac{b_0}{2} + \sum_{n=1}^{n=\infty} \left[a_n \sin\left(n\pi \frac{x}{l}\right) + b_n \cos\left(n\pi \frac{x}{L}\right)\right]. \tag{6.30}$$

The coefficients are given by,

$$a_n = \frac{1}{L} \int_{-L}^{L} u(x) \sin\left(n\pi \frac{x}{L}\right) dx, \tag{6.31}$$

and $b_n = \frac{1}{L} \int_{-L}^{L} u(x) \cos\left(n\pi \frac{x}{L}\right) dx$.

If $u(x)$ is an odd function, the cosine terms will vanish and

$$a_n = \frac{2}{L} \int_0^L u(x) \sin\left(n\pi \frac{x}{L}\right) dx. \tag{6.32}$$

If $u(x)$ is an even function, the sine terms will disappear and

$$b_n = \frac{2}{L} \int_0^L u(x) \cos\left(n\pi \frac{x}{L}\right) dx. \tag{6.33}$$

6.4.3 Laplace Transforms

For time-dependent boundary condition and semi-infinite domain problems Laplace transforms are very useful. Once the original boundary-value problem has been subjected to a Laplace transform, the transformed equation is often easily solvable. This solution is then back-transformed to get the required answer.

The Laplace transform of a function $f(x, t)$ is given by

$$Ł\ \{f(x,t)\} = \widehat{f}(x,p) = \int_0^\infty e^{-pt} f(x,t)dt. \tag{6.34}$$

"p" is the transformation variable. The key utility of the Laplace transform is the operation on time derivative.

$$Ł\left\{\frac{\partial f}{\partial t}\right\} = \int_0^\infty e^{-pt\frac{\partial f(x,t)}{\partial t}dt}. \tag{6.35}$$

Integration by parts leads to

$$Ł\left(\frac{\partial f}{\partial t}\right) = pŁ(f) - f(x,t=0). \tag{6.36}$$

From Eq. (6.24) the Laplace transform of a spatial derivative of f is equal to the spatial derivative of $\widehat{f}$.

In other words,

$$\left(\frac{\partial^n f}{\partial x^2}\right) = \frac{\partial^n f(x,p)}{\partial x^n}. \tag{6.37}$$

6.4.4 Example (i)

Diffusion into a semi-infinite body where the surface concentration c ($x = 0$, t) is fixed.

$$c(x=0,t) = 0; c(x=\infty\ ,t) = 0; c(x,t=0) = 0 \text{ for } 0 < x < \infty\ . \tag{6.38}$$

Apply the Laplace transform to both sides of the diffusion equation. Then,

$$Ł\left(\frac{\partial c}{\partial t}\right) = DŁ\left(\frac{\partial^2 c}{\partial x^2}\right); \frac{\partial^2 \widehat{c}}{\partial x^2} - \frac{p}{D}\widehat{c} = -\frac{c(x,t=0)}{D} = 0. \tag{6.39}$$

As a result, the Laplace transform removes the t-dependence and converts the partial differential equation into an ordinary differential equation.

$$\widehat{c}(x,p) = a_1 e^{\sqrt{\left(\frac{p}{Dx}\right)}} + a_2 e^{-\sqrt{\left(\frac{p}{Dx}\right)}}. \tag{6.40}$$

Transforming the boundary conditions, one gets,

$$\partial(x=0,p)=c_0\int_c^{\infty} e^{-pt}dt=\frac{c_0}{p};\ \frac{\partial\hat{c}}{\partial x}(x=\infty\ ,p)=0. \tag{6.41}$$

(p should be such that its real part must be positive and large and the integral should converge.)

Solving for a_1 and a_2 in Eq. (6.40), one obtains,

$$\hat{c}(x,p)=\frac{c_0}{p}e^{-\sqrt{\left(\frac{p}{D}\right)}x}. \tag{6.42}$$

Using the table of Laplace transforms (widely available) this solution is then inversely transformed to obtain the desired solution. For this case, the solution is:

$$c(x,t)=c_0\left[1-erf\left(\frac{1}{\sqrt{(4Dt)}}\right)\right]\equiv c_0 erfc\left(\frac{x}{\sqrt{(4Dt)}}\right), \tag{6.43}$$

where $erfc(z)\equiv[1-\ erf(z)]$ is the complementary error function.

6.4.5 Example (ii)

Time-dependent boundary conditions:

Here, for a constant flux of J_0 on the surface of a semi-infinite solid, the boundary conditions are,

$$\frac{\partial c}{\partial x}(x=0,t)=-\frac{J_0}{D}=\text{constant};\ c(x=\infty\ ,t)=c_0;\ c(x,t=0)=c_0,\ \text{for}\ 0\leq x\leq\infty\ . \tag{6.44}$$

When the rate is moderated by solute absorption due to the presence of some thin passive surface layer, the boundary condition might apply. Then, the surface concentration at $x=0$ must be a function of time for the constant-flux condition to be valid (see Fig. 6.5).

From Laplace transformation,

$$\frac{\partial^2 c}{\partial x^2}-\frac{p}{D}\hat{c}=\frac{c(x,t=0)}{D}=\frac{c_0}{D}. \tag{6.45}$$

This is an inhomogeneous ordinary differential equation. Its solution is the sum of the solution of its homogeneous form and the solution, $\hat{c}=\left(\frac{c_0}{p}\right)$. As before, using the boundary conditions and evaluating the coefficients, one reaches a conclusion that the surface concentration increases as

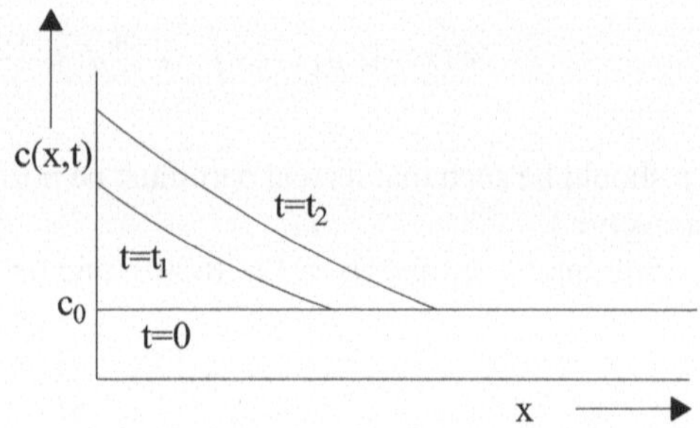

Fig. 6.5 Diffusion profiles necessary to maintain constant flux into a semi-infinite body. Note the fixed value of $\frac{dc}{dx}\big|_{x=0}$ for $t > 0$

$$c(x,t) = c_0 + 2J_0\sqrt{\frac{t}{\pi D}}. \tag{6.46}$$

6.5 Time of Approach to Steady State and Diffusion Depth

An approximate estimate of the diffusion penetration distance from a point source is the location where the concentration has fallen off by $\approx \left(\frac{1}{e}\right)$ of the concentration at $(x = 0)$. That is, when $x \approx 2\sqrt{Dt}$.

The penetration distance for the error-function solution is $c(x,t) = \left(\frac{c_0}{8}\right)$. That is, $\text{erf}\left(2\sqrt{Dt}\right) = -\left(\frac{3}{4}\right)$ or

$$x \approx 1.6\sqrt{Dt}. \tag{6.47}$$

Again, it follows that the penetration depth is approximately $2\sqrt{Dt}$.

If L represents the characteristic linear dimension of a body, steady state is reached at times $\tau \gg (L^2/D_{\min})$, where $D_{\min}$ is the smallest value of the diffusivity. In many physical problems steady-state conditions are not reached. For example, when the system is infinite or semi-infinite or when the boundary conditions are not independent of time.

Problems

1.	A thin sheet of steel initially contains 0.2% carbon. It is exposed to a carbon-rich atmosphere at 900 °C where the surface carbon content is maintained at 1.0%. Assuming the diffusion coefficient $D = 1.5 \times 10^{-11}$ $\mathrm{m^2/s}$ calculate the concentration at a depth of 0.5 mm after 10 h. (Use error function solution.) **Solution:** Given $C_0 = 0.2\%$ $C_s = 1.0\%$ $t = 10 \times 3600\ \mathrm{s} = 36{,}000\ \mathrm{s}$ $D = 1.5 \times 10^{-11}\,\mathrm{m^2/s}$ Using error function

(continued)

	$\frac{C_x - C_0}{C_s - C_0} = 1 - \text{erf}\left(\frac{x}{2\sqrt{Dt}}\right)$ $\frac{x}{2\sqrt{Dt}} = \frac{0.5 \times 10^{-3}}{1.4696 \times 10^{-3}} = 0.34$ From error function table: erf (0.34) = 0.370, Thus, $\frac{C_x - 0.2}{1.0 - 0.2} = 1 - 0.370 = 0.630$ $C_x = 0.704\%$ **0.704% carbon concentration at 0.5 mm depth after 10 h**
2.	A gas diffuses through a membrane 2 mm thick. The concentrations on either side are 0.05 and 0.02 kg/m^3. If the diffusion coefficient is 2 × 10–10 m^2/s, calculate the flux. **Solution:** Fick's first law (steady state): $J = -D\frac{\Delta C}{\Delta x}$ Where, $\Delta C = 0.02 - 0.05 = -0.03 \text{ kg/m}^3$ $\Delta x = 2 \times 10^{-3} \text{ m}$ Thus, $J = -(2 \times 10^{-10}) \times \left(\frac{-0.03}{2 \times 10^{-3}}\right)$ $J = (2 \times 10^{-10}) \times 15$ $J = 3 \times 10^{-9} \text{ kg/m}^2\text{s}$
3.	Given $D_0 = 2 \times 10^{-6}\ \text{m}^2/_\text{s}$ Q = 80,000 J, R = 8.314 J/Mol K and T = 1000, find D **Solution:** $D = 2 \times 10^{-6} \times e^{-80{,}000/(8.314 \times 1000)}$ $= 2 \times 10^{-6} \times e^{-9.62}$ $= 2 \times 10^{-6} \times 6.65 \times 10^{-5}$ $= 1.33 \times 10^{-10} \text{ m}^2/\text{s}$
4.	A carburizing operation maintains the surface concentration of carbon at 0.8% for 8 h. The initial carbon content is 0.2%. Given $D = 1.0 \times 10^{-11}\ \text{m}^2/_\text{s}$, find the depth where carbon concentration is 0.5% after 8 h. **Solution:** t = 8 × 3600 = 28,800 s $\frac{C_x - C_0}{C_s - C_0} = 1 - \text{erf}\left(\frac{x}{2\sqrt{Dt}}\right)$ $2\sqrt{Dt} = 2\sqrt{1.0 \times 10^{-11} \times 28{,}800} = 1.0734 \times 10^{-3} \text{ m}$ $\frac{C_x - C_0}{C_s - C_0} = 0.5$ $\text{erf}\left(\frac{x}{2\sqrt{Dt}}\right) = 0.5$ $\text{erf}(0.4769) = 0.5$

Reference

1. Balluffi, R. W., Allen, S. M., & Craig Carter, W. (2005). *Kinetics of materials*. Wiley.

Chapter 7
Diffusion in Multicomponent Systems

Contents

7.1 Introduction

A single interdiffusivity, $\tilde{D}$, is sufficient in a binary solution to describe composition evolution. In multi-component and ternary solutions, diffusion will have more (or at least two) independent variables in composition. Multiple cross terms and particle-particle chemical interaction terms appear and make the analysis complex. N^2 independent interdiffusivities for a solution with N independent components result. A variety of diffusional phenomena will be produced by multicomponent diffusion.

The general treatment for multicomponent diffusion results: A system of linear diffusion equations result from a general treatment of the problem. A linear transformation of the concentrations results in diffusion equations that are uncoupled and linear. General solutions for this situation can be developed by the techniques discussed earlier.

The assumptions made are that (a) only chemical potential gradients exist, and (b) thermal gradients and electric fields are absent. For one-dimensional diffusion, the general formulation will be

K. A. Padmanabhan et al., *Thermodynamics and Kinetics of Materials*, Engineering Materials, https://doi.org/10.1007/978-3-032-14527-7_7

$$\begin{aligned}\overrightarrow{J_1} &= L_{11}\frac{\partial \mu_1}{\partial x} - L_{12}\frac{\partial \mu_2}{\partial x} + \cdots - L_{1N}\frac{\partial \mu_N}{\partial x}.\\ \overrightarrow{J_N} &= -L_{N1}\frac{\partial \mu_1}{\partial x_1} - L_{N2}\frac{\partial \mu_2}{\partial x_2} + \cdots - L_{NN}\frac{\partial \mu_N}{\partial x_N}\end{aligned} \tag{7.1}$$

Rate of entropy production then is,

$$T\dot{\sigma} = -\sum_{i=1}^{i=N} J_i \frac{\partial \mu_i}{\partial x} \tag{7.2}$$

Let constant atomic volumes be present, and a V-frame is used to estimate the fluxes. Then,

$$\sum_{i=1}^{i=N} \Omega_i J_i = 0. \tag{7.3}$$

Then, the Nth flux will be

$$J_N = -\frac{1}{\Omega_N}\sum_{i=1}^{i=N} \Omega_i J_i. \tag{7.4}$$

and

$$T\dot{\sigma} = -\sum_{i=1}^{i=N-1} \frac{\partial}{\partial x}\left(\mu_i - \frac{\Omega_i}{\Omega_N}\mu_N\right). \tag{7.5}$$

Therefore, the force, the conjugate of flux, is

$$F_1 = -\left(\mu_i - \frac{\Omega_i}{\Omega_N}\right). \tag{7.6}$$

Then, the resulting (N – 1) general relations between the fluxes and the driving forces, each of which are independent, are linear. That is,

$$J_i = \sum_{i=1}^{i=N-1} L_{ij} F_j = -\sum_{j=1}^{j=N-1} L_{ij}\left(\frac{\partial \mu_j}{\partial x} - \frac{\partial \mu_N}{\partial x}\right). \tag{7.7}$$

The chemical potential gradients and Onsager coefficients (Eq. 7.7) are interchangeable with gradients in concentration and interdiffusivities. Table 7.1.

Each chemical potential is a function of the local concentration:

Table 7.1 Summary of definitions and relationships for diffusivities (adapted from Baluffi et al. [1])

Notation	Nomenclature	Relationships	Frame of reference
$^*D$	Homogeneous, pure substance self-diffusivity		C or V
$^*D$ applies to a single species, pure substance. Diffusion involving a radioactive isotope of an identical inert species is measured. There is no flow of mass. So, a V-frame or a C-frame can be used			
$^*D_i$	Component ***i*** self-diffuses in a two-component alloy	$D_i \approx \left[1 + \frac{\partial \ ln \ \gamma_1}{\partial \ ln \ c_1}\right] * D_i$	C or V
ith component in a two-component alloy diffusivity is denoted as D_i. The relation between D^* *and* $^*D_i$ given above. An isotope measures *D, *D. Zero transfer of mass. So, either a V-frame or a C-frame should be fine			
D_i	Diffusivity that is intrinsic	$\vec{J}_i^C = -D_i \nabla c_i$ $\vec{J}_i^V = -D_i \nabla c_i + \vec{\vartheta}_c^V C_i$	Local C or V
D_i is composition-dependent and is present in a chemically non-homogeneous alloy. The flux of component i in a two-component alloy and its gradient in concentration are correlated using the law due to Fick. The relation between D_i and $\tilde{D}$ is as given above.			
$\tilde{D}$	Interdiffusivity	$\vec{J}_i^V = -D_i \nabla c_i$ $\tilde{D} = c_1 \Omega_1 D_2 + c_2 \Omega_2 D_1$	V-frame V-frame
$\tilde{D}$ is the interdiffusivity; depends on composition in an inhomogeneous material. In a two-component system, the flux and gradient in concentration of component **1** or **2** are related via the law of Fick in a V - frame			

$$\mu_i = \mu_i(c_1, c_2, \ldots, c_{N-1}); i = 1, 2, \ldots, N - 1. \tag{7.8}$$

$(N - 1)$ is the number of concentrations that are independent since the relation $\sum_{i=1}^{i=N_c} \Omega_i dc_i = 0$ leads to a unique relation between atomic volumes and their concentrations. Assuming that equilibrium exists locally through the “Gibbs-Duhem relation” (which gives rise to an additional constraint on changes in chemical potential), one can conclude that of the different values of μ_i, $(N - 1)$ values can independently change. That is,

$$\frac{\partial \mu_i}{\partial x} = \sum_{j=1}^{j=N-1} \frac{\partial \mu_i}{\partial c_j} \frac{\partial c_j}{\partial x}. \tag{7.9}$$

$$J_i = \int_{i=1}^{j=N-1} L_{ij} F_j = -\sum_{j=1}^{j=N-1} L_{ij} \sum_{k=1}^{k=N-1} \left(\frac{\partial \mu_j}{\partial c_k} - \frac{\Omega_j}{\Omega_N} \frac{\partial \mu_N}{\partial c_k} \right) \frac{\partial c_k}{\partial x}. \tag{7.10}$$

Interdiffusivities, $\tilde{D}_{ij}$, are defined by Eq. (7.10). Then,

$$J_i = -\sum_{j=1}^{j=N-1} \tilde{D}_{ij}\frac{\partial c_j}{\partial x}, \tag{7.11}$$

when

$$\tilde{D}_{ij} = \sum_{k=1}^{k=N-1} L_{ik}\left(\frac{\partial \mu_k}{\partial c_j} - \frac{\Omega_k}{\Omega_N}\frac{\partial \mu_N}{\partial c_j}\right). \tag{7.12}$$

$\tilde{D}_{ij}$ is the product of two matrices,

$$\tilde{D}_{ij} = \sum_{k=1}^{k=N-1} L_{ik}\Upsilon_{kj}\ldots \tag{7.13}$$

where L_{ik} are known as "Onsager coefficients". Chemical potentials are coupled to concentrations, using the thermodynamic factors, Υ_{kj} as shown below.

$$\Upsilon_{kj} = \frac{\partial \mu_k}{\partial c_j} - \frac{\Omega_k}{\Omega_N}\frac{\partial \mu_N}{\partial c_j}. \tag{7.14}$$

Thus, the analysis of concentrations and diffusion involving N number of components will need independent concentrations that will number $(N - 1)$. The number of interdiffusivities will be $(N - 1)^2$. Therefore, for the ternary case, the interdiffusivities needed can be represented by a square matrix of order two. That is,

Eigenvalues, $\lambda_{\mp}$, *of the matrix, representing interdiffusivity should be positive and real.* When one considers a ternary alloy,

$$\lambda_{\mp} = \frac{Tr\left(\underline{\tilde{D}}\right) \mp \sqrt{Tr\left(\underline{\tilde{D}}\right)^2 - 4det\left(\underline{\tilde{D}}\right)}}{2} = \frac{Tr\left(\underline{\tilde{D}}\right) \mp \Delta}{2}. \tag{7.15}$$

Here $\left(\underline{\tilde{D}}\right) = \tilde{D}_{11} + \tilde{D}_{12}$, and $det(\tilde{D}) \equiv \tilde{D}_{11}\tilde{D}_{22} - \tilde{D}_{12}\tilde{D}_{21}$. The determinant Δ is given by,

$$\Delta \equiv \sqrt{Tr\left(\underline{\tilde{D}}\right)^2 - 4det\left(\underline{\tilde{D}}\right)} = \sqrt{[(\tilde{D})_{11} - (\tilde{D}_{22})]^2 + 4\tilde{D}_{12}D_{21}}. \tag{7.16}$$

The condition that the eigenvalues of the interdiffusivity matrix should be real and positive will restrict the values of the interdiffusivities. For a ternary case,

$$\tilde{D}_{11}+\tilde{D}_{22}>0;\ (\tilde{D}_{11}+\tilde{D}_{22})^2\geq 4\left(\tilde{D}_{11}\underset{22}{\check{D}}-\tilde{D}_{12}\tilde{D}_{21}\right);$$

$$(\tilde{D}_{11}\tilde{D}_{22}-\tilde{D}_{12}\tilde{D}_{21})\geq 0;\ \tilde{D}_{ij}\neq\tilde{D}_{ji}. \tag{7.17}$$

7.2 Multicomponent Diffusion Equations: Solutions

Now we consider the one-dimensional case of diffusion in a ternary alloy: Eq. (7.11) implies that,

$$\frac{\partial\vec{c}}{\partial t}=\frac{\partial}{\partial x}\left\{\begin{bmatrix}\tilde{D}_{11} & \tilde{D}_{12}\\ \tilde{D}_{21} & \tilde{D}_{22}\end{bmatrix}\frac{\partial}{\partial x}\begin{bmatrix}c_1\\ c_2\end{bmatrix}\right\}. \tag{7.18}$$

In general,

$$\frac{\partial\vec{c}}{\partial t}=\frac{\partial}{\partial x}\left[\underline{\tilde{D}}\right]\frac{\partial\vec{c}}{\partial x}. \tag{7.19}$$

Often, the $\tilde{D}_{ij}$ values are functions of the concentrations. Therefore, these equations are nonlinear. Hence, there is a need for numerical solutions. For a variety of special cases, solutions can be found.

7.2.1 *Case Where Diffusivity is Constant*

When the interdiffusivities are each constant and uniform, Eq. (7.19) forms a linear system,

$$\frac{\partial}{\partial t}\begin{bmatrix}c_1\\ c_2\end{bmatrix}=\begin{bmatrix}\tilde{D}_{11} & \tilde{D}_{12}\\ \tilde{D}_{21} & \tilde{D}_{22}\end{bmatrix}\nabla^2\begin{bmatrix}c_1\\ c_2\end{bmatrix}. \tag{7.20}$$

In general,

$$\frac{\partial\vec{c}}{\partial t}=\underline{\tilde{D}}\nabla^2\vec{c}. \tag{7.21}$$

Diagonalising the diffusivity matrix uncouples the expressions for the fluxes. The eigenvalues of λ_i, appear as the diagonal elements of the transformed interdiffusivity matrix. For a ternary alloy, the $\lambda_{\mp}$ values of Eq. (7.16) are the eigenvalues. It should be noted that for N independent components, N positive eigenvalues, λ_N will be there. In the general case, there will be N independent components. For a ternary alloy, the eigenvector for fast diffusion will be λ_+, say $\vec{f}$, and for slow diffusion it will be λ_-, say $\vec{s}$. Then,

$$\vec{s} = \begin{bmatrix} \dfrac{\tilde{D}_{11} - \tilde{D}_{22} - \Delta}{2\tilde{D}_{21}} \\ 1 \end{bmatrix}; \vec{f} = \begin{bmatrix} \dfrac{\tilde{D}_{11} - \tilde{D}_{22} + \Delta}{2\tilde{D}_{21}} \\ 1 \end{bmatrix}. \tag{7.22}$$

θ is the angle between the two (slow and fast) Eigen directions.

$$\cos\,(\theta) = \frac{\vec{s} \cdot \vec{f}}{|\vec{s}\,||\vec{f}\,|} = \frac{\tilde{D}_{21} - \tilde{D}_{12}}{(\tilde{D}_{11} - \tilde{D}_{22})^2 + (\tilde{D}_{12} + \tilde{D}_{21})^2}. \tag{7.23}$$

The transformation matrix, A, that diagonalises $\underline{\tilde{D}}$ leads to columns created by $\vec{s}$ and $\vec{f}$, the two eigenvectors considered above. Therefore, for a ternary alloy,

$$\underline{A} = \begin{bmatrix} \dfrac{\tilde{D}_{11} - \tilde{D}_{22} - \Delta}{2\tilde{D}_{21}} & \dfrac{\tilde{D}_{11} - \tilde{D}_{22} + \Delta}{2\tilde{D}_{21}} \\ 1 & 1 \end{bmatrix}; \underline{A}^{-1} = \begin{bmatrix} \dfrac{-\tilde{D}_{21}}{\Delta} & \dfrac{(\tilde{D}_{11} - \tilde{D}_{22} + \Delta)}{2\Delta} \\ \dfrac{\tilde{D}_{21}}{\Delta} & \dfrac{-(\tilde{D}_{11} - \tilde{D}_{22} - \Delta)}{2\Delta} \end{bmatrix}. \tag{7.24}$$

For the general case,

$$\underline{A} = [\tilde{x}_1, \tilde{x}_2, \ldots, \tilde{x}_{N-1}]. \tag{7.25}$$

Therefore,

$$\underline{A}^{-1}\underline{\tilde{D}}A = \begin{bmatrix} \lambda_- & 0 \\ 0 & \lambda_+ \end{bmatrix}. \tag{7.26}$$

Recall that for the ternary case,

$$\begin{bmatrix} J_1 \\ J_2 \end{bmatrix} = -\tilde{D}\nabla \begin{bmatrix} c_1 \\ c_2 \end{bmatrix}. \tag{7.27}$$

Therefore,

$$\underline{A}^{-1}\begin{bmatrix} J_1 \\ J_2 \end{bmatrix} = -\underline{A}^{-1}\underline{\tilde{D}}\underline{A}^{-1}\nabla \begin{bmatrix} c_1 \\ c_2 \end{bmatrix}. \tag{7.28}$$

Or

$$\underline{A}^{-1}\begin{bmatrix} J_1 \\ J_2 \end{bmatrix} = \begin{bmatrix} J_\alpha \\ J_\beta \end{bmatrix} = -\begin{bmatrix} \lambda_- & 0 \\ 0 & \lambda_+ \end{bmatrix} \nabla \begin{bmatrix} c_\alpha \\ c_\beta \end{bmatrix}. \tag{7.29}$$

Or,

$$J_\alpha = -\lambda_- \nabla c_\alpha; J_\beta = -\lambda_+ \nabla c_\beta. \tag{7.30}$$

Also,

$$\begin{bmatrix} c_\alpha \\ c_\beta \end{bmatrix} = \underline{A}^{-1} \begin{bmatrix} c_1 \\ c_2 \end{bmatrix}. \tag{7.32}$$

Steady-State Solutions: Eq. (7.27) holds good. (Recall: $\underline{\tilde{D}}\nabla^2 \vec{c} = 0$.) Solve as before.

Time-Dependent Solutions: In this case, the coupled diffusion equations are given by Eq. (7.21). Diagonalisation will uncouple them when the transformation matrix $\underline{A}$ is used. For the ternary case,

$$\frac{\partial}{\partial t}\left(\underline{A}^{-1}\vec{c}\right) = \underline{A}^{-1}\underline{\tilde{D}}\nabla^2\vec{c} = \underline{A}^{-1}\underline{\tilde{D}}\underline{A}\left(\nabla^2\underline{A}^{-1}\vec{c}\right), \text{or} \frac{\partial}{\partial t}\begin{bmatrix} c_\alpha \\ c_\beta \end{bmatrix} = \begin{bmatrix} \lambda_- & 0 \\ 0 & \lambda_+ \end{bmatrix} \nabla^2 \begin{bmatrix} c_\alpha \\ c_\beta \end{bmatrix}. \tag{7.33}$$

For the general case,

$$\frac{\partial}{\partial t}\begin{bmatrix} \check{c}_1 \\ \check{c}_2 \\ \cdots \\ \check{c}_{N-1} \end{bmatrix} = \begin{bmatrix} \lambda_1 & 0 & 0 & 0 \\ \cdots & \lambda_2 & \cdots & 0 \\ 0 & 0 & \cdots & 0 \\ 0 & 0 & 0 & \lambda_{N-1} \end{bmatrix} \nabla^2 \begin{bmatrix} \check{c}_1 \\ \check{c}_2 \\ \cdots \\ \check{c}_{N-1} \end{bmatrix} \tag{7.34}$$

where the concentrations $\check{c}_i$ are linear combinations given by the transformation of the eigensystem involving c_i. In the diagonal frame, every partial differential equation is independent. In general,

$$\begin{aligned} \frac{\partial \check{c}_1}{\partial t} &= \lambda_1 \nabla^2 \check{c}_1. \\ \frac{\partial \check{c}_2}{\partial t} &= \lambda_2 \nabla^2 \check{c}_2. \\ \cdots &= \cdots \\ \frac{\partial \check{c}_{N-1}}{\partial t} &= \lambda_{N-1} \nabla^2 \check{c}_{N-1} \end{aligned} \tag{7.35}$$

In general, for one-dimensional diffusion,

$$\vec{\tilde{c}}\,(x=L,t)=\underline{A}^{-1}\vec{c}\,(x=L,t);\quad \vec{\tilde{c}}\,(x=R,t)=\underline{A}^{-1}\vec{c}\,(x=R,t);\\ \vec{\tilde{c}}=(x,t=0)=\underline{A}^{1}\vec{c}\,(x,t=0). \tag{7.36}$$

The net result is a set of diffusion equations that are uncoupled. The diffusivities are constructed from the interdiffusivities of the components obtained from an algorithm. The solutions to these equations are generated using procedures that rely on Laplace transforms.

7.2.2 Concentration-Dependent Diffusivities

If the diffusivities are functions of concentration, for a three-component system the diffusion equations are

$$\frac{\partial c_i}{\partial t}=-\nabla.\vec{J}_1=\frac{\partial}{\partial x}\left(\tilde{D}_{i1}\frac{\partial c_1}{\partial x}\right)+\frac{\partial}{\partial x}\left(\tilde{D}_{i2}\frac{\partial c_2}{\partial x}\right). \tag{7.37}$$

If $\eta\equiv\left(\frac{x}{\sqrt{t}}\right)$, the equations become

$$-\frac{\eta}{2}\frac{dc_i}{d\eta}=\frac{d}{d\eta}\left(\tilde{D}_{i1}\frac{dc_1}{d\eta}+D_{i2}\frac{dc_2}{d\eta}\right). \tag{7.38}$$

Analytical solutions of these equations are difficult and time-consuming. Usually, these are solved numerically, and the solutions are compared with experimental results. Such solutions are available in the literature. Due to strong particle-particle interactions, *uphill diffusion* is exhibited by quite a few systems. In some cases, thermodynamic activity data are interpreted using diffusivity behaviour. This has led to some particle-particle interaction models. In others, a satisfactory explanation is lacking. Diffusion behaviour is explained in terms of single-body or many body models. They are all based on statistical mechanics and the concept of thermally activated complexes and thermally assisted atom jumps. This topic is beyond the scope of the present book.

7.2.3 Diffusion and Random Walks

It is assumed that a series of displacements is responsible for the motion of a particle, each independent of the one preceding it. Then it is described as a *random walk*. The displacements could vary in length and direction or be fixed. This approach predicts the positions, in the form of a distribution, of the particles. They can then be compared with predictions based on macroscopic models of diffusion. Random

Table 7.2 Solutions assuming instantaneous, 1D, 2D, and 3D models of diffusion in an infinite medium (reproduced from Baluffi et al. [1])

1D Diffusion		
Source of 0D in 1D 1D source in 2D Plane source (2D) in 3D	$\frac{d^2}{dx^2}$	$c\left(\vec{r},t\right)=\frac{n_d}{(4\pi Dt)^{(1/2)}}\cdot e^{-x^2/(4Dt)}$
2D Diffusion		
Source of 0D in 2D 1D source in 3D	$\frac{1}{r}\frac{d}{dr}r\frac{d}{dr}$	$c\left(\vec{r},t\right)=\frac{n_d}{(4\pi Dt)}\cdot e^{-r^2/(4Dt)}$
3D Diffusion		
Source of 0D in 3D	$\frac{1}{r^2}\frac{d}{dr}r^2\frac{d}{dr}$	$c\left(\vec{r},t\right)=\frac{n_d}{(4\pi Dt)^{\frac{3}{2}}\cdot e^{-r^2/(4Dt)}}$

walk solutions also lead to an understanding of diffusive processes that are nonrandom.

For a random walker, the distribution of positions can be easily formulated on a one-, two-, and three-dimensional lattice. Balluffi et al. [1] have elegantly summarised such a distribution, comparing it with diffusion that is interpreted in terms of macroscopic assumptions. In these analyses, the particles are assumed to migrate independently. Their views are collated in Table 7.2.

Let a particle be located at the origin. It can then jump either along the +z direction (p_R would the probability), or along the (-z) direction for which p_L is the probability. Here, $0 < p_R < 1$; $p_R + p_L = 1$. In addition, let unit length define a displacement. Then, when $N_r >> 1$, the concerned particle could reach $-N_r$, $-N_r + 1, \ldots - 1, 0, 1, \ldots N_r - 1, N_r$. The concerned particle is not likely to jump positively all the time to reach site N_r. It is also not likely to jump negatively all the time to reach $-N_r$. If $p_R = p_L = \left(\frac{1}{2}\right)$ (equal probability), the particle, on average, will be present at the origin, with jumps in the positive and the negative directions cancelling each other. To have the particle end up in position n after N_r jumps, let the positive displacements (number) be ($+N_R$). Let the negative displacements be (N_L),

$$N_R - N_L = n. \tag{7.39}$$

$$N_R + N_L = n. \tag{7.40}$$

The "binomial coefficient" gives the different sequences/trajectories in which the walker can reach site n. That is,

$$\Omega(n, N_R) = \frac{N_r}{N_R!N_L!} = \frac{N_r!}{[(N_R + n)/2]!}\,\frac{N_r!}{[(N_R - n)/2]!}. \tag{7.41}$$

Then the probability that one can get to site n after N_r jumps will be given by,

$$p(n, N_r) = \frac{N_r!}{[(N_r + n)/2]![(N_r - n)/2]!} p_R^{N_R} p_L^{N_L}. \tag{7.42}$$

The probability that one can get to site n after N_r jumps, p_L, will be,

$$p(n, N_r) = \frac{N_r!}{[(N_r + n)/2]![(N_r - n)/2]!} \left(\frac{1}{2}\right)^{N_r}. \tag{7.43}$$

From Stirling's approximation, one gets,

$$Q! = \sqrt{2\pi Q}\, Q^Q e^{-Q}. \tag{7.44}$$

In the limit, when

$$n/N_r \ll 1, p(n, N_r) \alpha\, e^{-n^2/(2N_r)}. \tag{7.45}$$

That is, a one-dimensional distribution of a point source is Gaussian (Eq. 7.45). If R $= n(r^2)^{1/2}$, the probability distribution for a displacement R is,

$$p(R, N_r) \alpha\, e^{-R^2/2\langle r^2 \rangle \Gamma \tau}. \tag{7.46}$$

Normalizing the probability distribution, one can write,

$$\int_{-\infty}^{\infty} p(R, \tau) dR = 1. \tag{7.47}$$

Therefore, the probability distribution becomes

$$p(R, N_r) = \frac{1}{\sqrt{2\pi N_r \langle r^2 \rangle}} e^{-R^2(2\langle r^2 \rangle \Gamma \tau} \tag{7.48}$$

From Table 7.2, it follows that 1D diffusion problem solution for a point source based on macroscopic diffusion resembles Eq. (7.48). From Eq. (7.48), one gets that its first and second moments are:

$$\langle R \rangle = \int_{-\infty}^{\infty} p(R, N_r) R dR = 0. \tag{7.49}$$

$$\langle R^2 \rangle = p(R, N_r) R^2 dR = N_r \langle r^2 \rangle. \tag{7.50}$$

It is evident from Eq. (7.49) that every jump associated with the walk is random, with zero displacement and no change in the centre of mass, on average, due to the random jump. Eq. (7.50) is the random walk problem which gives rise to mean-square displacement, that is, $N_r\langle r^2\rangle$. The last two equations have been derived for a one-dimensional random walk. This solution applies to the 2D and 3D cases also.

Comparing Eq. (7.50) and Eq. (7.48), one can get that

$$D = \frac{\Gamma\langle r^2\rangle}{2}. \tag{7.51}$$

Equation (7.51) links 1D uncorrelated diffusion on a macroscopic scale to parameters associated with microscopic jumps.

7.2.4 Correlated Jumps and Diffusion

In a general sequence of jumps for calculating the root mean square displacement, two terms are important. (1) $N_r\langle r^2\rangle$ in Eq. (7.50) is associated with ideal random walk. (2) The second term is due to correlation effects that could be present when successive jumps are not entirely random. A correlation factor, f, takes care of correlated walks. D, the isotropic diffusivity associated with the macroscopic scale, and the jump events on the microscopic scale, all on 3D level are correlatable. Following a similar derivation as in the previous case, one obtains for isotropic diffusion $D = \frac{\Gamma\langle r^2\rangle}{2d} f$ where d is the dimensionality of the analysis.

There are several processes in which a random walk is observed ($f = 1$). But diffusion processes with different values for f are also present. An important application area of the theory is diffusion in crystals. Topics like atomic mechanisms in different materials, e.g., metallic, ionic materials, and nonstoichiometric materials, are discussed in this connection. Other important topics are isotropic and anisotropic diffusional anelasticity, the presence of hysteresis loops, etc. Many textbooks on Solid State/Condensed Matter Physics discuss these and other similar topics.

7.2.5 Measurement of Diffusivities

The above analysis provides a powerful technique to determine the diffusivities of species that exhibit anelastic relaxation. The frequency, ω_p, corresponding to the Debye peak can be known using a torsional pendulum. Relaxation time is obtained from the relation $\tau = \frac{1}{\omega_p}$. The diffusivity is derived using the relationships involving the relaxation time, the diffusivity and the jump frequency. The interstitial diffusivity is given by $D = \left(\frac{\Gamma a^2}{12}\right)$. Therefore, $D = \left(\frac{a^2}{18\tau}\right)$. This technique is useful in determining the diffusivities of diverse interstitial atoms, especially at low temperatures, where measurable (by a torsional pendulum), but low jump frequency is found. In this connection, it is well-known that the diffusivity of Carbon (an interstitial atom) in BCC iron is important.

Reference

1. Balluffi, R. W., Allen, S. M., & Carter, W. C. (2005). *Kinetics of materials*. Wiley.

Chapter 8
Influence of Crystal Imperfections on Diffusion

Contents

8.1 Introduction

It has been experimentally observed that in crystals containing only point defects, the rate of diffusion is significantly slower than in crystals containing imperfections, such as dislocations, internal interfaces, and free surfaces. The short-circuit diffusion paths provided by these linear and planar defects contribute dominantly to diffusion in such crystalline material.

Rapid diffusion transpires along line and plane flaws within a thin region. The region is cylindrical for dislocations, roughly of diameter that is twice the distance between two atoms, which represents "defective material" within dislocation core. The grain boundaries consist of a narrow slab, around two to three interatomic distances thick, which is referred to as the core of the grain boundary. At a surface that is free diffusion means that is prevalent in the initial layers of atoms, which are far thinner than typical diffusion distances. To understand diffusion caused by these defects, thin slabs or cylinders of effective thickness, δ, are used as substitutes. These locations demonstrate markedly greater diffusivity than the crystalline material. Table 8.1 lists the notations, sourced from Baluffi [1], that define the diffusivities in several regions characterized by linear and planar flaws.

Figure 8.1 illustrates self-diffusion data for several diffusion-types in FCC metals. The data, without that of surface diffusion, are displayed in an Arrhenius sketch, signifying that the different activation energies remain constant, not dependent on

K. A. Padmanabhan et al., *Thermodynamics and Kinetics of Materials*, Engineering Materials, https://doi.org/10.1007/978-3-032-14527-7_8

Table 8.1 Notation for short-circuit diffusivities [1]

D^D*(undissociated)*	Along the core of an undissociated dislocation diffusivity is present, it is a cylindrical structure or 'pipe' with a diameter of δ
D^D*(dissociated)*	Core of a dislocation that is dissociated displays diffusivity. It has a cylindrical structure with a diameter of δ, and appears like a pipe
D^B	A grain boundary experiencing diffusivity (rectangular slab of thickness δ)
D^S	A free surface diffusivity (again of thickness δ)
D^{XL}	Bulk crystal without 1D and 2D defects exhibiting diffusivity
D^L	A liquid displaying diffusivity

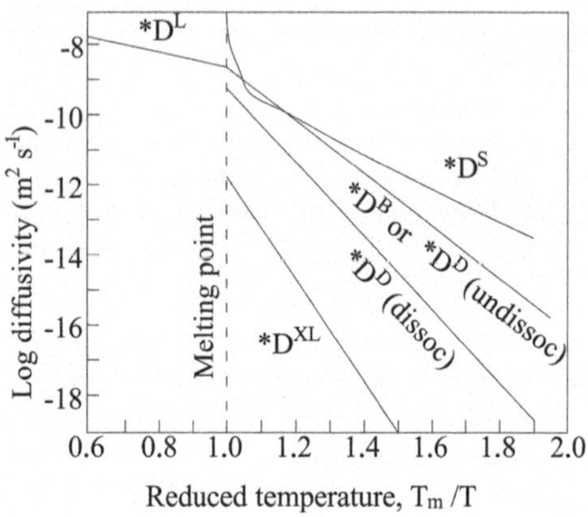

Fig. 8.1 Arrhenius diagram for $^*D^{xL}$, $^*D^D$*(dissociated)*), $^*D^D$*(undissociated)*, $^*D^B$, $^*D^s$, and $^*D^L$ in FCC metal [1, 2]

temperature. Indecisively accurate and non-comprehensive data do not allow an inference that a curvature does not exist in certain diffusion types.

Figure 8.1 presents the Master Arrhenius plot illustrating many types of diffusion in face-centered cubic (FCC) metals. The picture has normalized data pertaining to metals that have FCC structure. A reduced scale using a reciprocal of the temperature has been used, i.e., $(1/T)/(1/T_m) = (T_m/T)$. The diffusivities were calculated from experimental data under the assumption that $\delta = 0.5$ nm. This data originates from the predominant locations of high-angle grain boundaries.

Dislocations, grain boundaries, and surfaces exhibit significantly distinct structures, resulting in markedly variable diffusivities. If the defective core region exhibits lower density than a region where defects are absent, or the region has a structure in its core that has "open" channels, faster diffusion will be present via these open pathways.

To describe a crystalline interface, a minimum of five geometric parameters is necessary: three for the misorientations between crystals, and two for the axis of rotation and the angle of rotation—the last two indicate the plane tilt. For free interface, surface inclination in terms of its unit normal is necessary. The specific parameter values at which interfacial energy reaches extreme levels are determined by crystal symmetries. Certain parameters may be restricted while others are

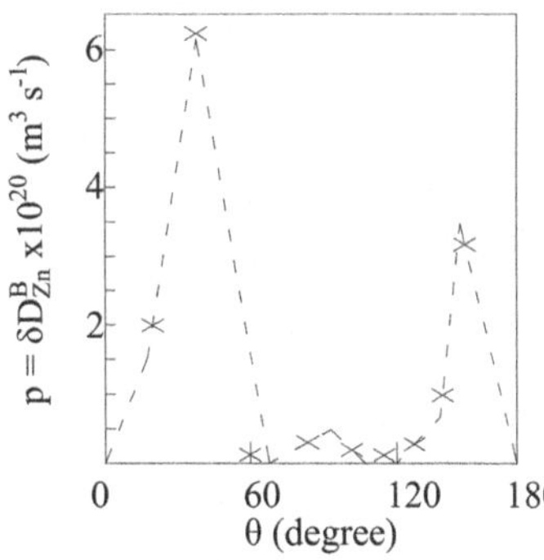

Fig. 8.2 The diffusivity of Zn at grain boundaries in Al along [3] tilt axes of (tilt) symmetric boundaries. Presented against the tilt angle, θ [4]

permitted to pursue a lower-energy state, contingent upon their individual characteristics.

Low-angle grain boundaries exhibit a misorientation of around 15° or less between crystals and consist of arrays of discrete dislocations. For misorientations exceeding around 15°, the dislocation spacing is significantly reduced. The cores overlap, rendering the boundary comparable to a continuous slab of "bad" material. These are called the high-angle grain boundaries.

High-angle boundaries are further classified into three types: "singular" boundaries, "vicinal" boundaries, and "general" boundaries. A singular boundary corresponds, for a given degree of freedom, to a minimum in energy, making it resistant to changes in that degree of freedom. A "vicinal" boundary shows a slight difference in one or more geometric parameters compared to a singular interface. Therefore, it can lower its energy by adopting a fit-misfit structure consisting of patches of minimum energy singular boundary delineated by arrays of discrete interfacial dislocations or steps. A general boundary is not energy-minimized with respect to any of its degrees of freedom and deviates considerably from the values that determine the degrees of freedom of a singular boundary. This type of interface cannot reduce its energy through structural changes and therefore they cannot support localized dislocations or steps.

With misorientations, the structures of boundaries change significantly. The minima in diffusivity (Fig. 8.2) are seen at singular and vicinal boundaries. Crystal misorientations towards zero are small-angle boundaries. The regions centered around the maxima represent the general grain boundaries. Polycrystalline materials mostly consist of general boundaries. Their diffusivities (large) are close to average normalised values.

If an atomic jump is increasingly constrained, rates of jump, Γ, decrease as follows, $\Gamma^s > \Gamma^B \simeq \Gamma^D(undissoc) > \Gamma^D(dissoc) > \Gamma^{XL}$. These diffusion processes exhibit activation energies in an inverse fashion. The self-diffusivities are as follows,

$$*D^s > *D^B \simeq *D^D(undissoc) > *D^D(Dissoc) > *D^{XL}. \tag{8.1}$$

Diffusivities of metals at free surfaces and grain boundaries at 0.5 T_m are 7–8 orders of magnitude higher than diffusivities of crystals. At higher defect densities, crystals

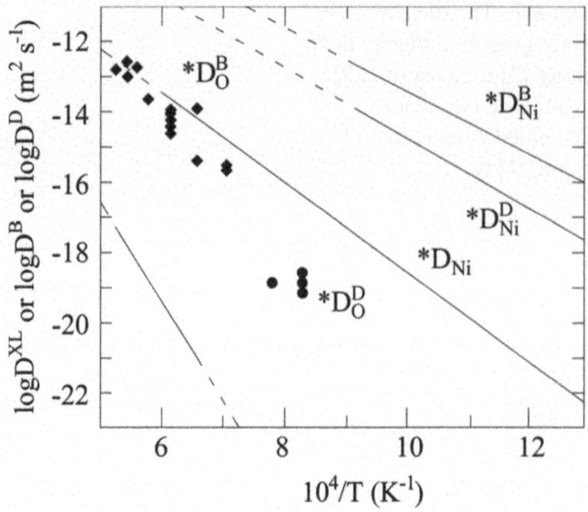

Fig. 8.3 Oxygen and nickel self-diffusion on their individual sublattices. A single crystal nickel oxide that does not have many dislocations. Grain boundaries are the diffusion paths in a polycrystalline structure. In (semiconductor) nickel oxide, grain boundary diffusivities of both nickel and oxygen are more than the respective bulk diffusivities [5]

undergo at 0.5 T_m significant diffusion through grain-boundaries and interfaces, even though the diffusion flux is across a limited cross-sectional area. At lower temperatures, short-circuit diffusion is more important. Overall, a similar behaviour is seen in ionic crystals, Fig. 8.3.

At boundaries between grains and free interfaces low temperature short circuit diffusion frequently governs diffusional transport and influences kinetic processes. For dislocation diffusional short-circuiting also similar conclusions to a lesser extent are valid because the high diffusivity pipes are very narrow.

8.2 Atomic Transport Along Grain Boundaries in Polycrystalline Materials

Atoms can move throughout both the interiors of grains and the grain boundaries (assumed to be "slabs"). Various phenomena may manifest based on grain size, temperature, diffusion duration, and the mobility of the border network. As grain size diminishes, diffusion accelerates due to the increased rate of diffusion at the grain boundaries.

At higher temperatures the ratio of boundary to bulk diffusivity diminishes, indicating reduced boundary diffusion. An atom in the bulk will be overtaken by a boundary that is mobile, rapidly diffusing into the grain boundary and subsequently depositing back into the crystalline substance behind it.

In stationary grain boundaries, when the moving time, t, is accessible, each diffusing atom disperses through the bulk and multiple boundaries, specifically, $*D^{XL}t > L^2$, where the grain size is L. The proportion of time allocated to diffusion for every atom in the boundaries of grains equals atom sites within the boundaries of grains)/(atomic sites present in specimen, expressed as η= $\left(\frac{3\delta}{L}\right)$. The traverse,

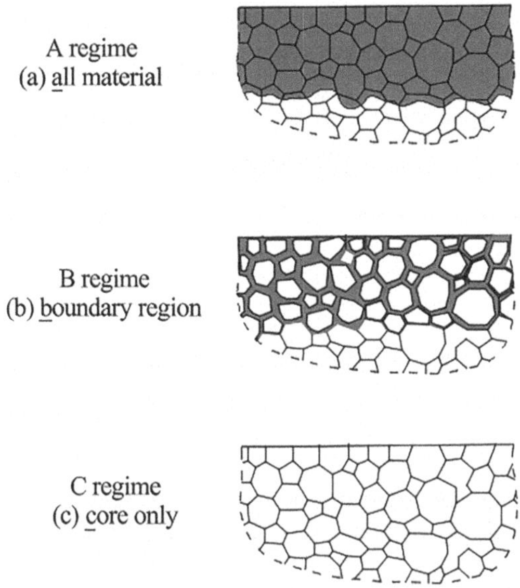

Fig. 8.4 "Self-diffusion" regimes in a polycrystalline sample (*A*, *B*, *C* regimes). The semi-infinite sample is diffused by the tracer atoms from the surface, and diffuses into the semi-infinite specimen. High concentration domains of tracer isotopes are shaded. (**a**) Regime *A*: Average grain size is far less than diffusion length within grains. (**b**) Regime *B*: Grain size is greater than diffusion length. (**c**) Regime *C*: Within grains diffusion length is minimal. Significant grain boundary diffusion is present. Preferential penetration with the grain boundaries cannot be represented pictorially because it is too narrow [1]

expressed as a "mean square", of an atom from grain boundary diffusion is represented as $D^B\eta t$. Within a grain "mean-square displacement" is $*D^{XL}(1 - \eta)t$. Its overall value is given by,

$$\langle *D \rangle t = *D^{XL}(1-\eta)t + *D^B\eta t. \tag{8.2}$$

$$\text{As } \eta \ll 1, \langle *D \rangle = *D^{XL} + \left(\frac{3\delta}{L}\right) + \left(\frac{3\delta}{L}\right) * D^B; \quad *D^L t > L^2. \tag{8.3}$$

The quantity $\langle *D \rangle$ represents the average effective diffusivity; in other words, it defines the overall diffusion. In a material that is homogeneous, macroscopic uniform diffusion behaves similarly—Eq. (8.3). A schematic illustration is available in Fig. 8.4a, and experimental data on Fig. 8.5. As the diffusion field overlaps many boundaries, this is known as multiple boundary diffusion. Fast grain-boundary diffusion causes preferential diffusion along the grain-boundary cores (Fig. 8.4a). This effect cannot be illustrated because very few atoms are involved.

When minimal diffusion is present within grains, grain boundary diffusion is substantial. Total diffusion will solely result from boundary diffusion—refer to Fig. 8.4c. This transpires at low temperatures or brief durations under the criteria $(*D^{XL}t) < \lambda^2$ and $(*D^Bt) > \lambda^2$, where λ represents the interatomic distance.

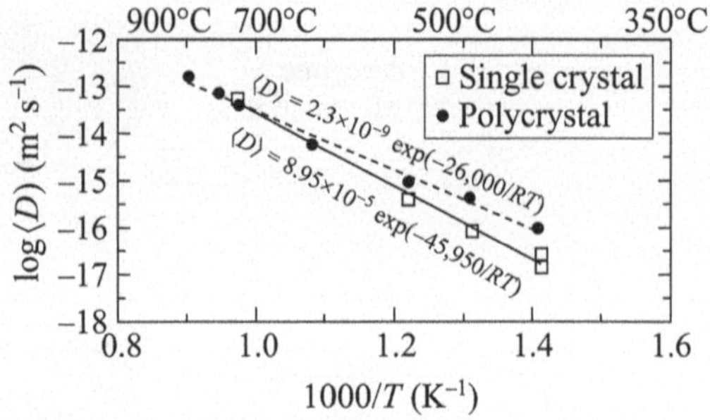

Fig. 8.5 Mean self-diffusivity values ($*D$) at reduced temperatures. Monocrystalline and polycrystalline silver. Grain-boundary diffusion substantially influences the assessed overall average diffusivity [1]

Numerous intermediate situations may exist inside grain boundaries and grains. Grain size, L, is more than diffusion length within grains (Fig. 8.4b). Criteria for this kind of diffusion are $\lambda^2 < *D^{XL}t < L^2$ and $*D^{B}t > \lambda^2$. These two instances are referred to as isolated-boundary diffusion regimes, in which diffusion field overlap with individual boundary segments is absent. This is in contrast with regimes characterized by multiple boundaries. These are presented as domains A, B, and C—see Fig. 8.4.

When the boundaries are in motion, like during grain growth or recrystallisation, the the situation is changed. If v is the average speed of the boundary, nearly stationary grain boundaries will be present when $v_t < \lambda$, and above regimes will once more become relevant. However, if $\left[\sqrt{*D^{XL}t} + vt\right] > L$, the prevalent regime will pertain to diffusion along multiple-boundaries. Equation (8.3) remains valid even if $\sqrt{*D^{XL}t}$ is insignificant. In this scenario, the grain boundaries approach the atoms and not the other way round. Conversely, when $\left[\sqrt{*D^{XL}t} + vt\right] < L$, the relevant regime will be diffusion along isolated boundaries.

The potential behaviour during diffusion in different regions are illustrated approximately here—see Fig. 8.6. It is a $log(*D^{XL}t)$ vs. log (vt) plot. Logarithmic scales elucidate the features around the origin since $\left(\frac{L}{\lambda}\right)$ is generally 10^3 or greater. The stationary-boundary regimes $\left(\frac{vt}{\lambda}\right)$ appear at left and encompass regimes A, B, and C. The regions where diffusion is along isolated boundaries encompass the origin and continue to include the ordinate and the abscissa respectively till $(*D^{XL}t) = L^2$ and $vt = L^2$. Outside the domain of diffusion along boundaries that are isolated, the region where diffusion along multiple boundaries is present will prevail. Figure 8.6 displays boundaries moving in a region where isolated boundaries move and it is categorised into two sub-regimes, contingent upon crystal diffusion being fast enough to let atoms into the grains before the boundaries advance.

When a mobile boundary is attached to the coordinate system, under quasi-steady state conditions, to advance grain flow by diffusion the available flux is $J = \left[- *D^{XL}\left(\frac{dc}{dx}\right) - vc\right]$ and the diffusion equation is

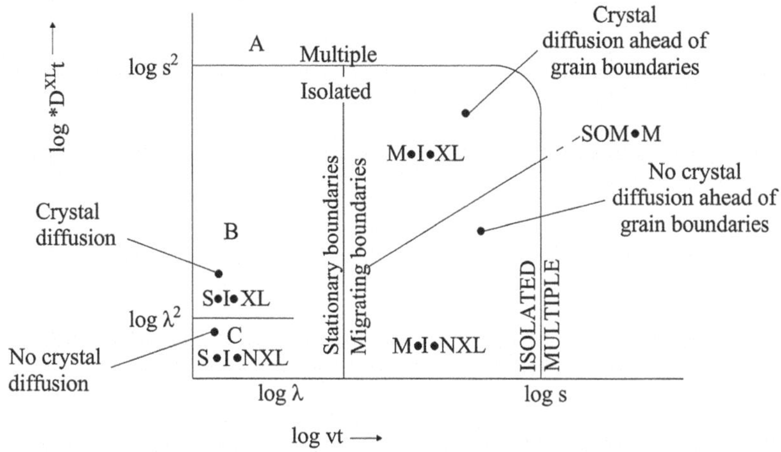

Fig. 8.6 In a polycrystalline material both bulk and grain boundary diffusion can be present. λ denotes interatomic separation, and d represents the grain size. The boundaries remain fixed on the left side. S.I.XL = Stationary boundaries. Diffusion across an isolated border. Diffusion penetration of crystal (XL) into neighbouring grains. S.I.NXL is equivalent to stationary boundaries. Diffusion across an isolated border. No diffusion penetration of crystal (XL) into neighbouring grains. M.I.NXL is equivalent to moving boundaries. Diffusion across an isolated border. Crystal (XL) diffusion ahead of boundaries. SOM.M ≡ Boundaries that are either stationary or in motion. Multiple boundary diffusion [6]

$$-\nabla\vec{J} = -\frac{d}{dx}\left[- * D^{XL}\frac{dc}{dx} - vc\right] = 0 \tag{8.4}$$

$$\text{whose solution is } * D^{XL}\frac{dc}{dx} + vc = A. \tag{8.5}$$

At a considerable distance beyond the boundary, $\frac{dc}{dx} \Rightarrow 0$ and $c \Rightarrow 0$. Therefore, $A = 0$. On integration, one gets

$$c = c^0 e^{(-vx/*D^{XL})}. \tag{8.6}$$

c^0 is the concentration sustained at the boundary. Figure 8.7b illustrates the concentration profile. At the boundary front negligible concentration will be present if $\left(\frac{*D^{XL}}{v}\right) < \lambda$ (Eq. (8.6). Consequently, domains M.I.XL (moving boundaries) and M.I. NXL (stationary boundaries), separated by a curve in Fig. 8.6, adheres to a linear relationship $*D^{XL}t = \lambda vt$.

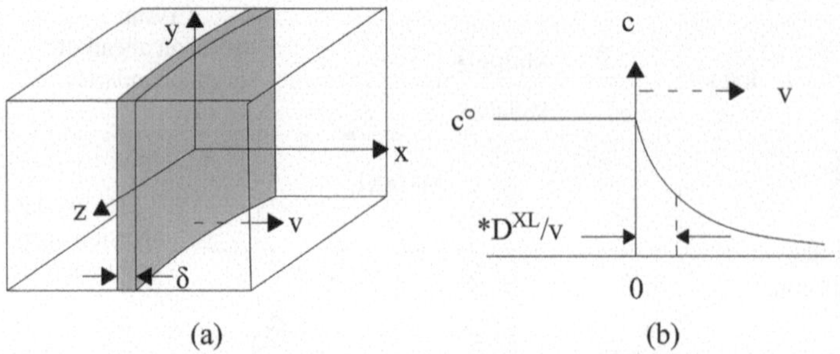

Fig. 8.7 (**a**) illustrates that v represents the constant velocity of the moving grain boundary. Tracer atoms diffuse in the transverse direction from the edges into the boundary slab. They also diffuse from the boundary into the grain behind the boundary to deposit into it. Equilibrium conditions prevail during diffusion. The boundary and coordinate system move together. (**b**) Near the boundary the tracer concentration obeys Eq. (8.6). If diffusion prevails long in both stationary and moving boundaries, it will approach the multiple-boundary regime [1]

8.3 Fast Grain Boundary Diffusion

Rapid diffusion along grain boundaries is not well understood, despite the availability of sufficient experimental evidence. In defect-mediated grain-boundary diffusion, atomic movement within the core is present between several places in the core. They have different activation energies. Then, the overall diffusion rate cannot be accounted for by a single activation energy. So, a curvature is present in Arrhenius plots for grain-boundary diffusion. Such a curvature cannot be detected with insufficient data obtained in a limited temperature range.

8.4 Diffusion Along Dislocations

The dislocation structure influences the rate of dislocation diffusion. Evidence suggests that the rate is greater along an edge dislocation compared to a screw dislocation. Dissociating into partial dislocations is a way of dislocation relaxation in close-packed structures—Fig. 8.8. The stacking fault energy controls the degree of dissociation.

High stacking fault energy materials do not dissociate, e.g., Al. Those of low stacking fault energy easily dissociate, e.g., in FCC metals diffusion rate along dislocations decreases as the dislocations dissociate into partials. This happens because the core region of relaxed partial dislocations lacks the necessary openness or disorder to enable rapid diffusion, in contrast to the cores of complete dislocations (Fig. 8.9). $^* D^D$ for non-dissociated dislocations is $\simeq {}^* D^B$, which suggests that diffusion processes in cores of non-dissociated dislocations and boundaries of high misorientation are rather similar. The net self-diffusion-induced-transport of atoms observed supports this inference. That is, in both undissociated dislocations and

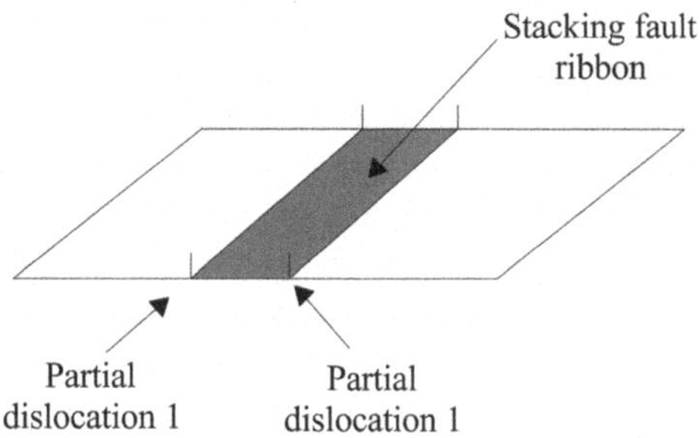

Fig. 8.8 Dissociated lattice dislocation in FCC metal. A ribbon of stacking fault separates the partial dislocations

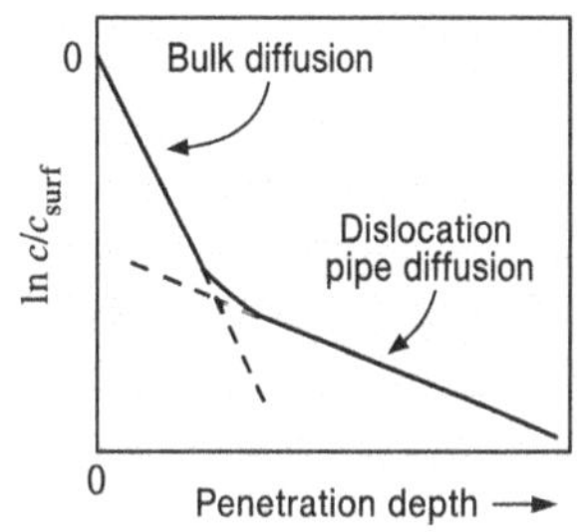

Fig. 8.9 Illustrates tracer isotope penetration by tracer self-diffusion from a free surface into a single crystal that contains dislocations. The tracer concentration is c_{surf}. Near-surface movement is dominated by bulk diffusion, whereas dislocation pipe diffusion prevails at greater depths

high-angle grain boundaries the structure is rather similar and a defect-mediated mechanism is present. In the entire volume of displaced crystal containing dislocations self-diffusion is uniformly distributed. The categorization of these regions is like what is seen in polycrystalline materials. The diffusion can either involve multiple boundaries or solitary boundaries, with or without lattice diffusion. Stationary or moving dislocations may be involved. The important parameters should include $* D^D$ and not $* D^B$. The dislocation density is more relevant than the grain size.

In a dislocated single crystal diffusion from source at surface predominates the crystal surface. Diffusion fields of neighbouring dislocations do not overlap. This behaviour aligns with the dislocation counterpart of regime B in Fig. 8.4.

8.5 Diffusion Along Free Surfaces

Fast diffusion along free surfaces has features like diffusion along grain boundaries. A species capable of diffusing can undergo surface or bulk diffusion and enter/exit either. A species that diffuses rapidly along a surface may enter a contiguous grain. The analytical treatments are similar. At high temperatures, all surfaces experience a "roughening" transition and so become "general interfaces/grain boundaries". Several details of atomic-level surface diffusion are unknown.

At low temperatures mechanisms involving jumps of surface vacancies and adatoms predominate "special" ("vicinal" or "singular") boundaries. The formation energies of these defects depend on surface inclination, i.e., *(hkl)*. Surface migration energies reveal that the adatom moves more slowly. Moreover, the diffusion is anisotropic (due to low two-dimensional symmetry). In surface formations in which atoms are arranged in closely spaced parallel rows and there is larger gap between rows, diffusion occurs more readily along the dense rows than transversely across them. Transverse diffusion proceeds by a substitution mechanism. Between atoms arranged in dense rows another atom, by diffusion, supplants an atom in that row, moving the atom that is vacated into a neighboring gorge in the middle of the atom rows. A repetition of this pattern is analogous to the "bulk interstitialcy" mechanism. The diffusion rate along and over ledges at vicinal boundaries differs from that in unique regions. High temperature diffusion mechanisms involve transitions to sites that are more remote and are more complicated. Jumps, that are collective, across many faults, and similar phenomena may occur. At temperatures that are raised, delocalization of adatoms may occupy a large part of the duration of this process. The Arrhenius plot frequently exhibits curvature as a result of supplementary mechanisms. Consequently, diffusion is more isotropic.

Problems

1.	A Metal has a diffusivity of $D = 1.2 \times 10^{-10}\ \mathrm{m^2/s}$ at 500 °C. Find the average distance travelled by an atom in 10 h. **Solution:** Use $x = \sqrt{2Dt}$ Given $D = 1.2 \times 10^{-10}\mathrm{m^2/s}$ $t = 10 \times 3600 = 36{,}000\ \mathrm{s}$ $x = \sqrt{2 \times 1.2 \times 10^{-10} 36{,}000}$ $x = 2.94$ mm
2.	Calculate the time needed for an atom to diffuse an average distance of 0.5 mm in a material with diffusivity $5 \times 10^{-11}\ \mathrm{m^2/s}$. **Solution:** Use $t = \frac{x^2}{2D}$ $x = 0.5 \times 10^{-3}\mathrm{m}, D = 5 \times 10^{-11}\mathrm{m^2/s}$ $t = 41.67\ \ \mathrm{min}$
3.	If $D_{gb} = 2 \times 10^{-12}\ \mathrm{m^2/s}$, $\delta = 5$ mm, and grain size $d = 0.5$ mm, find D_{eff} **Solution:** $\delta = 5 \times 10^{-9}\mathrm{m}; d = 0.5 \times 10^{-3}\mathrm{m}$ $D_{\mathrm{eff}} = 2 \times 10^{-12} \times \dfrac{2 \times 5 \times 10^{-9}}{0.5 \times 10^{-3}}$ $D_{\mathrm{eff}} = 2 \times 10^{-12} \times 2 \times 10^{-5}$ $D_{\mathrm{eff}} = 4 \times 10^{-17}\mathrm{m^2/s}$
4.	If $D_s = 3x10^{-8}\ \mathrm{m^2/s}$ and $\frac{dC}{dx} = 1 \times 10^5\ \mathrm{mol/m^4}$, calculate surface diffusion flux. **Solution**

(continued)

$$J = -D_s \frac{dC}{dx}$$
$$J = -\left(3 \times 10^{-8}\right) \times \left(1 \times 10^{5}\right)$$
$$J = -3 \times 10^{-3} \,\mathrm{mol}/_{\mathrm{m^2 s}}$$

References

1. Balluffi, R. W., Allen, S. M., & Craig Carter, W. (2005). *Kinetics of materials*. Wiley.
2. Gjostein, N. A. (1973). Short circuit diffusion. In *Diffusion* (pp. 241–274). American Society for Metals.
3. ThermoCalc. (n.d.). *ThermoCalc databases*. Retrieved from https://thermocalc.com/products/databases/
4. Herbeuval, I., & Biscondi, M. (1974). Diffusion of zinc in grains of symmetric flexion of aluminum. *Canadian Metallurgical Quarterly, 13*(1), 171–175.
5. Atkinson, A. (1994). In R. W. Cahn, P. Haasen, & E. Kramer (Eds.), *Diffusion in ceramics. Materials Science and Technology-A comprehensive treatment* (Vol. 11). VCH Publishers.
6. Cahn, J. W., & Balluffi, R. W. (1979). Diffusional mass-transport in polycrystals containing stationary r migrating grain boundaries. *Scripta Metallurgica et Materialia, 13*(6), 499–502.

Chapter 9
Diffusion in Non-crystalline Materials

Contents

9.1 Introduction

A variety of non-crystalline materials exist—liquids (e.g., mono-atomic), amorphous materials, glass networks, and long-chain polymers. Diffusion in these materials is through different mechanisms. Often, quantitative analyses are difficult. Diffusion in non-crystalline materials is poorly understood. Opinions differ on the operating mechanisms in many cases. Some examples are examined here.

9.2 Free Volume Model for Self-Diffusion in Liquids

Regions that contain free volume in monoatomic liquids are caused by fluctuations that lead to displacements, which at times can cause large enough fluctuations that give rise to diffusive displacements. "The hard-sphere model of liquid" is a good approximation. The inter-atomic potential energy is taken as constant till atoms touch each other, when the energy reaches infinity. Inter-particle distance, on average, in liquids is more than if the atoms were in contact. The difference between the volumes of the solid and the liquid is known as "free volume". Inter-atomic potentials are nearly constant. The volume traversed by a particle is confined and small. The average velocity of a particle in a region of near-constant potential and that of a gas particle are the same. Predominantly, a particle remains on average in a fixed position. But fluctuations can create a large enough space to displace a particle

K. A. Padmanabhan et al., *Thermodynamics and Kinetics of Materials*, Engineering Materials, https://doi.org/10.1007/978-3-032-14527-7_9

significantly. If another particle occupies that space before the original particle comes back, the result is a diffusive-type jump. That is, redistribution at constant energy of free volume causes diffusion.

The hard-sphere gas model, the basis of the kinetic theory of gases, reveals that the self-diffusivity is given by $* D^G = \left[\frac{2}{5}\langle u\rangle L\right]$, where $\langle u\rangle$ is the average velocity, and L is the mean free path. The diameter of the confining volume nearly equals the mean free path of a confined particle in the liquid. So, the self-diffusivity of the liquid, assuming that it is very small, is,

$$*D(V) = C_{\text{geom}} a(V)\langle u\rangle. \tag{9.1}$$

$a(V)$ is the diameter of the confining volume, V is the free volume associated with the particle, $\langle u\rangle$ is the average velocity of the particle, and C_{geom} is a geometrical constant. Unless V exceeds a critical value, V^{crit}, $*D(V) = 0$. And so, the overall diffusivity is

$$*D^L = \int_{V^{\text{crit}}}^{\infty} * D^V p(V) dV \tag{9.2}$$

where probability of the presence of free volume, $p(V)dV$, is between V and $V + dV$. If $N_i(V_i)$ is the number of particles with free volume V_i and υ^{free} is the total free volume,

$$\upsilon^{\text{free}} = \gamma \sum_i N_i V_i \tag{9.3}$$

is valid. γ represents the free-volume overlaps between neighbor particles. Its value is approximately one. That is, N, the total number of particles, is given by

$$N = \sum_i N_i. \tag{9.4}$$

The number of ways the free volume is distributed at constant energy (entropy) is expressed as

$$S^{\text{conf}} = k \ln\left(\frac{N!}{\Pi_i N_i!}\right). \tag{9.5}$$

Here N_i are the bin populations. Its equilibrium probability distribution (Eq. 9.2) is the continuum limit of the bin populations N_i maximizes S^{conf}. It is constrained by Eqs. (9.3) and (9.4). For the total free volume and fixed-number of constraints Lagrange multipliers β and λ are included. Using Sterling approximation ($N_i\,! \simeq N_i \ln N_i$), the extremal conditions become $N_i \gg 1$. After some complex computations, we get,

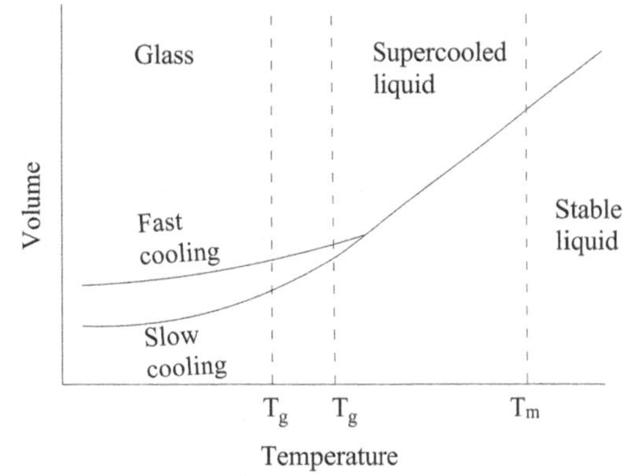

Fig. 9.1 Metallic glass volume vs. T during rapid and relatively slow cooling. T_m is the melting temperature, T_g is the glass-transition temperature (values for both rapid and relatively slow cooling) [1]

$$*D^L \simeq C_{\text{geom}} a\left(V^{\text{crit}}\right)\langle u\rangle e^{-\gamma V^{\text{crit}}/\Omega^{\text{free}}}. \tag{9.6}$$

This equation reveals the diffusivities seen in simple liquids. If $*D^L$ is not thermally activated, there is no Arrhenius dependence for the diffusivity of crystals. Roughly, $\langle u\rangle \alpha T^{0.5}$ and $\langle \Omega^{\text{free}}\rangle$ rise linearly with T.[1]

9.3 Diffusion in Amorphous Metals

Rapid cooling (quenching) from the liquid phase causes metallic glasses (amorphous metallic alloys) formation. In fact, it first becomes a supercooled liquid. At still lower temperatures, it experiences a glass transition—Fig. 9.1.

The cooling rate determines the transformation temperatures. An abrupt change in the rate at which the volume and other properties change with decreasing temperature characterizes the glass transition. This temperature is given by the intersection of the extrapolated cooling curves above and below the transition. As this temperature is higher for fast rather than slow cooling, the free volume in glasses formed at low temperatures is less. The combined effects of the low temperature and less free volume leads to a loss of fluidity of the metallic glass and the glass becomes a frozen liquid.

9.3.1 *Self-Diffusion*

On reheating and isothermally annealing at a T below T_g, a metallic glass loses the frozen-in excess free volume and the system relaxes. And this leads to equilibration

[1] As this is a first course on this topic, complicated mathematical calculations which do not add to the physical knowledge have been omitted.

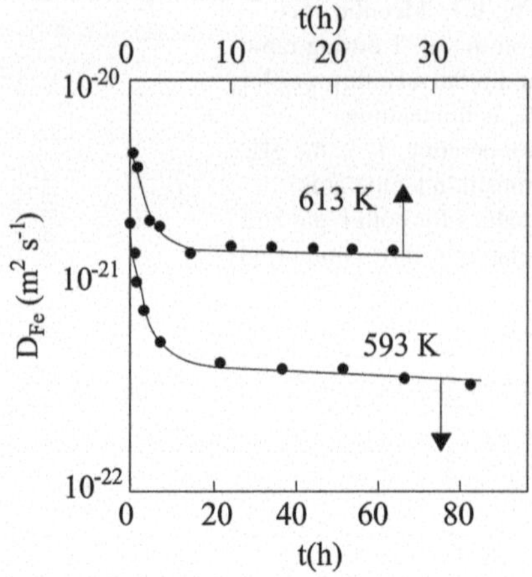

Fig. 9.2 Self-diffusivity of Fe in a rapidly cooled amorphous $Fe_{40}Ni_{40}B_{20}$ due to isothermal annealing below T_g. Arrows indicate the different time scales [2]

without crystallizing. In the regions of higher-than-average atomic density, the free volume is lost. In the beginning, the self-diffusivity decreases during annealing but reaches an asymptotic, time-independent value later—Fig. 9.2.

This structure is dense and is randomly packed. The atoms are arranged for the highest density of the hard-sphere radii and exhibit no translational symmetry. Locally, several polyhedral units in different proportions are formed. Large holes and micro-crystallites are not present. This structure is metastable vis-a-vis the crystalline state. The time dependence is present in the self-diffusivities. The temperature dependence follows the Arrhenius equation, i.e., Ln* D_i vs. 1/T plots are linear.

The mechanism of self-diffusion in relaxed metallic glasses is controversial.

Diffusion in small interstitials, polymer chains of different types and oxide glasses is not discussed here.

Problems

1.	In an amorphous materials, the diffusivity is $D = 7 \times 10^{-12} \mathrm{m^2/s}$. Find the average diffusion distance after 5 h. **Solution:** Use $x = \sqrt{2Dt}$ $D = 7 \times 10^{-12} \mathrm{m^2/s}$ $t = 5 \times 3600 = 18,000 \ \mathrm{s}$ $x = \sqrt{2 \times 7 \times 10^{-12} \times 18,000}$ $x = \sqrt{2.52 \times 10^{-7}}$ $X = 0.502 \ \mathrm{mm}$
2.	If the free volume fraction $f = 0.04$, atomic jump frequency $\nu = 10^{13}$ Hz, and jump distance $l = 2 \times 10^{-10}$ m, find the self-diffusion coefficient. **Solution:**

(continued)

	$D = \nu l^2 f$ $D = 10^{13} \times (2 \times 10^{-10})^2 \times 0.04$ $= 1.6 \times 10^{-7} \, \mathrm{m^2/s}$
3.	At 600 °C, the self-diffusion coefficient of an amorphous solid is $3 \times 10^{-13} \, \mathrm{m^2/s}$. Calculate the diffusion time needed to achieve an average displacement of 10 μm. **Solution:** $X = 10 \ \mu\mathrm{m} = 10^{-5} \ \mathrm{m}$ $t = \frac{x^2}{2D} = \frac{(10^{-5})^2}{2 \times 3 \times 10^{-13}} = 0.167 \ \mathrm{s}$

References

1. Balluffi, R. W., Allen, S. M., & Craig Carter, W. (2005). *Kinetics of materials*. Wiley.
2. Horvath, J., & Mehrer, H. (1986). Tracer diffusion of Fe-59 in amorphous $Fe_{40}Ni_{40}B_{20}$. *Lattice Defects Amorphous Materials, 13*(1), 1–14.

Chapter 10
Motion of Dislocations and Interfaces

Contents

Motion of dislocations and interfaces is key to several kinetic processes. The fact that dislocations and interfaces can have different structures complicates the matter further. This can lead to varying responses. The mechanisms of dislocation glide and climb are taught in Courses on the Theory of Dislocations. In view of its importance a short summary is provided first.

10.1 Driving Forces on Dislocations

External forces move dislocations in crystals. When a dislocation displacement reduces the energy of a system an external driving force is exerted. Such forces may arise in different ways.

K. A. Padmanabhan et al., *Thermodynamics and Kinetics of Materials*, Engineering Materials, https://doi.org/10.1007/978-3-032-14527-7_10

10.1.1 Mechanical Forces

The effective force on a dislocation ("mechanical force") is $f = \frac{\delta W}{\delta r}$. According to the Peach-Koehler equation,

$$\overrightarrow{f_\sigma} = \left(\overrightarrow{b^T} \cdot \sigma\right) \times \widehat{\zeta} \equiv \vec{d} \times \widehat{\zeta}, \tag{10.1}$$

where $\overrightarrow{f_\sigma}$ is the mechanical force exerted on the dislocation (per unit length), σ is the stress tensor at a dislocation, and $\widehat{\zeta}$ the unit vector tangent to the dislocation along its positive direction. Both $\vec{d}$ and $\widehat{\zeta}$ have three components in the cartesian coordinate system. That is,

$$\vec{d} = \begin{bmatrix} d_x \\ d_y \\ d_z \end{bmatrix} = \begin{bmatrix} b_x\sigma_{xx} & b_y\sigma_{xy} & b_z\sigma_{xz} \\ b_x\sigma_{xy} & b_y\sigma_{yy} & b_z\sigma_{yz} \\ b_x\sigma_{xz} & b_y\sigma_{yz} & b_z\sigma_{zz} \end{bmatrix}. \tag{10.2}$$

This allows the computation of the mechanical force exerted on a straight dislocation by a stress field. A shear stress, σ_{xy}, acting on an edge dislocation $\vec{b}$ = [b00] ($\widehat{\zeta}$ = [OO1]) (Fig. 10.1) enables it to glide in the x-direction on its slip plane. A tensile

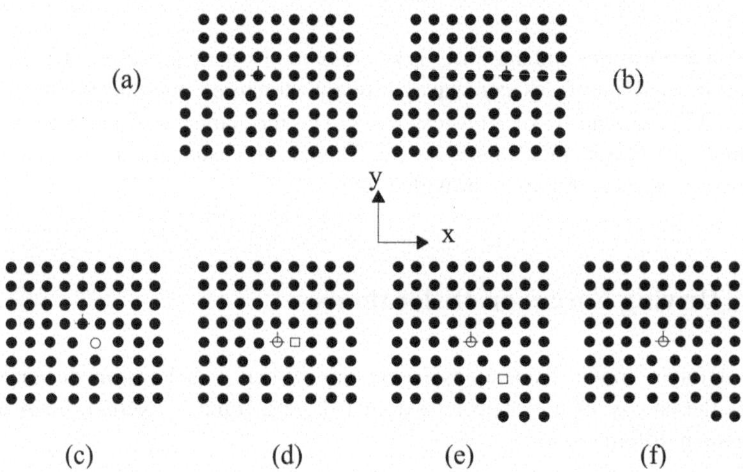

Fig. 10.1 Edge dislocation in cubic crystal ($\vec{b}$ = [b00]. ($\widehat{\zeta}$ = [OO1])—glide and climb. (**a**) and (**b**) Glide from left to right. (**c**–**f**) Downward climb along $-y$. (**d**) the substitutional atom is shown in lighter shade near the dislocation core in (**c**) has joined the extra half plane to create a vacancy. (**e**), diffusion has caused the migration of the vacancy away from the dislocation core. In (**f**), at the surface step the vacancy is eliminated (equivalent to removing an atom from the surface and transporting it to the dislocation at its core). This creates a new site at the dislocation, i.e., it became a vacancy source. Subsequently, at the surface this step is removed, i.e., it is an atom source [1]

stress, σ_{xx}, (Eq. 10.1), on the other hand, subjects the dislocation to a force $\overrightarrow{f_\sigma} = -\hat{j}b\sigma_{xx}$ (i.e., a force that causes climb in the $-y$ direction). The force (always normal to the dislocation line) in a general stress field can have a component in the glide plane of the dislocation also. Then, both glide and climb result from the overall force. But at low temperature profuse diffusion is absent. Only glide will be present.

This can be extended to treat the motion of curved dislocations using traditional mathematics.

10.2 Dislocation Glide

It is essential to know the driving force and glide rate of a dislocation in a crystal. As the velocity of dislocations approach that of sound in the given medium, relativistic effects appear in perfect crystals. At high temperatures, dissipative phonon effects cause frictional drag forces. At the core, shuffling of atoms necessary for motion, is difficult in some crystals. This slows down glide. Point, line, and planar defects and inclusions present in imperfect crystals function as additional constraints and restrict dislocation glide. Glide in a perfect single crystal is assumed to be a linear elastic continuum.

10.2.1 Glide in Perfect Single Crystals

In Fig. 10.2, the elastic displacements, u_1, u_2, and u_3, around a screw dislocation that moves at a constant velocity, $\vec{v}$, along the x axis can be found by solving the Navier-Stokes equations as it applies to isotropic linear elasticity. The only nonzero displacements for a screw dislocation are along z. Therefore, the Navier equations reduce to

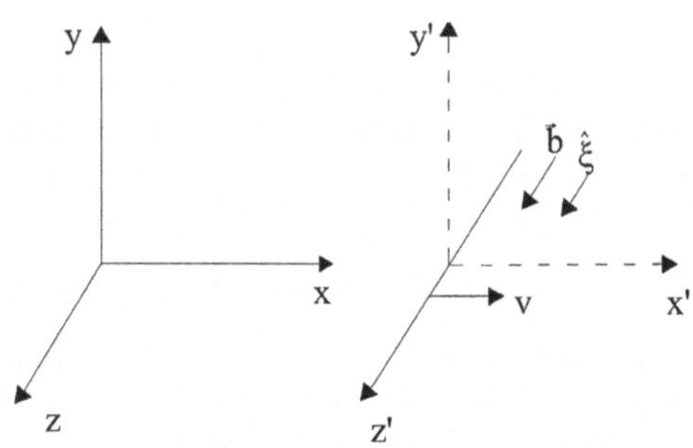

Fig. 10.2 Screw dislocation with $\vec{b} = [00b]$, $\hat{\zeta} = [001]$ in the +x direction which moves at a constant velocity $\vec{v}$. The origin of the (x', y', z') coordinate system moves with the moving dislocation [2]

$$\rho \frac{\partial^2 u_3}{\partial t^2} = \mu \left(\frac{\partial^2 u_3}{\partial x^2} + \frac{\partial u_3}{\partial y^2} \right), \tag{10.3}$$

where ρ is the medium density, μ is the shear modulus, and the term on the LHS is the inertial term due to the acceleration of mass due to the moving dislocation. This equation is solved by changes variables as,

$$x' = \frac{x - vt}{\gamma_L}, \quad \gamma_L \equiv \sqrt{1 - \left(\frac{v}{c}\right)^2}, \quad y' = y, \quad z' = z, \quad t' = \frac{t - vx}{c^2 \gamma_L}. \tag{10.4}$$

Here, $c = \sqrt{\frac{\mu}{\rho}}$ is the transverse shear sound wave velocity in the elastic medium. One sees from Fig. 10.2 that the origin of the (x', y', z') coordinate system moves with the moving dislocation.

From Eqs. (10.3) and (10.4),

$$\frac{\partial^2 u_3}{\partial x'^2} + \frac{\partial^2 u_3}{\partial y'^2} = 0. \tag{10.5}$$

(u_3 is not a function of t'. Therefore, $\left(\frac{\partial u_3}{\partial t'}\right) = 0$.) The form of Eq. (10.5) corresponds to the Navier equation for a static screw dislocation. After transformation back to x, y, t space the solution is,

$$u_z(x, y, t) = \frac{b}{2\pi} \tan^{-1}\left(\frac{\gamma_L y}{x - vt}\right). \tag{10.6}$$

The shear stress of the dislocation in cylindrical coordinates: The shear stress, in cylindrical coordinates, σ_{θ_z}, is known from $\sigma_{xz} = \mu\left(\frac{\partial u_z}{\partial x}\right)$, $\sigma_{yz} = \mu\left(\frac{\partial u_z}{\partial y}\right)$, $\sigma_{\theta z} = \sigma_{yz} \cos\theta - \sigma_{xz} \sin\theta$. Thus, the result is,

$$\sigma_{\theta z} = \frac{\mu b}{2\pi} \left[\frac{\gamma_L (x^2 + y^2)^{1/2}}{x_0^2 + \gamma_L^2} \right]. \tag{10.7}$$

Here the distances x_0 and y_0 (measured from the moving dislocation) are newly presented. The stress field in Eq. (10.7) continuously contracts along the x_0 axis and extends along the y_0 axis with increasing dislocation velocity. This distortion resembles the Lorentz contraction and expansion of the electric field around a moving electron. The quantity γ_L resembles the Lorentz-Einstein term $\left(1 - \frac{v^2}{c^2}\right)^{1/2}$ in the theory of relativity of the electron. Here, c is the transverse shear wave. The velocity of light in the theory of relativity is replaced by the sound velocity here. As $v \to c$ and $\gamma_L \to 0$, the stress around the dislocation becomes zero everywhere except along the y'-axis, where it is infinite.

Thus, the energy of the dislocation is velocity dependent. The energy density around the dislocation, w, is the sum of the elastic strain-energy and the kinetic-energy density. That is,

$$w = 2\mu\varepsilon_{xz}^2 + 2\mu\varepsilon_{yz}^2 + \frac{1}{2}\rho\left(\frac{\partial u_z}{\partial t}\right)^2 = \frac{\mu}{2}\left[\left(\frac{\partial u_z}{\partial x}\right)^2 + \left(\frac{\partial u_x}{\partial y}\right)^2 + \frac{1}{2}\left(\frac{\partial u_z}{\partial t}\right)^2\right]. \quad (10.8)$$

In the first two terms in the above equation on the RHS each expression is the elastic strain-energy density, and the third term is the kinetic-energy density. When the total energy is obtained by integrating the energy density over the volume around the dislocation it becomes

$$W = \frac{W^0}{\gamma_L} = \frac{W^0}{\left(1 - v^2/c^2\right)^2} \quad (10.9)$$

where W^0 is the elastic energy of the dislocation per unit length at rest. That is,

$$W^0 = \frac{\mu b^2}{4\pi}\, ln\left(\frac{R}{R_0}\right). \quad (10.10)$$

In this case, R_0 is the cutoff radius at the core. R is the dimension of the crystal containing the dislocation. From Eq. (10.9), the moving dislocation energy tends to infinity as its velocity tends towards the sound velocity, c.f., as in a relativistic particle the energy/mass tends to infinity as the velocity tends to that of light. So, relativistic effects are important only when the velocity of the dislocation approaches that of sound, c.

An edge dislocation behaves in a more complicated manner. (Its displacement field experiences both shear and normal stresses.) Superposition of two terms, each behaving relativistically with limiting velocities corresponding to the speed of transverse shear waves and longitudinal waves, respectively, leads to the solution. In both the terms the relative magnitudes are velocity-dependent.

Drag Effects: Dislocation glide encounters dissipative frictional forces in real crystals. Several sources contribute to these frictional forces, e.g., Scattered phonon and electron drag restricts the dislocation velocity to values well below c. Quantitative analyses are difficult to achieve.

Peierls Force/Peierls Energy Barrier: In some crystals (e.g., covalent crystals) the Peierls force may be so large that the driving force due to the applied stress cannot drive the dislocation forward. The dislocation is immobile under such a situation. At elevated temperatures, the Peierls energy barrier may be crossed by the dislocation through thermal activation aided by stress. If the applied stress is σ_A and too small to overcome the barrier, the swept-out area is A_1, and the energy that must be supplied by thermal activation is then the shaded area (Fig. 10.3). Therefore, the activation energy, E, is,

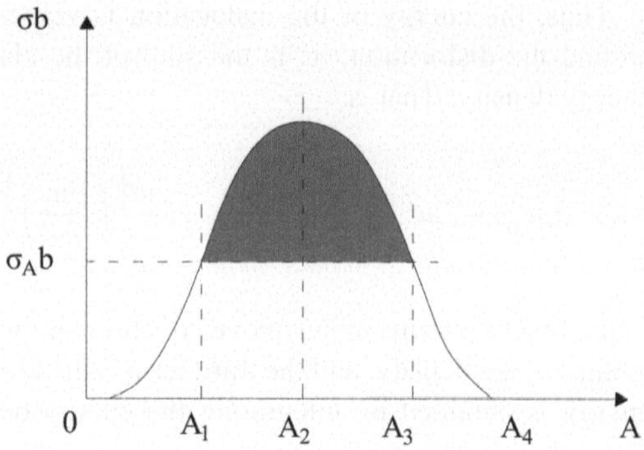

Fig. 10.3 Applied force, σb, vs. area swept out, A, when a dislocation overcomes an obstacle to glide [2]

$$E = b\int_{A_1}^{A_3} (\sigma - \sigma_A)dA. \tag{10.11}$$

b is the Burgers vector of the dislocation, σ is the Pierls stress, $(A_3 - A_1)$ (=A) is the area swept out through the thermal activation event.

The dislocation velocity will be,

$$v = v_0 e^{-E/(kT)}, \tag{10.12}$$

where v_0 is a constant proportional to the attempt frequency. The area $(A_3 - A_1)$, called the activation area, is one that is swept out during the activation event. From a kinetics standpoint Eq. (10.11) is of interest. It says that the activation energy decreases with increasing applied stress. Therefore, the process is known as "stress-aided thermal activation".

10.2.2 Glide Overcoming Many Obstacles in Imperfect Crystals

Different types of defects, e.g., point, line, and planar crystal defects, solute atoms, embedded particles which interact with dislocations, can hinder glide. Solute atoms dispersed in a solution will prevent a dislocation from moving through as a rigid line but will lead to segments that bulge in and out as the dislocation experiences close encounters with nearby solute atoms. In effect the dislocation experiences a uniform motion. Superimposed rapid forward or backward localized bulging is also present. Rapid bulging motion dissipates energy and so exerts a drag force. At adequately high temperatures, solute atoms migrate in the dislocation stress field. Such induced diffusion can dissipate energy and produce a drag force, particularly if the dislocation is slow moving. Moreover, solute atoms with anisotropic displacement fields can change orientations because of the stress field of a moving dislocation. This causes a macroscopic strain increment, and a dissipative drag force. Solute atom segregation to the cores of dislocations is a possibility and they form atmospheres around dislocations and hinder/pin them. If the binding is very strong, motion gets impeded.

Dislocations can attract or repel each other, e.g., during the plastic deformation of crystals which leads to work hardening. In this case, numerous dislocations are generated during the deformation, many are retained in the crystal, and they prevent the further passage of dislocations, and give rise to material hardening/strengthening. The glide of dislocations, held up at obstacles in their slip planes, can climb around them by thermal activation at elevated temperatures. This will allow the continuation of glide.

Grain boundaries can block slip. At a boundary a discontinuity in the slip plane and the Burgers vector are experienced by a gliding dislocation. It then attempts to pass through an adjoining grain. The kinetic processes involved in the movement of dislocations through materials containing various obstacles are many and are discussed in specialized books.

10.2.3 Experimental Observations

The velocity of edge and screw dislocation segments in LiF single crystals as a function of applied force/stress is displayed in Fig. 10.4. Stresses above a yield threshold stress are necessary for motion to be observed. With increasing stress, the velocity increases but as the sound velocity is approached, it levels off as the velocity of sound is approached. The significantly relativistic range was not reached in these

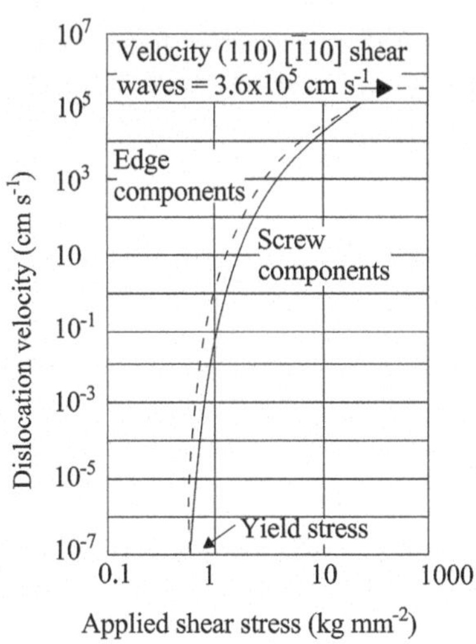

Fig. 10.4 Velocity vs. resolved shear stress for dislocation motion in LiF single crystals [3]

experiments (for all measurements $\gamma_L \simeq 1$). At the lower stresses (where the results are impurity sensitive), obstructions arising from dislocation-dislocation and dislocation-defect interactions restrict the velocities. At the higher stresses (Fig. 10.4), where impurity insensitivity exists and the stress decreases with increasing temperature, the higher velocities are limited by phonon-viscosity drag. Even during low-strain-rate deformation in the beginning high dislocation velocities may be present if the mobile dislocation concentration in the beginning is very low. Then the movement of a limited number of dislocations can attain the required strain. Experimental evidence exists for the presence of strong frictional drag as the velocities approach the relativistic range. Mechanical energy is converted to heat during viscous damping of dislocations. This leads to internal friction when a dislocation network is subjected to oscillating stress.

10.2.4 Glide at Supersonic Velocities

If in a crystal a dislocation is inserted and it moves at supersonic speed, *sound waves*, like the electromagnetic (Cherenkov) radiation that enters a material at a velocity greater than the speed of light in that medium. This leads to a fast deceleration of the dislocation. Steady state supersonic displacement of dislocations is allowed when dislocation motion in its glide plane gives rise to a sufficiently large energy reduction in the system. This reduction in energy supplies the energy needed for radiation during supersonic motion, e.g., a partial dislocation that moves to remove its associated fault or motion in a glissile martensitic interface, which changes the higher free-energy parent phase to the lower-energy martensitic phase. Models for the motion of such dislocations are totally different, see, for example, Nabarro [4, 5] and Hirth and Lothe [6]. Clear evidence is not available for the supersonic motion of martensitic interfaces, perhaps due to the drag forces. There is some evidence that supersonic dislocations are present in shock-wave fronts (Fig. 10.5). Models for the

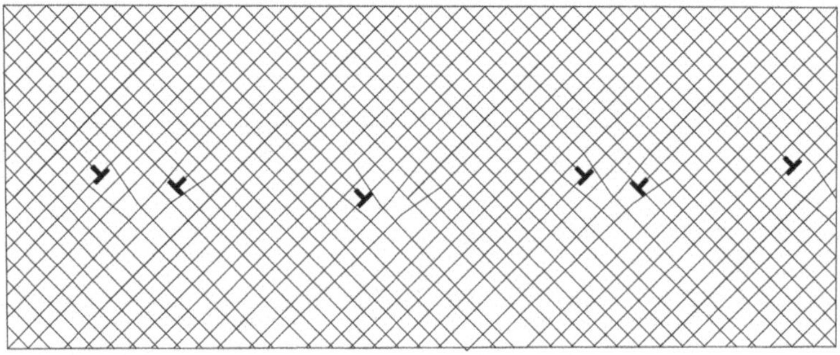

Fig. 10.5 A shock-wave front in a possible interface between normal (lower) and compressed (upper) lattice [2]

motion of such fronts and some evidence generated by computer simulation are available.

10.2.5 Dislocation Motion and Anelasticity

Dislocation segments in dislocation networks can produce anelastic strains by stress-induced dislocation glide. If a pinned dislocation segment is subjected to oscillating stress, periodically it will bow in and out like a vibrating string—Fig. 10.6. The dislocation velocity is finite, dissipative drag forces exist, and the material will exhibit a small oscillating strain which will lag the stress. Internal friction results.

10.3 Climb of Dislocation

Vacancies are created at the jogs (steps at the edge of an extra plane) present in the dislocation core. A jog is reached by either a vacancy jumping into it or by first jumping into a dislocation core and then diffusing along it to a jog. The jogs are eliminated there. The elementary processes here are:

- A vacancy jumping into a jog and getting eliminated in the process.
- The jumping of a vacancy into the core, where it gets attached.
- A vacancy attached at a jog getting destroyed.
- An attached vacancy diffusing along the core.

Often an attractive binding energy binds vacancies to a dislocation and the vacancies diffuse along the dislocation quicker than in the crystal. So, fast diffusion enables many vacancies to reach a jog by moving along a dislocation core than by diffusing through the crystal. The nucleation and growth of strings of attached excess vacancies along the core allows the required formation of the number of jogs needed for dislocation climb. If a string is sufficiently long, it will collapse and produce a fully formed jog pair, as in the region along a core. The continued elimination increases the jog pair distance till the excess vacancies at the jogs disappear and from the edge of an extra plane a complete row of atoms is eliminated. Then, this process repeats itself (steady-state climb).

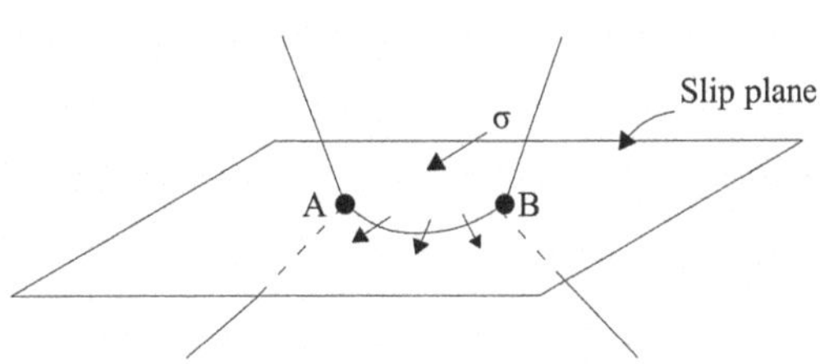

Fig. 10.6 Dislocation segment pinned at A and B bowing out in the slip plane due to the applied shear stress, σ [2]

10.3.1 Diffusion vs. Interface-Reaction-Controlled Climb

The climb rate of this process has been described using a kinetic model. Equilibrium concentration in the vicinity of the jogs is assumed. Due to the strongly attractive binding energy between vacancies and the core they diffuse rapidly along it through a huge mean distance (Z) before jumping back into the crystal. Z increases with the vacancy binding energy between the dislocation and the relative rate of diffusion of the vacancy along the dislocation core. Therefore, each jog can maintain the concentration at equilibrium value on either side of a dislocation for roughly to a distance, $\langle Z \rangle$. Two segments of high-diffusivity core on either side aids every jog to serve as an ellipsoidal sink of semi-axes b and $\langle Z \rangle$. The concentration is maintained in local equilibrium with the jog. The effectiveness of a dislocation as a sink depends on the magnitude of $\langle Z \rangle$ and the spacing of the jogs along the dislocation $\langle S \rangle$. When the vacancy concentration is small and the system is near equilibrium, the jog spacing is nearly equal to the Boltzmann equilibrium expression $\langle S \rangle \approx$ b exp $\left(\frac{-E_j}{kT}\right)$, where E_j is the formation energy of a jog. For excess supersaturations the vacancies quickly nucleate jog pairs along the dislocation. The number of jogs will cross the equilibrium value. Their spacing will get less. The dislocation sink efficiencies in such situations can widely vary. The jog sinks along the dislocation will overlap if $2\langle Z \rangle / \langle S \rangle < 1$. The efficiency of the sink then is high. Local vacancy equilibrium everywhere along its length is ensured. The vacancy elimination rate, will be decided by the diffusion rate at which the vacancies can reach the dislocation. The dislocation will be considered to be an "ideal" sink. (i) A strong binding energy between the jogs and the vacancies, (ii) a fast diffusion rate in the core, (iii) a low formation energy for a jog, and (d) a large vacancy supersaturation will ensure ideal behavior.

If diffusion of vacancies attached to the jogs is obstructed and $\langle Z \rangle$ is small, i.e., $\langle Z \rangle \approx b$, each jog is analogous to an isolated spherical sink of radius b. If in addition, $\langle S \rangle$ is significant, the jog sinks will be separated, and the sink efficiency of the dislocation is very limited. Then the vacancy destruction rate will be dependent on the rate of destruction of vacancies along the dislocation line. In the extreme case, the vacancy elimination rate is nearly independent of the rate at which the vacancy can transported over long distances to the dislocation line by diffusion. Then the sink limitation will determine the reaction rate, the so called "reaction rate control".

If the dislocation acts as a sink for vacancy flux by diffusion, or as a source of atoms or a source for a vacancy flux, the sink efficiency, η, is expressed as,

$$\eta = \frac{\text{flux of atoms created at actual source}}{\text{flux of atoms created at corrresponding ideal source}}. \tag{10.13}$$

If the climb is sufficiently fast to reach ideal diffusion-limited conditions, the operating efficiency is one. Slow acting sources reduce efficiency to 0.

The climb of mixed dislocations can operate by the same jog-diffusion mechanism, which applies to screw dislocations. A pure screw dislocation climbs forming

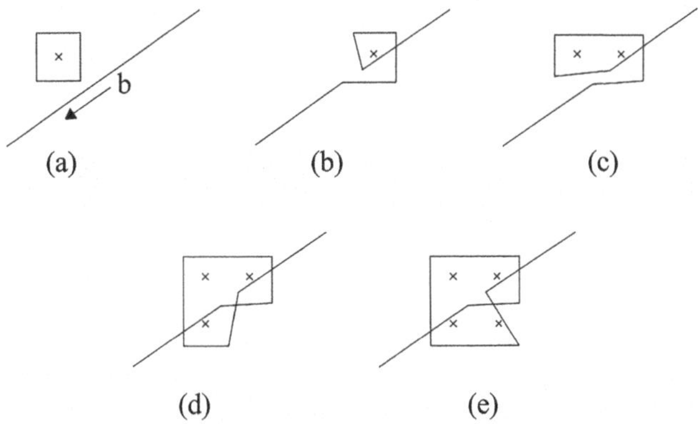

Fig. 10.7 A helical segment formation on an initially straight screw dislocation lying along [100] in a primitive-cubic crystal by progressive addition of vacancies to the core. (Each vacancy is represented by a vacancy-type prismatic loop of atomic size.) (**a**) Vacancy in a crystal with a straight screw dislocation nearby. (**b**) Configuration after a vacancy has joined the dislocation. (**c–e**) Configurations after two, three, and four vacancies are added [7, 8]

helixes using the excess vacancies (Fig. 10.7). As the helical dislocation turns have edge components, the overall behavior is that of mixed dislocation segments. Further, the jogs may nucleate at steep curvature or heterogeneous nodes. The dislocation may have split into partials in materials of low stacking fault energy (high formation energy). Then, (dislocation) climb is difficult.

10.4 Experimental Observations

Many semi-quantitative results for metals only are available. For example,

- Metals quenched from high temperatures have been used to measure the annihilation rate of supersaturated vacancies by dislocation climb.
- Dislocation loop shrinkage due to dislocation climb has been measured.
- The rate of removal of vacancies from voids has been measured in sintering experiments.

When the driving force increases (i.e., more energy is available for climb), dislocation climb becomes more efficient.

All dislocations (dissociated dislocations and non-dissociated dislocations) are efficient sources, e.g., quenched metals. Much lower efficiencies are present for lower-stacking-fault-energy metals. In the higher-stacking-fault-energy metals the efficiency gets lowered less rapidly with energy change, $|g_s|$. This is because the dislocations contain more jogs as $|g_s|$ increases. Forming a jog on a dissociated dislocation is more than in an undissociated dislocation because the forming energy of the former is more.

Problems

1. Uniform shear stress $\tau = 30$ MPa acts on a screw dislocation with Burgers vector $\mathbf{b} = 0.286$ nm along the line direction. Calculate Force per unit length on the dislocation.

   Solution

$$f = \tau b$$

$$b = 0.286 \times 10^{-9} \text{ m}$$

$$f = 30 \times 10^{6} \times 0.286 \times 10^{-9} = 8.58 \times 10^{-3} \text{ N/m}$$

   Answer $f = 8.6 \times 10^{-3}$ N/m along the glide direction.

2. Given that $= 30\ \text{GPa}, \rho = 7800 \text{kg}/_{\text{m}^3}$. At a point, $u_{z,\ x} = 1.0 \times 10^{-3}$, $u_{z,y} = 2.0 \times 10^{-3}, u_{z,t} = 1.0 \times 10^{-2} \text{m}/_{\text{s}}$, find w.

   Work:

$$(u_{z,x})^2 + (u_{z,y})^2 = 1 \times 10^{-6} + 4 \times 10^{-6} = 5 \times 10^{-6}$$

   Strain part: $\frac{\mu}{2}(\cdot) - 1.5 \times 10^{10} \times 5 \times 10^{-6} - 7.5 \times 10^{4} \text{ J/m}^3$.

   Kinetic: $\frac{1}{2}\rho u_{z,t}^2 = 0.5 \times 7800 \times 10^{-4} = 0.39 \text{ J/m}^3$.

   Answer: $\omega \approx 7.5000 \times 10^4 + 0.39 \approx 7.5 \times 10^4$ J/m^3 (kinetic is negligible here).

References

1. Allen, S. M., & Thomas, E. L. (1999). *The structure of materials*. Wiley.
2. Balluffi, R. W., Allen, S. M., & Carter, W. C. (2005). *Kinetics of materials*. Wiley.
3. Johntson, W. G., & Gilman, J. J. (1959). Dislocation velocities, dislocation densities and plastic flow in lithium fluoride crystals. *Journal of Applied Physics, 30*(2), 129–144.
4. Nabarro, F. R. N. (1967). *Theory of crystal dislocations*. Clarendon Press.
5. Nabarro, F. R. N. (Ed.). (1979–2004). *Dislocations in solids (series)* (Vol. 1–12). Elsevier North-Holland.
6. Hirth, J. P., & Lthe, J. (1982). *Theory of dislocations* (2nd ed.). Wiley.
7. Yang, M., Zhang, Y., Dong, J., Huang, Y., Lu, Z., Wang, L., et al. (2024). Developing centimeter-sized CuZr-based metallic glass composites via multi-element microalloying. *Intermetallics, 164*, 108114.
8. Balluffi, R. W., & Granato, A. V. (1979). Dislocations, vacancies and interstitials. In F. R. N. Nabarro (Ed.), *Dislocations in solids* (Vol. 4, pp. 1–133).

Chapter 11
Crystals, Boundaries Between Crystals And Their Motion

Contents

Crystals and the boundaries between crystals display line defects, including interfacial dislocations, dislocation ledges, and pure ledges. The boundaries between crystals exhibit more complex dynamics compared with vapour-crystal and liquid-crystal interactions. Movements at crystal and crystal interfaces occur through various atomic mechanisms, ranging from glide (rapid) to thermally triggered flow (slow).

11.1 Thermodynamics of Crystalline Interface Motion

For the motion of crystal/crystal interfaces, there are several sources available, e.g., recrystallisation. The interfaces between a deformed matrix and growing recrystallised grains exhibit a differential in the value of free energy per atom in the bulk, the ΔG value across two sides of the interface. This leads to compatibility stresses in polycrystals. Therefore, the individual crystals encounter different stresses. The force that drives the process is due to the difference in the energy

K. A. Padmanabhan et al., *Thermodynamics and Kinetics of Materials*, Engineering Materials, https://doi.org/10.1007/978-3-032-14527-7_11

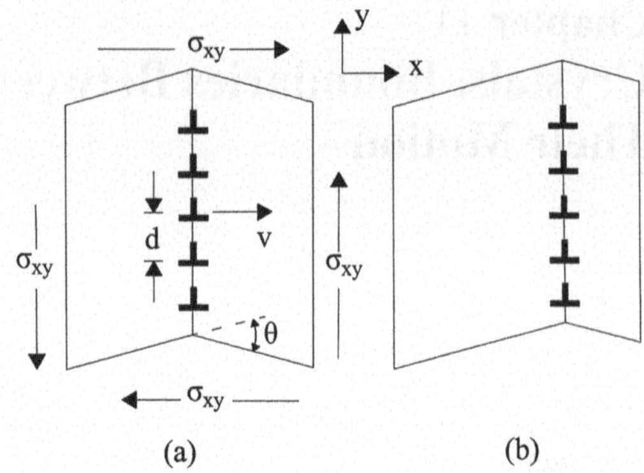

Fig. 11.1 Forward glissile motion under an applied stress of an edge dislocation (which comprises a boundary along *x* and a shear in the *y* direction) of a small-angle symmetric tilt boundary. Dislocation spacing, *d*, equals (b/θ), with *b* the Burgers vector and θ the misorientation between the two grains [1]

arising from the elastic strain per atom across the contact. Crystal/crystal interfaces may also move due to pressure and a change in the body shape (Fig. 11.1).

When this transpires, due to the work performed by the imposed stresses, the system's potential energy diminishes. If δx is the displacement of the interface, the applied stress, σ_{xy}, does work equal to $\sigma_{xy}\theta\delta x$. That is, the interface experiences pressure,

$$P = \frac{\sigma_{xy}\delta x}{\delta x} = \sigma_{xy}\theta. \tag{11.1}$$

This is comparable to the mechanical force applied to a dislocation by stress. The estimated values of the pressures experienced by crystal/crystal interfaces extend over about six orders of magnitude—the highest, 10^7–10^9 Pa, during phase transformations and only 10^3–10^5 Pa at interface curvatures.

11.2 Conservative and Nonconservative Motion of Interfaces

When a diffusion flux is not present, conservative motion is present. Nonconservative motion occurs due to fluxes caused by diffusion over a long range in a species or in many of them.

Steady-state conservative motion is possible when the atomic fraction on either side of the interface is the same. Atoms are locally moved across the interface between two crystals for precise interactions. Long-range diffusion to the boundary does not exist. This local transfer is by atom shuffling across the interface and/or by crystal defect formation, e.g., vacancies/interstitials. Diffusion across the boundary, getting annihilated in the adjacent grain, becomes possible. (i) The motion (glissile) of martensitic interfaces, (ii) the thermally activated motion of grain boundaries, are examples of conservative motion. The boundary serves as a source of flux in nonconservative motion. The addition or removal of atoms from one or both crystals at the interface results in the growth or shrinking of either or both adjacent crystals. This causes the motion of the interface relative to either of the crystals will be the result. Point defects needed at the interface to support long-range diffusional fluxes, causing substitutes shuffling to add or remove interstitial atoms, is a result.

Instances where interfaces exhibit non-conservative movement and act as sources and sinks of diffusional flux occur in several phenomena, such as diffusional creep and polycrystalline material sintering, when atoms migrate to grain boundaries that serve as sources of point defects. In multicomponent-multiphase systems, lattice sites at the boundary are annihilated or created by long-range diffusion. In such a case, the hetero-phases serve as sources of diffusing component sources.

11.3 Motion That Is Preserved

11.3.1 Migration of Neatly Defined Boundaries Through Gliding of Dislocation

Clearly defined interfaces can shift conservatively by interfacial dislocation glide. This motion frequently happens throughout extensive temperature ranges, and this phenomenon is evident even at low temperatures with restricted potential for thermal activation.

Grain boundaries of small misorientation are semi-coherent. Dislocations in well-spaced arrays are present inside the crystal lattice. Along these boundaries, dislocations can glide without the boundary changing its structure, e.g., in Fig. 11.1, a boundary that is tilted symmetrically glides through dislocations that have an edge orientation. A change in the shape of the two crystals that together form the boundary between them is seen. Through an intricately connected operation, atoms from grain 1 are transferred to grain 2. An atom in a grain that contracts moves to a pre-determined site in an expanding crystal/ grain. Such a shuffling replaces the arrays of glissile dislocations already present. In other words, all the atomic positions in the bicrystal are correlated in their positions at the interface. Boundary motion changes the macroscopic shape of the sample, the so-called "military motion" to separate it from the chaotic, "civilian" movement of boundaries present when a general boundary that is incoherent moves. The sample shape does not change in the second case. Strong experimental evidence for the glide of small-angle boundaries is available. General small-angle boundaries frequently exhibit many families of dislocations, characterised by distinct Burgers vectors. Change will occur without structural alteration only when segments of dislocations in all the glide planes align inside a common zone, with their axes perpendicular to the boundary plane. If such a condition is not met, conservative motion of the boundary is possible only if a combined glide-climb of the dislocations is present.

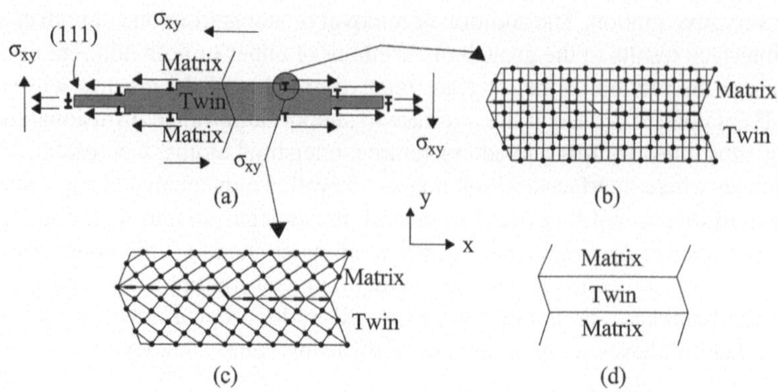

Fig. 11.2 (**a**) The glissile interfaces of a twin (of lenticular shape) in an FCC structure comprise dislocations with a ledge when viewed along the sheet normal. (**b**) A magnified perspective of the dislocation-step area. This is a semi-coherent interface that corresponds to a bicrystal obtained by a 60° rotation around z axis. A vector representing a translation in a DSC-lattice of the concerned bi-crystal corresponds to the Burgers vector of the dislocation—see the fine grid in **c**. Now the boundary becomes coherent vis-à-vis the reference structure of the FCC crystal matrix. The dislocation, then, is referred to as a coherency dislocation. (**d**) The shape change across the entire specimen cross-section was produced by the twin formation. (Such interpretation using the DSC lattice has been abandoned in recent times) [1]

General/ High-Angle Grain Boundaries: High-angle, semi-coherent interfaces, which have localised linear imperfections—dislocations and ledges—can advance through the movement by glide of the dislocations, such as motion at the boundary on a (111) plane in a FCC lattice, i.e., a mechanical twin (Fig. 11.2).

A twin boundary is seen as a large-angle grain boundary that results from rotation by 60° about the direction [3]. The plane (111) of the matrix parallels the surface that defines the twin. The island grain formed by the twin is lenticular in shape and has lower elastic energy. The curved top as well as the bottom regions of the boundary reveal a parallel arrangement of dislocations comprising ledges and dislocations—refer to Fig. 11.2b. Relative to a reference structure, the twin surface exhibits semi-coherence, resembling a bicrystal that is parallel to the (111) plane and in addition, a twin. Its motion makes the interface upper and lower segments shift perpendicularly, thereby augmenting the thickness of the lens-shape. Consequently, specialised locations exist, and atoms can be shuffled locally over a boundary via glissile line defects and a process that enables the barrier to glide. The result is a macroscopic alteration of the sample shape. If a dislocation of a Burgers vector $\vec{b}$ intersects a boundary, the boundary experiences parallel shearing at the surface of the Burgers vector, $\vec{b}$.

If h is the height of a ledge in the line defect, the interface advances by a distance of h.

These effects, when combined, produce a change of shape. The strain on the interface parts induces twin expansion and results in a shape alteration that can be accomplished with the application of shear stress, σ_{xy}—Fig. 11.2a. The ratio of ΔG

(change in per-atom free energy) to R (associated with atomic volume) yields this pressure. Glissile dislocations experience a (force/ length) given by, $f = b\sigma_{xy}$, derived in Eq. (11.1). The applied force does work to displace unit boundary area by δx, $((\delta x/h)b$.Thus, the pressure is,

$$P = -\frac{\partial G}{\partial x} = \frac{(\partial x/_h) b\sigma_{xy}}{\partial x} = \frac{b\sigma_{xy}}{h}. \tag{11.2}$$

When twins form in a matrix due to applied shear stresses, glissile boundary motion causes mechanical twin formation. The lateral movement of line faults can occur at velocities approaching the speed of sound. The numerous line defects on parallel (111) planes can be formed in many ways, e.g., a dislocation "pole" mechanism. Whether such a basic mechanism exists is not clear.

As a result of glissile line defects, sharp hetero-phase interfaces can move in military fashion, e.g., displacive martensitic transformations. The interface, in this instance, is semi-coherent and is devoid of a stress field that is of long range. The parallel arrangement of interfacial dislocations can then move by glide by atom shuffling. This moving interface will change the parent phase in military fashion into the martensite phase and cause a macroscopic shape change.

11.3.2 Glide-Triggered Thermally Activated Motion of Sharp Interfaces and Interfacial Dislocation Climb

Combined glide-climb of interfacial dislocations is required in the motion of several interfaces. This is feasible solely at extreme temperatures, as adequate thermal activation for ascent is a requisite. As previously stated, a grain boundary of low misorientation can migrate by dislocation glide when all segments in its dislocation structure are aligned along a zone with an axis perpendicular to the boundary plane. This condition is frequently unmet. Movement of the interface boundary then has glide and climb components, see Fig. 11.3. A shear stress σ_{zy} oriented at 45° to the boundary plane is met by arrays of edge dislocations of two kinds.

A stress that acts tangentially and causes a shear leads to a climb force, i.e., $f = b\sigma_{xy}$ on every dislocation (Eq. (11.1)). Then, dislocation climbs result. The forces among the dislocations in the array maintain a consistent spacing indicative of the boundary structure with minimum energy. Subsequently, all dislocations will progress in a steady condition along the +x direction by a combined glide-climb mechanism. Consequently, the structure of the mobile boundary remains unchanged. This motion induces a transformation in the specimen's morphology, as illustrated in Fig. 11.1. A sequential arrangement of dislocations exhibits positive climb, facilitated by the establishment of atomic diffusion currents between them, as depicted in Fig. 11.3. Each current parallels the transition from a diminishing to an expanding crystal.

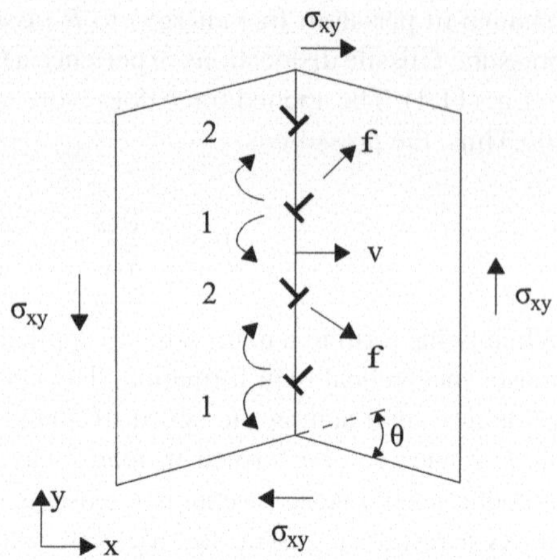

Fig. 11.3 A very low misoriented boundary, that is tilted symmetrically (possessing edge dislocations of two arrays of orthogonal Burgers vectors), exhibiting movement that is conserved, is presented. This is facilitated by thermal activation. The imposed stress gives rise to *f* which acts on every dislocation. Atom flows between dislocations are represented by arrows [1]

Assuming that the rate of dislocation climb is diffusion-limited, a simple model for the boundary motion can be arrived at. Ignoring the effects of dislocation motion and the local dislocation stress fields on diffusion, a flux equation can be written down as,

$$\overrightarrow{J_A} = -\frac{*D}{\Omega kTf}\nabla\Phi_A = -\frac{*D}{\Omega kTf}(\mu_A - \mu_V). \tag{11.3}$$

The vacancies at the dislocations are assumed to remain in equilibrium. They are ideal sources and so, at the dislocations $\mu_V = 0$. If at a dislocation of type 2 acting as a sink, an atom is inserted (Fig. 11.3), the dislocation will cover a distance of $((\sqrt{2}\Omega)/b)$ along +*x*. In that direction, a force $\sigma_{xy}b/\sqrt{2}$ acts on it, and the energy expended $= (\sqrt{2}\Omega/b)(\sigma_{xy}b/\sqrt{2}) = (\sigma_{xy}\Omega)$. Then, the potential due to diffusion, Φ_A, has a "boundary value" at the dislocation cores,

$$\Phi_A^D(sink) = \mu_A^0 - \sigma_{xy}\Omega. \tag{11.4}$$

where μ_A^0 denotes the chemical potential of atoms in an unstressed material. Comparably, for type 1 dislocations, Φ_A^D (source) $= \mu_A^0 + \sigma_{xy}\Omega$.

The mean potential gradient between neighbouring dislocations is $(\nabla\Phi_A) = \frac{2\Omega\sigma_{xy}}{d}$, where the distance between dislocations is *d*. Let the area to length ratio be ***A***, across which the flux due to diffusion is A ≃ d. Based on the values (Eq. 11.3) and considering that *D correlates with the local vacancy concentration fluctuations, which are negligibly small, the per unit length atom current in total passing through a dislocation of type 2 is given by,

$$I_A = 2AJ_A = \frac{4 * D\sigma_{xy}}{fkT}. \tag{11.5}$$

Here, the equilibrium self-diffusivity is ***D**. Then, the volume of atoms per unit length and time giving rise to dislocation climb is $I_A\Omega$, and the rate of dislocation climb is $v_c = \frac{I_A\Omega}{b}$. Each dislocation progresses along the x-axis through a combination of climb and glide, with the climb rate being $\sqrt{2}$ times higher. Then, the speed of the boundary can be expressed as $v = \sqrt{2}v_c$. That is,

$$v = \frac{4\sqrt{2}\Omega * D}{bfkT}\sigma_{xy}. \tag{11.6}$$

As $d = \frac{b}{\theta\sqrt{2}}$, the boundary experiences a pressure of $\mathrm{P} = \theta\sigma_{xy}$.

Then, Eq. (11.6) becomes

$$v = \frac{8\Omega * Dd}{fkTb^2}P = M^B P \tag{11.7}$$

That is, pressure and velocity/ speed are directly proportional. M^B, the mobility of the boundary and *D, are also involved (Eq. (11.7). Boundary motion requires an energy of activation of magnitude similar to that of self-diffusion in crystals, which will determine the activation energy. That is, this is a crystal diffusion-limited (self-diffusion limited) process.

Dislocation motion in high-angle grain boundaries is a controversial topic, which brings into question many of the derivations presented earlier in this and the previous chapters. The reader can find this advanced topic discussed in, e.g., [9, 10].

11.3.3 Thermally Activated Sharp Interface Movement

The rearrangement of individual atoms or tiny, diminishing atom clusters onto an expanding grain at the inflexions known as "ledges" will facilitate conservative movement. Such a motion involves no specimen shape change. An elementary investigation indicates that the interface velocity is approximated using a straightforward calculation as follows:

$$\mathrm{v} = \left[\frac{N_k N_s^2 \Omega^2 v_0}{kT} e^{S^s/k} e^{-E^s/(kT)}\right] P = M^B P \tag{11.8}$$

where N_k is per unit area and concerns the kink sites present. N_s corresponds to atoms shuffled per every transfer (average value). v_0 signifies the periodicity due to thermal vibration. M^B is the boundary mobility, and S_s and E_s are the activation entropy and energy needed for atom shuffling.

11.3.4 Self-Diffusion Leading to Thermally Activated Motion

Diffuse interfaces can migrate through the movement of diffuse anti-phase boundaries that delineate two ordered regions situated on distinct sublattices. In alloys that are ordered, self-diffusion facilitates migration between sub-lattices of different atoms, thereby altering the order locally.

11.3.5 Conservative Interface Motion Limiting Impediments

Solute atoms, either by alloy design or as impurities, are responsible for segregation to surfaces in many instances. This is like solute-atom atmospheres at dislocations. The solute atmosphere must either be dragged by the interfaces with them, or the interface must pull itself away. Solute drag implies that the solute atoms should be able to diffuse along with the moving interface due to the attractive interaction between them. Often, the diffusion of the solute atmosphere is slower compared with that of the interface movement with solute impeding motion. The movement of a boundary is constrained by the drag due to the interaction of a solute and an atom. Under substantial applied pressure, the interface readily detaches itself from the solute environment.

A patch of interface can be pinned by an embedded second phase particle. Interfacial energy can hold up an interface at a particle. So, it will bulge around it. When all the interfacial energies are equal (isotropic case), $F_{max} \approx \pi R\gamma^{12}$ (11.9). Here, R is the sphere radius and γ^{12} is the interfacial energy. With immobile particles and low driving force, the interfaces may be held stationary by the particles. The interface may become free at higher pressures. This unpinning may be thermally activated at high temperatures, and the particle size is small. If the particles are mobile, the interface and its attached particles may move forward together.

11.3.6 Thermally Activated Motion of Grain-Boundary—Experimental Observations

Solute-atom drag effects affect boundary mobilities severely, even in minute solute atom concentrations. The extent of solute segregation and drag depends on the structure of the interface, the solute species, in addition to its concentration.

If solute drag is rate-controlling, boundary mobility depends on boundary structure. Interpreting boundary-motion experiments only in terms of the boundary structure is not recommended. For general boundaries, experimentally, a linear relationship exists between velocity and pressure. The shuffling model gets support from the experimental results.

11.4 Sources and Sinks Action at the Interfaces of Atomic Fluxes—Nonconservative Motion

The inquiry into whether an interface that shifts during nonconservative motion can be swift enough to uphold local equilibrium among all species at the interface is presently being analysed. If the interface kinetics (in local equilibrium) are governed by the atomic diffusion rate (to or from the interface) rather than the flux rate, the process is classified as diffusion-limited. When the rate of accommodation at the interface governs the kinetics, it is source-limited (i.e., interface reaction control).

Discrete lattice dislocation climb in small-angle grain boundaries renders them sources of point defects, such as vacancies. The dislocation segments at the boundary of the array will endeavour to ascend. The need to maintain a balanced configuration in the boundary array constitutes the constraint. In Fig. 11.1 (a symmetric tilt boundary), an edge dislocation can easily climb. A boundary that just has a twist orientation, in contrast, can adopt helical climb with the aid of screw dislocations (Fig. 11.5). But the boundary structure will be disturbed. This can happen only when the driving forces are large, i.e., point defects are large, which could lead to supersaturation or sub-saturation. Figure 11.4 illustrates small-angle boundaries in gold that function as effective sinks for supersaturated vacancies, driven by a substantial force. Subsequently, vacancies in the bulk, distant from the boundary, arise from the collapse of supersaturated vacancies. But the annihilation of supersaturated vacancies will ensure the presence of a precipitate-denuded zone by the boundary sink action.

When large-angle singular or vicinal grain boundaries contain dislocations with ledges, they may emit vacancies through climb and potentially accompany glide across their faces. The coherent interface patches that are situated between the line defects are comparatively stable, resulting in their inactivity. Next, source efficiency is contingent upon the capacity to disperse/disperse/climb/collect the vacancies by

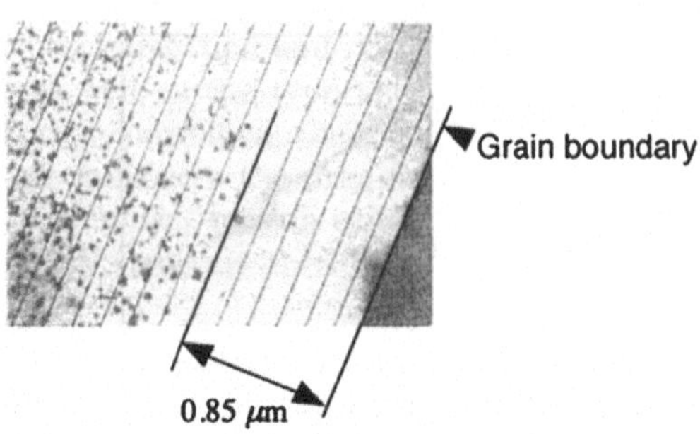

Fig. 11.4 Vacancy precipitate denudation in quenched and subsequently annealed gold adjacent to a small-angle grain boundary. The border (bottom right) functioned as a sink for the supersaturated vacancies. Precipitation of vacancies constitutes diminutive dislocation arrangements that arise from the collapse of a vacancy aggregation [4]

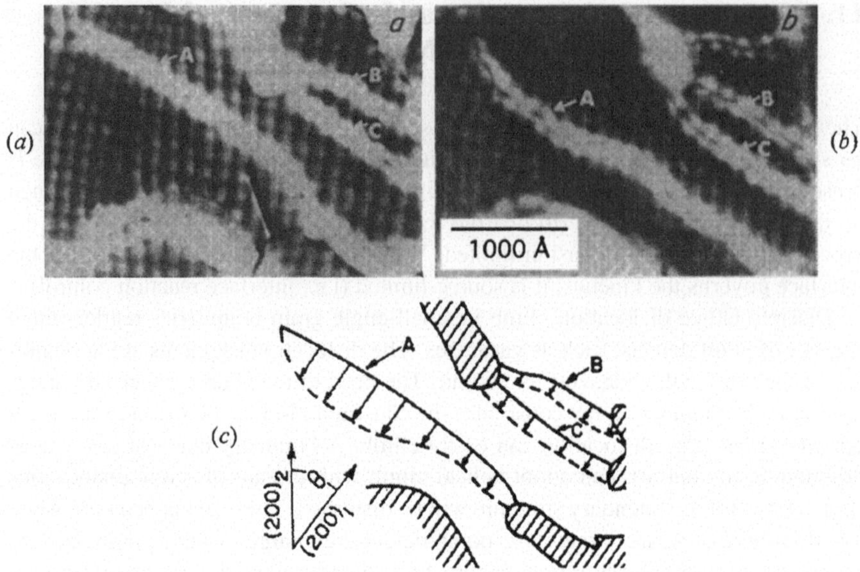

Fig. 11.5 Illustrates boundary dislocations, A, B, C, that climb in an extrinsic manner in a near-CSL boundary of gold in the (001) twist. The background dislocations that are evident from the misorientation vis-à-vis the adjacent one that is singular and is of twist character (the vicinal boundary is near-CSL to this), accommodate the twist deviation. Excessive interstitial defects were generated by fast ion irradiation, and they were subsequently absorbed by ascent at the boundary dislocations. Consequently, the boundary functioned as a defect receptacle. (**a**) Before irradiation; (**b**) the identical area, after irradiation, as in (**a**). (**c**) a sketch that illustrates the ascent [5, 6]

lengthwise diffusion and within the grain-boundary core. A vicinal boundary that functions as a sink for supersaturated interstitial defects is illustrated in Fig. 11.5.

The interfacial line defects needed for the source action are often produced by the impinged lattice dislocations. However, at sufficiently high driving forces, dislocation-like defects may be homogeneously nucleated in the boundary by forming small loops with ledge character. When vacancies available are supersaturated, the energy needed for the nucleation of a loop can be approximately written as,

$$\Delta F = \frac{\mu b^2 R}{2(1-\nu)}\left[ln\left(\frac{4R}{R_0}\right) - 1\right] + 2\pi R f^L - \frac{\pi R^2\left(\vec{b}\,.\dot{n}\right)}{\Omega} E^{defect}. \qquad (11.9)$$

R represents the loop radius (the first term is elastic loop energy), f is the core energy, i.e., the unit length energy. E^{defect} is needed to precipitate the defect. For vacancy supersaturation, e.g., during quenching from a high temperature, Eq. (11.9) is used to evaluate ΔF. Subsequently, loop nucleation occurs at high speeds. Boundaries can act as sinks when a material is highly irradiated, by nucleation and subsequent

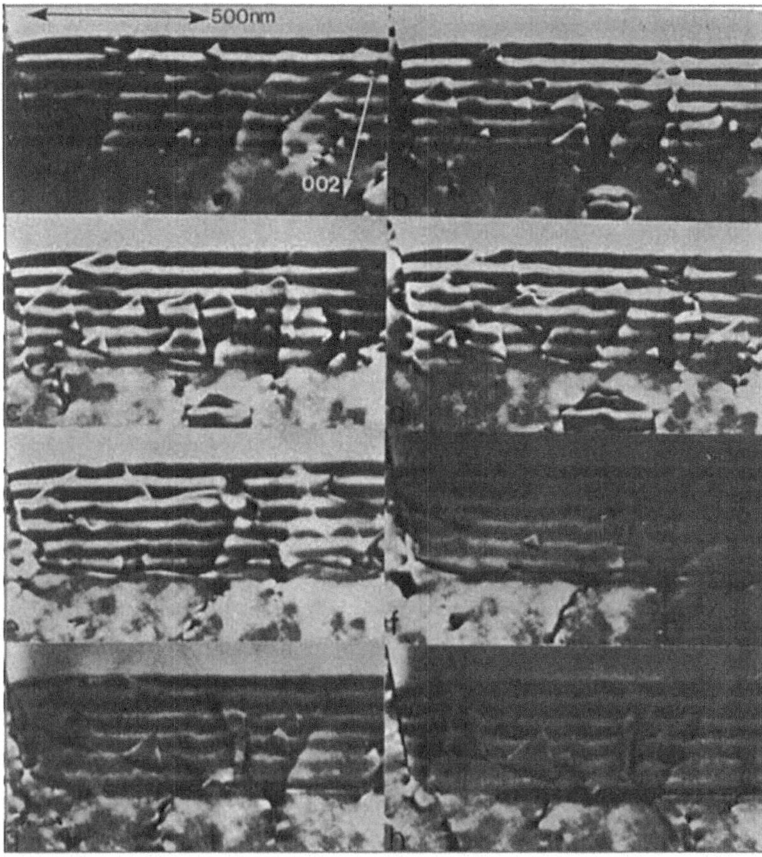

Fig. 11.6 (111) twin boundary in copper functioning as a sink for surplus interstitial defects after one million electron volt irradiation. Coalescence eliminates the defects in boundary regions. Subsequently, they converge to form triangular grain-boundary dislocation loops. Through climb and expansion, the loops, once established, destroy other imperfections [6]

formation of boundary dislocation loops. Figure 11.6 displays irradiated copper twin boundaries containing loops that are triangular.

Vicinal boundaries can be effective vacancy sinks when, as seen in experiments, the climb of dislocations is feasible and the forces that drive the boundaries are substantial. Results in large-angle boundaries cannot be interpreted in terms of dislocation movements. The efficiency of high-angle boundaries is low when the driving forces are low. Vicinal hetero-phase boundaries may allow flux of solute atoms by acting as either sinks or sources. Then, they will exhibit both dislocation and ledge characteristics. The line defects are the sources. Shuffling of lattice sites from one crystal to the next causes lateral motion. The interface shifts in relation to both stages. When two adjacent phases at an interface possess distinct compositions,

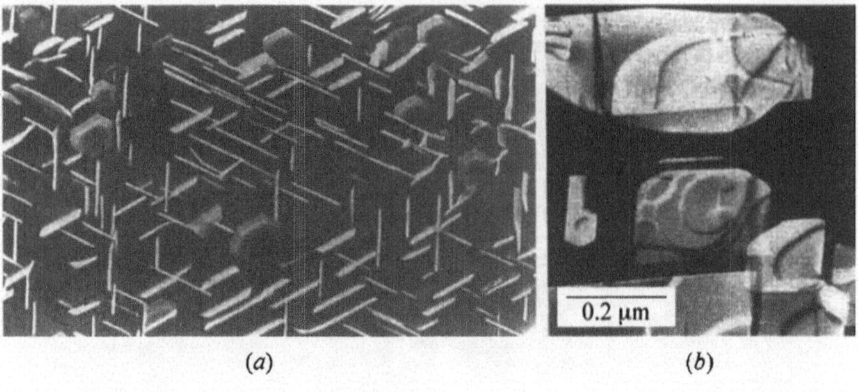

Fig. 11.7 (**a**) Ag_2Al precipitates in the form of thin platelets in Al-Ag alloy. The broad faces of the platelets are parallel to (111) planes of the matrix, which lie at different angles to the normal to the paper. (b) Precipitate platelets in Cu-A1 alloy. Line defects that possess both dislocation and ledge character are present on the broad faces. Platelets grow in thickness by climb of these line defects across their faces [7, 8]

diffusion at the ledge should either provide or extract the solute atoms. Consequently, the ledge facilitates the shuffling and long-range diffusion of solute atoms, allowing them to bond.

Figure 11.7 illustrates the growth and thickening of platelet precipitates, by the movement of the line defects. The density of the ledge and its capacity to accommodate solute atoms will determine the efficacy of the developing precipitate platelet (as a possible sink for solute atoms). Experimental evidence indicates that the effectiveness of hetero-phase interfaces is diminished in the absence of suitable ledge defects.

Incoherent interfaces are characterized by homo-phase and hetero-phase configurations that are general with respect to all degrees of freedom. They do not sustain localized line defects. In many cases they can act as highly efficient sources of fluxes or solute atoms by atom shuffling in the interface core. Some models assume that the dislocation core is a slab of "bad" material that contains a density of sites where point defects can be created and destroyed, or where atomic sites containing solute atoms can be transferred across the interface by the uncorrelated local atom shuffling. Some experimental evidence for such a possibility perhaps exists in the case of homo-phase (grain) boundaries. However, it is not be the case for hetero-phase boundaries.

11.5 Diffusion-limited and Source-Limited Kinetics

The former has been discussed here in detail. By making several unverified assumptions about a source-limited process, some mathematical expressions have been derived to account for the latter case. Notwithstanding the mathematical elegance, in the application, the idea of source-limited behaviour is explained by deducting a constant stress (called the threshold stress) from the total applied stress to determine the effective stress that drives the kinetic process. The only justification for the deduction is that it allows one to fit the experimental data. For such treatments, one is referred to, among others, to reference [9].

Problems

1	Determine the work performed by the imposed stress on a crystalline interface with an area of 2 cm^2 when the interface is displaced by 1 μm under an applied force of 50 N **Solution:** $W = F \times d$ $W = 50 \times 10^{-6} = 50\ MJ$
2	A crystal interface moves under the influence of an applied shear stress of 150 MPa. If the area of the interface is 5 × 10^{-4} m^2, calculate the force applied on the interface **Solution:** $F = \tau \times A$ $F = 150 \times 10^6 \times 5 \times 10^{-4} = 75N$
3	A sharp interface in a crystal progresses at a velocity of 10^{-5} m/s due to a shear stress of 120 MPa. Calculate the interface velocity when the applied stress is increased to 200 MPa, given that the interfacial dislocation glide velocity is exactly proportional to the applied stress **Solution:** $\frac{v_1}{v_2} = \frac{T_1}{T_2}$ Where $V_1 = 10^{-5}$ m/s $T_1 = 120$ MPa $T_2 = 200$ MPa Solving for V_2 $v_2 = \frac{v_1 \times T_2}{T_1} = \frac{10^{-5} \times 200}{120} = 1.67 \times 10^{-5} m/s$
4	A sharp interface in a material undergoes thermally activated motion, and the dislocation velocity is found to increase by a factor of 3 when the temperature increases from 600 K to 800 K. Calculate the activation energy for the motion $\ln\left(\frac{v_2}{v_1}\right) = \frac{Q}{k}\left(\frac{1}{T_1} - \frac{1}{T_2}\right)$ **Solution:** *Given*: $\frac{v_2}{v_1} = 3,$ $T_1 = 600K$ $T_2 = 800K$ Solving for Q: $\ln(3) = \frac{Q}{8.617 \times 10^{-5}}\left(\frac{1}{600} - \frac{1}{800}\right)$ $Q = 1.85eV$
5	An interface acts as a source for atomic fluxes with a flux of 3 × 10^{20} atoms/m2s. If the interface area is 2 cm^2, calculate the total number of atoms entering the interface per second **Solution:** $N = flux \times A$

(continued)

where:
$flux = 3 \times 10^{20} atoms/m^2 s$,
$A = 2 \times 10^{-4} m^2$,
$N = 3 \times 10^{20} \times 2 \times 10^{-4} = 6 \times 10^{16} atoms/s$

References

1. Sutton, A. P., & R.WBalluffi. (1996). *Interfaces in crystalline materials*. Oxford University Press.
2. ThermoCalc. (n.d.). *ThermoCalc databases*. Retrieved from https://thermocalc.com/products/databases/
3. Dornelas, P., Filho, J., Farias, F., Moraes e Oliveira, V., Moraes, D., & Zumpano Júnior, P. (2020). FEM-thermodynamic simulation methodology to predict the influence of t8/5 on the coarse grain heat-affected zone of a Cr-Mo low-alloy steel pipe. *Journal of Manufacturing Processes, 60*, 520–529. https://doi.org/10.1016/j.jmapro.2020.10.082
4. Siegel, R. W., Chang, S. M., & Balluffi, R. W. (1980). Vacancy loss at grain-boundaries in quenched polycrystalline gold. *Acta Metallurgica, 28*(3), 249–257.
5. Komem, Y., Petroff, P., & Balluffi, R. W. (1972). Direct observation of grain boundary dislocation climb in ion-irradiated gold bicrystals. *Philosophical Magazine, 26*, 239–252.
6. King, A. H., & Smith, D. A. (1980). On the mechanisms of point-defect absorption by grain and twin boundaries. *Philosophical Magazine A, 42*(4), 495–512.
7. Rajab, K. E., & Doherty, R. D. (1989). Kinetics of growth and coarsening of faceted hexagonal precipitates in an fcc matrix.1. Experimental-observations. *Acta Metallurgica, 37*(10), 2709–2722.
8. Weatherly, G. C. (1971). The structure of ledges at plate-shaped precipitates. *Acta Metallurgica, 19*(3), 181–192.
9. Balluffi, R. W., Allen, S. M., & Craig Carter, W. (2005). *Kinetics of materials*. A John Wiley & Sons.
10. Padmanabhan, K. A., & Davies, G. J. (1980). Superplasticity, Springer Verlag, Heidelberg, Germany.

Chapter 12
Evolution of Surface and Volume Morphology: The Role of Capillary and Mechanical Forces

Contents

"Morphology" includes both the external macroscopic shape and the internal microstructure of a material. Morphology can change when suitable driving forces are present. These driving forces result from differential decreases in total free energy that gives rise to the driving forces.

Shapes and positions of interfaces mostly determine morphology. A real surface has a minimum of one interface distinct from the environment. This is known as a free interface. Interfaces that exist within a condensed phase or between condensed phases are termed as internal interface. 2D (surface) and 3D (volume) evolution represent morphological changes. The areas and crystallographic degrees of freedom contribute to the excess free energy of interfaces. Mechanical forces can produce a response in them. So, two independent forces drive the interfaces which then move and experience morphological change. If the interface motion decreases If the total excess free energy associated with the interface is decreased by interface motion, i.e., the area and energy per unit area, the resulting force causing motion is known as a capillary force. The capillary force in a curved isotropic interface arises from the pressure difference across the interface caused by its local mean curvature. Such capillary forces are important in the development of fine-scale microstructures with interfaces of large curvatures. Then, the pressure differences can become large. In addition, if an interface allows an applied force to do work, a force for interface motion is produced. Such a force is an applied force. This chapter does not consider chemical driving forces and phase transformations, which were discussed in an earlier chapter.

First, the morphological evolution of free surfaces influenced by capillary forces, i.e., surface smoothing and faceting, surface grooving at intersections with grain

K. A. Padmanabhan et al., *Thermodynamics and Kinetics of Materials*, Engineering Materials, https://doi.org/10.1007/978-3-032-14527-7_12

boundaries, and incipient evolution of a cylinder into a row of spheres, is examined. Surface and volume diffusion and vapor transport are examined. Morphological change due to the capillarity-driven grain growth in polycrystal and the coarsening of a fine distribution of second-phase particles are taken up next. Next in line is morphological change involving free and internal interfaces and capillary and applied mechanical forces. Surface smoothing and porosity removal which occurs during sintering of porous bodies and body shape change by diffusional creep are technologically important.

12.1 Surface Evolution

The integral of the interfacial free energy per unit area γ, over all interfaces constitutes the total interfacial contribution to the system free energy. The geometric degrees of freedom of an interface decides the value of γ (assumed to be isotropic for a simple analysis). Two sources, interfacial area reduction and interfacial energy per unit area due to changes in the geometrical/crystallographic attributes of the interfaces, drive interface motion. The rate of interfacial area reduction per unit volume transferred across the interface equals the local geometric mean curvature. These local driving forces caused by mean curvature variations lead to models for the capillarity-induced morphological evolution of isotropic interfaces. The kinetics of surface smoothing, the formation of surface grooves on free surfaces at intersections with grain boundaries, and the incipient evolution of a cylinder into a row of isolated spheres are derived for bodies with isotropic surfaces. Surface and volume diffusion and vapor transport are investigated.

Methods for the evolution of anisotropic surfaces are available. Faceting and crystal growth are examined. Here, the surface velocity and the surface inclination are related. A driving force is there for transport from particles enclosed by isotropic surfaces of larger constant mean curvature to particles of smaller constant mean curvature. In the case of isotropic surface undulations, the driving force caused by differences in curvature will lead to a constant mean curvature. Material below a bump of smaller mean curvature has a lower diffusion potential than a bump of smaller mean curvature. Capillarity-induced evolution flattens the irregularities on a uniform flat surface. A cylinder with surface irregularities can either evolve back to a uniform cylinder or a row of isolated spheres.

The rate of transport mechanism for surface motion decides its evolution. The processes to be considered are surface diffusion, diffusion through the bulk, or vapor transport. Kinetic models of capillarity-induced interface evolution are mainly due to Mullins, 1957 [1]. Here, the models involve surface diffusion (which relates interface velocity to fourth-order spatial derivatives of the interface), and vapor transport (which relates velocity to second-order spatial derivatives).

12.1.1 Flattening of Free Surface

Consider a system of only one type of atom. Let its equilibrium point defects and an initial surface profile be

$$y - y_0 = h(x); \quad \text{or} \quad F(x, y) = y - h(x) = constant. \tag{12.1}$$

$h(x)$ indicates deviations from a flat surface ($y = constant$) (Fig. 12.1). Local surface curvature determines the diffusion potential. The curvature can be determined as a function of x. The boundary conditions produced local diffusion potential decides the boundary conditions. Fluxes needed for surface diffusion are caused by potential gradients. A thin surface slab of thickness δ (an efficient source of point defects assumed to be present in their local equilibrium numbers) controls migration.

The force-flux equation and the diffusion potential at a curved interface, when combined leads to,

$$\overrightarrow{J^S} = c(v) = -cM^S \nabla_{surf} \Phi^S = -c\Omega\gamma^S M^S \nabla_{surf} k = -\frac{\gamma^S * D^S}{kT} \nabla_{surf} k. \tag{12.2}$$

where $\left(\frac{1}{\Omega}\right)$ is the concentration of atoms in a surface layer of thickness δ. The surface mobility M^S and surface diffusivity ($*D^S$) are connected. D_{surf} is the surface (2D) gradient operator in the interface tangent plane and has derivatives within the interface. For a 2D interface inside a 3D ($x(u, v)$), $y(u, v)$, $z(u, v)$)—interfacial coordinates (u, v) and curvature k—the surface gradient of k is,

$$\nabla_{surf} \kappa(u, v) = \nabla|_{x(u,v),y(u,v),z(u,v)} = \left(\frac{\partial \kappa}{\partial u}, \frac{\partial \kappa}{\partial v}\right). \tag{12.3}$$

This vector is normal to the surface that points out the direction of fastest increase in κ. For the surface described by Eq. (12.1)

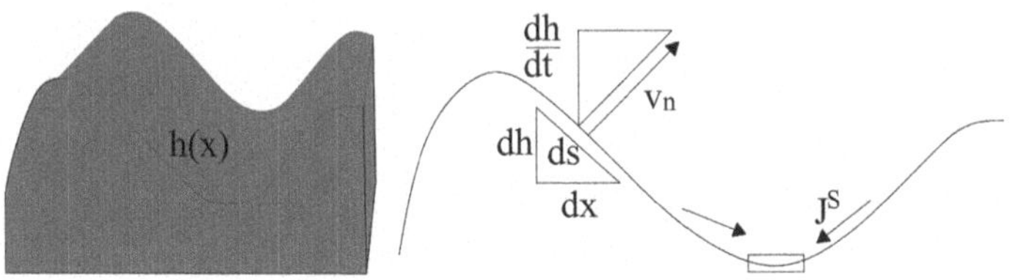

Fig. 12.1 Surface diffusion flattens a surface $h(x)$. Flux accumulation is proportional to the normal velocity. Vertical motion $\left(\frac{dh}{dt}\right)$ rate is a function of normal velocity v_n and the local surface geometry [1, 2]

$$J^S = \frac{\gamma^S * D^S}{kT} \frac{\partial \kappa}{\partial s} \tag{12.4}$$

where s is the arc length along the surface curve $y - h(x) = \text{constant}$ (Fig. 12.1). The particle accumulation rate at a given surface site is obtained by the integral of the flux over the surface that encloses each surface site. When each surface site has a square base of area $\left(\frac{1}{c^S}\right)$ (a box) (c^S is the number of surface sites per area) and height δ, the rate of particle accumulation is

$$N^S = -\int_{box-walls} \overrightarrow{J^S}.\overrightarrow{dA} = -\int_{box-volume} \nabla \overrightarrow{J^S} dV \approx -\nabla \overrightarrow{J^S} \frac{\delta}{c^S}. \tag{12.5}$$

The atom addition per unit area is $N^S c^S = -\delta \nabla . \overrightarrow{J^S}$, and so the normal surface velocity is

$$v_n = \Omega \dot{N}^S c^S = -\Omega \delta \nabla . \overrightarrow{J^S} = \frac{\Omega \gamma^S \delta * D^S}{kT} \nabla^2_{surf} \kappa = B^S \frac{d^2 \kappa}{ds^2} \tag{12.6}$$

where B^S contains the kinetic and material coefficients that multiply the surface Laplacian, $\nabla^2_{surf} \equiv \nabla_{surf} . \nabla_{surf}$; i.e., the 2D Laplacian operator with derivatives with respect to coordinates in the interface as in Eq. (12.3). Hence the relation between the change rate of the surface height, h, and v_n is,

$$\frac{\partial h}{\partial t} = \frac{v_n}{\sqrt{1 + \left(\frac{\partial h}{\partial x}\right)^2}} = \frac{B^S}{\sqrt{1 + \left(\frac{\partial h}{\partial t}\right)^2}} \frac{\partial^2 \kappa}{\partial x^2}. \tag{12.7}$$

The derivatives with respect to arc length, s, in Fig. 12.1 can be converted to derivatives with respect to x using $ds^2 = dh^2 + dx^2$. From the expression for κ,

$$\frac{\partial h}{\partial t} = \frac{B^S}{1 + \left(\frac{\partial h}{\partial t}\right)^2} \frac{\partial}{\partial x} \left[\frac{1}{\left[1 + \left(\frac{\partial h}{\partial x}\right)^2\right]^{1/2}} \frac{\partial \kappa}{\partial x} \right]; \quad \frac{\partial h}{\partial t} = \frac{B^S}{1 + \left(\frac{\partial h}{\partial t}\right)^2} \frac{\partial}{\partial x}$$
$$\times \left[\frac{1}{\left[1 + \left(\frac{\partial h}{\partial x}\right)^2\right]^{1/2}} \frac{\partial}{\partial x} \left\{ \frac{\frac{\partial^2 h}{\partial x^2}}{\left[1 + \left(\frac{\partial h}{\partial x}\right)^2\right]^{3/2}} \right\} \right]. \tag{12.8}$$

In the limit of small slopes, as $\left|\left(\frac{\partial h}{\partial x}\right)\right| << 1$,

$$\frac{\partial h}{\partial t} = -B^S \frac{\partial^4 h}{\partial x^4} + 3B^S \left(\frac{\partial^2 h}{\partial x^2}\right)^3. \tag{12.9}$$

The second term on the right-hand side is mostly negligible as local regions of rapidly changing slopes disappear quickly due to the first term. Therefore, the linearized surface diffusion equation is

$$\frac{\partial h}{\partial t} = -B^S \frac{\partial^4 h}{\partial x^4} = \frac{-\Omega \gamma^S \delta * D^S}{kT} \frac{\partial^4 h}{\partial x^4}. \tag{12.10}$$

The dependence of h on the surface roughness length scales can be analyzed with independent Fourier components

$$h(x,t) = A(t)\, sin\left(\frac{2\pi x}{\lambda}\right). \tag{12.11}$$

Substitution in Eq. (12.11) into Eq. (12.10) leads to

$$\frac{dA(t)}{A(t)} = \frac{-\Omega \gamma^S \delta * D^S}{kT} \left(\frac{2\pi}{\lambda}\right)^4 dt \tag{12.12}$$

so that the Fourier amplitude with a time amplitude, $A(t)$, is

$$A(t) = A(0) e^{B^S (2\pi/\lambda)^{4t}}. \tag{12.13}$$

The amplitude decay exponent is linearly related to $\left(\frac{1}{\lambda^4}\right)$. Or, fine-scale roughness is removed faster than longer-wavelength surface roughness. General roughness is split into a sum of Fourier terms. In linear Eq. (12.10), each term dies down independently. So, the longest wavelength controls the isotropic surface flattening rate by surface diffusion. Equation (12.13) is an approximation (valid for small slopes). The numerical solutions of the evolution equation, Eq. (12.8), for large-slope conditions indicate that the small-slope approximation is valid for the general case (no significant error was seen). Equation (12.13) (wavelength dependence) can be used for experimental surface measurements and kinetic coefficients that lead to B^S. If a surface has an array of parallel grooves, the amplitude-decay factor dependence on the spacing can be measured.

Surface evolution by vapor transport is not discussed because it is beyond scope of this book.

12.2 Anisotropic Surfaces

12.2.1 Geometrical Aspects

The inclination decides the anisotropic surface energy per unit area, $\gamma(\hat{n})$. For isotropic surfaces, the surface energy varies linearly as the area, but for anisotropic surfaces two additional degrees of freedom are involved, viz., the parameters necessary to define the inclination. Anisotropic surface can decrease its energy at constant area by changing its normal. The treatment of morphological anisotropic surface evolution is analogous to that for an isotropic surface-kinetics, that is, $\int \gamma(\hat{n})dA$, replaces $\gamma \int dA$ for an isotropic surface, and must decrease monotonically. Local surface-energy density is connected to the local surface-energy density through $\hat{n}$, the inclination to the surface. But the treatment is more involved which for anisotropic surfaces of large magnitude inclinations $\hat{n}$, become unstable. They cannot be in local equilibrium. The missing inclinations will make surface derivatives discontinuous. Facets may appear on planes. Such a surface can have both smoothly curved and flat potions. They are separated by edges or corners with discontinuous derivatives.

Figure 12.2 presents a three-grain junction on a high temperature annealed polycrystalline Al_2O_3 surface. The surfaces of each grain and facet morphology are different. Grain 1 is flat. Grain 2 has two facet inclinations. And grain 3 displays three facet inclinations. Wulff construction finds the shape of a body of fixed volume with minimal total free energy.

12.2.2 Morphological Evolution Rate

Morphological evolution kinetics of anisotropic interfaces is obtained as an extension of the isotropic case. An isotropic interface evolves from a diffusion potential

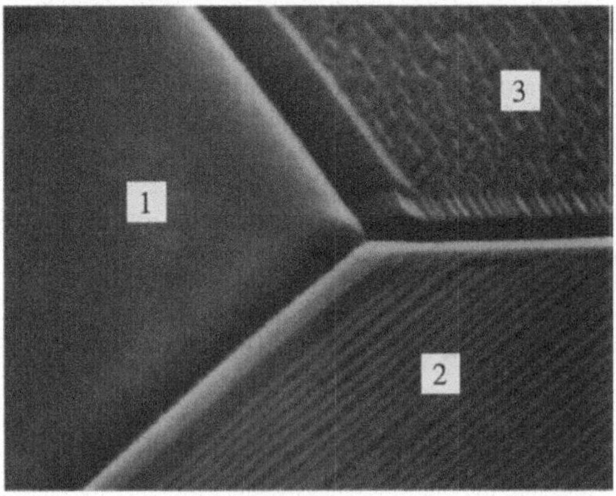

Fig. 12.2 Three faceted grains in annealed Alumina surface morphology [3]

which is linearly related to the product of the local geometric curvature (mean curvature) and the surface energy per unit area. The local geometric curvature equals the interface area change, δA, with the volume addition, δV, i.e., $\kappa = \frac{\delta A}{\delta V}$. Hence, the increase in local energy due to an atom volume increase, Ω, is $\Omega\gamma\kappa$. The anisotropic analogous increase in energy is the weighted mean curvature $\kappa_\gamma = \delta(\gamma A)/\delta V$. That is, the diffusion potential for the anisotropic case is increased by the local adatom local energy increase, $\Omega\kappa$. Then it is derived that,

$$\kappa_\gamma = \nabla_{surf} \overrightarrow{\xi(\widehat{n})}. \tag{12.14}$$

where $\vec{\xi}$ is the capillarity vector and ∇_{surf} is the surface divergence operator (like a surface gradient). This expression involves two types of derivatives to express κ_γ, wherein the first produces $\vec{\xi}$ from a derivative in γ-space. The derivative evolving interface is used to obtain the second derivative.

Surface Evolution: Morphological evolution calculations can be tedious. But the kinetic equations follow as extensions of the isotropic case. For anisotropic surface movement by surface diffusion, the normal interface velocity follows from Eq. (12.6) (the isotropic case). For the anisotropic case,

$$v_n = N^S \Omega \dot{c}^S = \frac{\Omega * D^S}{kT} \nabla^S_{surf} \kappa_\gamma. \tag{12.15}$$

If anisotropic surface diffusivity exists, its surface derivatives also must be present.

For movement by vapor transport of an anisotropic interface that is exposed to a vapor in equilibrium with a very large particle (this equation is assumed in this book), the normal interface velocity is:

$$v_n = -\frac{K\Omega^2 p^{eq}(\kappa = 0)}{kT} \kappa_\gamma. \tag{12.16}$$

The weighted mean curvature for a surface in local equilibrium is simple if the Wulff shape is fully faceted. The current models ignore edge and corner energies, nucleation barriers for ledge and step creation, ledge-ledge interactions, and elastic effects associated with edges and corners. Growth rate for inclination-dependent interface velocity also involves similar assumptions.

Problems

1	Consider a free surface of a material subjected to a tensile stress of 200 MPa. The initial curvature of the surface is 1 mm^{-1}. Calculate the rate of flattening of the surface if the surface tension is 1.5 N/m **Solution:** $\frac{dR}{dt} = \frac{\gamma}{E}.curvature$ *where*: $\gamma = 1.5 N/m$ (Surface tension) E = 200 MPa (Young's modulus)

(continued)

	Curvature = 1 mm^{-1} The rate of flattening is: $\frac{dR}{dt} = \frac{1.5}{200 \times 10^6} \times 1 = 7.5 \times 10^{-9} m/s$
2	In a polycrystalline material, the surface evolves such that the radius of curvature changes by 0.1 mm in 10 min. If the change in curvature is due to the geometrical aspects of the material, calculate the evolution rate **Solution:** $Rate = \frac{\Delta Curvature}{\Delta t}$ $Where:$ $\Delta Curvature = \frac{1}{0.1mm} = 10mm^{-1}$ $\Delta t = 10\ \text{min} = 600s.$ The rate of curvature evolution is: $Rate = \frac{10}{600} = 1.67 \times 10^{-2} mm^{-1}$
3	Consider a grain boundary with an initial surface area of 5 sq.cm The morphological evolution due to surface diffusion results in a change of the surface area by 0.2 sq.cm over 24 h. Calculate the rate of morphological evolution **Solution:** $Rate = \frac{\Delta Area}{\Delta t}$ Where: ΔArea = 0.2 cm^2 Δt = 24 h = 86,400 s The rate of morphological evolution is $Rate = \frac{0.2}{86400} = 2.31 \times 10^{-6} cm^2/s$

References

1. Mullins, W. W. (1957). Theory of thermal grooving. *Journal of Applied Physics, 28*(3), 333–339.
2. Mullins, W. W. (1962). Solid surface morphologies governed by capillarity. In N. A. Gjostein (Ed.), *Metal surfaces: Structure, energies and kinetics* (pp. 17–66). American Society for Metals.
3. Dynys, J. M. (1982). Sintering mechanisms and surface diffusion for aluminium oxide. PhD thesis, Department of Materials Science and Engineering, Massachusetts Institute of Technology.

Chapter 13
Capillary Forces and Microstructure Coarsening

Contents

Two types of capillary force-driven cases were considered earlier: (a) those that led to surface area reduction, and (b) those that altered surface inclinations. Here the microstructure length scale changes are treated.

Characteristic length scale increase due to microstructure evolution leads to coarsening. Coarsening of particle distribution results from a reduction in total interfacial energy. The two effects of microstructure coarsening are: (i) a decrease in total hetero-phase interfacial energy due to a flux from smaller to larger particles, and (ii) the increase in the size of the particle. The larger particles retain the solutes. The fine precipitates, if present, can lower many material properties. Larger grains grow instead of smaller grains in single phase polycrystals, reducing the total grain-boundary free energy (a competitive process). Undesired coarse-grained structures could result.

13.1 Particle Distribution Coarsening

13.1.1 Classical Mean Field Theory of Coarsening

Some essential elements of this theory worked out in 1961 were known earlier. At an elevated temperature let a binary system composed of A and B atoms containing a distribution of spherical β-phase particles of pure B be embedded in an A-rich matrix phase, α. The B atom concentration in the vicinity of each β-phase particle has an

K. A. Padmanabhan et al., *Thermodynamics and Kinetics of Materials*, Engineering Materials, https://doi.org/10.1007/978-3-032-14527-7_13

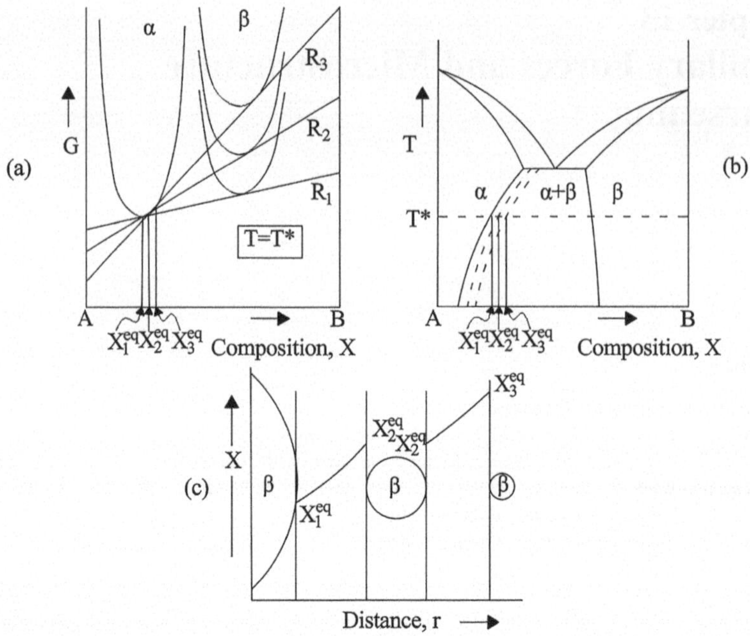

Fig. 13.1 β-phase particle size at temperature T^* affects the concentration, x^{Eq}, of component B in the α phase in equilibrium with a β-phase particle in a binary system. It is assumed that β is pure B. (**a**) presents schematic free-energy curves for α phase and three β-phase particles of different radii, $R_1 > R_2 > R_3$. A smaller radius decreases the free energy (per mole) of the particles due to interfacial energy contribution which increases with the interfacial area to volume ratio. (**b**) The corresponding phase diagram. Common tangent construction in (α) decides the concentration of B in the α phase in equilibrium with the β-phase particles. As R decreases this concentration increases (exaggerated for clarity). (**c**) Concentration profiles in α matrix between the three β-phase particles concentration profile [1]

equilibrium value, which increases with decreasing particle radius, see Fig. 13.1. Concentration differences in the matrix causes a flux of B-atoms from smaller to larger particles. Due to this flux smaller particles shrink, and larger particles are enlarged.

Here diffusion- or source-limited kinetics for growth is assumed. The growth law for particle distribution in each case is different. The distribution of particle sizes at any time t can be quantified as, $f(R, t)$ [units, $(length)^{-4}$], so that the number of particles per unit volume with radii between R and $(R + dR)$,

$$n(R, R + dR; t) = f(Rt)\, dR. \tag{13.1}$$

If during coarsening the total number of B atoms in solution is assumed to be constant, a particle volume by increasing R as it absorbs a B atom from the solution. The particle total volume remains constant.

That is,

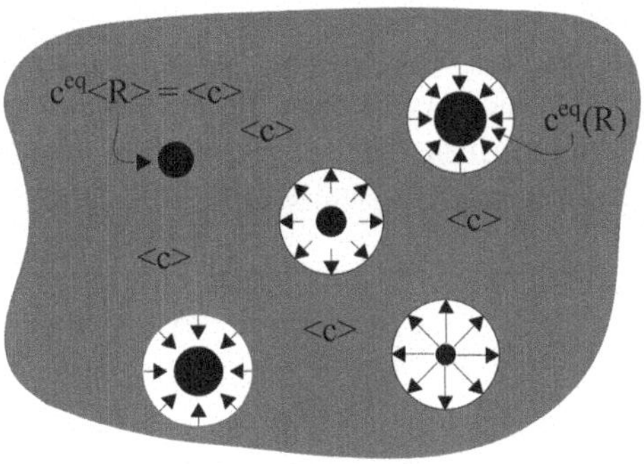

Fig. 13.2 Diffusion-limited coarsening (a mean field approximation). A diffusion field of spherical symmetry surrounds every particle. Arrows indicate the fluxes. The matrix concentration at each particle interface is fixed by Eq. (13.4). The concentration far-removed from particles is given by ⟨c⟩. Near particles of average size, ⟨R⟩, the flux is zero [2, 3]

$$\frac{d}{dt}\sum_{part}\frac{4}{3}R^3 = 0. \tag{13.2}$$

Thus,

$$\sum_{part} R^2 \frac{dR}{dt} = 0. \tag{13.3}$$

Here all particles in the distribution are involved in the summation.

Coarsening Limited By Diffusion: In this case, the hetero-phase interfaces around the particles serve as very effective point defect sources and sinks. The concentrations of B in the α phase in their direct vicinities are maintained in this case at the equilibrium values. The rate of diffusion between the particles then controls the coarsening rate. It is assumed that α and β are fluid phases, β is pure component B and is an incompressible spherical particle of radius, R, and α is a dilute solution. Then the equilibrium concentration as a function of particle radius can be arrived at. In such a case, B atom concentration in α at the α/β interface is

$$c^{eq}(R) = c^{eq}(\infty)e^{\frac{2\Omega\gamma}{kTR}} \approx c^{eq}(\infty)\left[1 + \frac{2\gamma\Omega}{(kTR)}\right] \tag{13.4}$$

where $c^{eq}(\infty)$ is the solubility of B in α for a system with a planar α/β interface.

A random distribution of different sized particles is present. An exact model coupling diffusion and particle size evolution cannot be easily derived. As diffusion near a spherical sink has a short transient and a steady state is characterized by steep concentration gradients near the surface, a mean field approximation is reasonable to adopt. The particles are taken as independent sinks in contact with a mean-field (Fig. 13.2).

A result of mean-field approximation is that a diffusion field of spherical symmetry surrounds every particle. The far-field boundary condition is fixed by the mean concentration, (c), which is less than the equilibrium concentration of the smallest particles and more than the equilibrium concentration for the largest particles. Thus, large particles grow, and small particles shrink (Fig. 13.2). The particle growth rate is,

$$I = 4\pi D_B R\left(\langle c\rangle - c^{\alpha\beta}\right) \tag{13.5}$$

where I is the diffusion current. Then, the particle growth rate with radius R is,

$$\frac{dR}{dt} = -\tilde{D}\frac{\tilde{D}^{eq}(R)\langle c\rangle}{R}\Omega. \tag{13.6}$$

Combining Eqs. (13.3) and (13.6),

$$\sum_{part} R(c^{eq}(R) - \langle c\rangle) = 0. \tag{13.7}$$

Substitution of the expression for $c^{eq}(R)$ into the above equation (Eq. 13.7) leads to,

$$\sum_{part} R\left[\langle c\rangle - c^{eq}(\infty)\left(1 + \frac{2\gamma\Omega}{kTR}\right)\right] = 0. \tag{13.8}$$

Rearranging one gets

$$[\langle c\rangle c^{eq}(\infty)]\sum_{part} R = \frac{2\gamma\Omega c^{eq}(\infty)}{kT} \text{ and } \langle c\rangle - c^{eq}(\infty) = \frac{2\gamma\Omega c^{eq}(\infty)}{kT\langle R\rangle} \tag{13.9}$$

where $\langle R\rangle$, the average particle size (radius), is given by,

$$\langle R\rangle = \frac{\sum_{part} R}{N^{tot}}. \tag{13.10}$$

Here $N^{tot} = \sum_{part} 1$, the total number of particles.

Comparing Eqs. (13.4) and (13.9) one gets that when the matrix concentration, $\langle c\rangle$, is in equilibrium with particles of size $\langle R\rangle$, i.e.,

$$\langle c\rangle = c^{eq}(\langle R\rangle). \tag{13.11}$$

Subtracting Eq. (13.4) from Eq. (13.9),

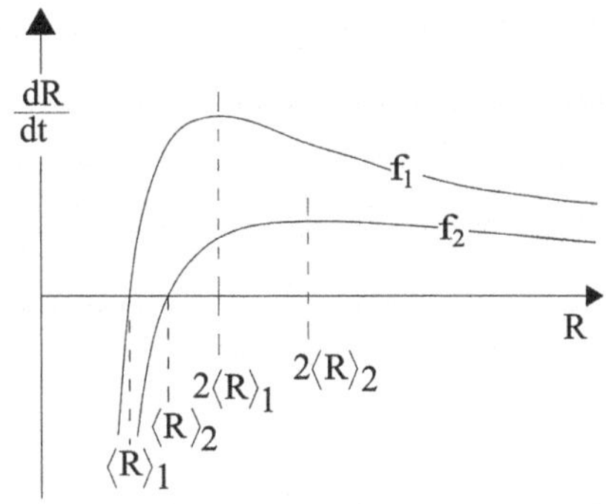

Fig. 13.3 Particle growth rate vs. particle size in Eq. (13.13). f_1 and f_2 are for different growth rates for different particle-size distributions, e.g., $\langle R\rangle_2 = 1.5\langle R\rangle_1$ [4]

$$\langle c\rangle c^{eq}(R) = \frac{2\gamma\Omega c^{eq}(\infty)}{kT}\left(\frac{1}{\langle R\rangle} - \frac{1}{R}\right). \tag{13.12}$$

Combining Eqs. (13.4) and (13.12),

$$\langle c\rangle - c^{eq}\langle R\rangle = \frac{2\gamma\Omega c^{eq}(\infty)}{kT}\left(\frac{1}{\langle R\rangle} - \frac{1}{R}\right). \tag{13.13}$$

Combining Eq. (13.6) with Eq. (13.13), one obtains,

$$\frac{dR}{dt} = \frac{2\tilde{D}\gamma\Omega^2 c^{eq}(\infty)}{kTR}\left(\frac{1}{\langle R\rangle} - \frac{1}{R}\right). \tag{13.14}$$

When R $< \langle R\rangle$, $\left(\frac{dR}{dt}\right)$ is negative. When R $> \langle R\rangle$, $\left(\frac{dR}{dt}\right)$ is positive. Eq. (13.14) is an example of a mean-field theory solution, which suggests that any particle behavior depends only on its size compared with the mean particle size $\langle R\rangle$. Figure 13.3 presents a schematic plot of this equation for two different particle-size distributions, f_1 and f_2, so that $\langle R\rangle_2 = 1.5\langle R\rangle_1$. The particle size for which the rate of particle-size growth is a maximum will obey,

$$\frac{d}{dt}\left(\frac{dR}{dt}\right) = 0 = \frac{2\langle R\rangle - R_{max}}{R_{max}^3\langle R\rangle}. \tag{13.15}$$

Thus, the particle size increase rate is maximum at $R_{max} = 2\langle R\rangle$.

It follows from Eq. (13.15), that the largest particles grow slowly and particles of smaller radii (but greater than $\langle R\rangle$) will grow quicker. Then, the distribution of particles shifts toward $2\langle R\rangle$. The average radius, $\langle R\rangle$, increases with time. R $>> \langle R\rangle$ will seldom be seen because the large R tail will diminish with time. The expected shift in distribution is presented in Fig. 13.4.

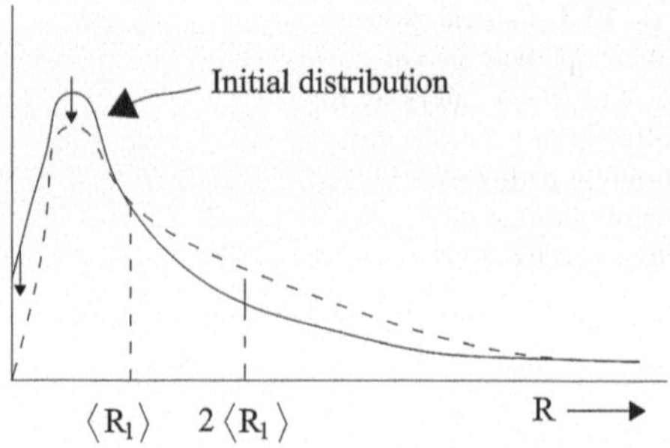

Fig. 13.4 $f_1(R)$, change with time (Eq. 13.14). The smallest particles have large shrinkage rates. The distribution lower end collapses to zero [4]

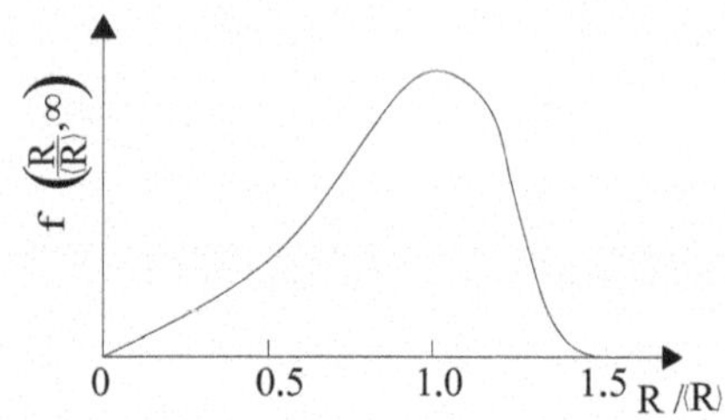

Fig. 13.5 Normalized particle-size distribution (for the final steady-state) for particle coarsening limited by diffusion [2, 3]

A differential equation involving the flux of particles in particle-size space can predict the particle-size distribution as the distribution evolves with time. The particle flux density passing the size R is,

$$J = (R, t)\frac{\partial R}{\partial t}. \tag{13.16}$$

The accumulation is given by,

$$\frac{\partial f(R,t)}{\partial t} = -\frac{\partial R}{\partial t} = -\frac{\partial}{\partial R}\left[f(R,t)\frac{\partial R}{\partial t}\right]. \tag{13.17}$$

The change rate of f in the interval $(R, R + dR)$, is given by Eq. (13.1). The difference between the rates of particles entering from below and exiting the interval at $R + dR$ is related to the earlier mentioned quantity. Combining Eqs. (13.14) and (13.17) one gets,

$$\frac{\partial f(R,t)}{\partial t} = -\frac{2\tilde{D}\gamma\Omega^2 c^{eq}(\infty)}{kT\langle R\rangle}\frac{\partial}{\partial R}\left[\frac{(R-\langle R\rangle)f(R\ t)}{R^2}\right]. \tag{13.18}$$

If the form of $f(R, t = 0)$ is assumed to be Gaussian, the form of $f(r, t)$ at later times can be known using Eq. (13.18). Solutions yield the results given below.

- As $t \to \infty$, steady-state (normalized) distribution function is approached asymptotically (see Fig. 13.5). All initial distributions follow this path. The most frequent particle size (mode) in the steady-state distribution is $1.13\langle R\rangle$. No particles larger than $1.5\langle R\rangle$ (the cut-off size) will be present.

- The mean particle size increases with time during annealing. N^{tot}, the total number of particles decreases because the larger ones grow at the expense of the smaller ones.
- Equation (13.14) (the growth law) and the continuity equation (Eq. 13.18) (for the evolution of the mean particle size), results in

$$\langle R(t)\rangle^3 - \langle R(0)\rangle^3 = \frac{8\tilde{D}\gamma\Omega^2 c^{eq}(\infty)}{9kT} = K_D t. \tag{13.19}$$

Here coarsening, limited by diffusion, has a rate constant, K_D.

These are the experimental findings. Normally, experimental studies measure two particle coarsening quantities:

1. the functional dependence of mean particle size. For volume-diffusion-limited coarsening, the $t^{1/3}$-law is often observed consistent with theoretical predictions (see Eq. 13.19), e.g., coarsening of semi-solid Pb-Sn alloys. But some predictions of the classical theory are not obeyed in these alloys (see below).
2. the particle-size distribution, along with its time dependence. The latter aspect does not always match predictions. The distributions seen are often broader than predicted by the classical theory. Moreover, the particles are many times larger than the predicted cut-off size, $1.5\langle R\rangle$.

During the occurrence of coarsening limited by the source (also known as "interface reaction control"), the interfaces surrounding the particles behave as poor sources and sinks, and the coarsening rate is decided by the diffusion fluxes between the particles present at the particle interfaces. In a simple model, the same assumptions can be made about the source action at the particles. The rate of particle growth is then,

$$\frac{dR}{dt} = K[c(R) - c^{eq}(R)]\Omega. \tag{13.20}$$

Here K is a rate constant. If the matrix diffusion between the particles is very rapid, or K is very small, all composition gradients are eliminated on a short time scale compared with the particle growth tome, $c(R) = \langle c\rangle$. Source action is the rate-limiting step. The average composition $\langle c\rangle$ can be obtained by combining the growth law, Eq. (13.20), with Eq. (13.3) and assuming the presence of a uniform concentration field, i.e., =c(R). c(R) as in (Fig. 13.6), the average composition, <c>, is obtained.

As the matrix concentration field is uniform, $c = \langle c\rangle$ everywhere, including regions adjacent to the particles where $c(R) = \langle c\rangle$. Therefore,

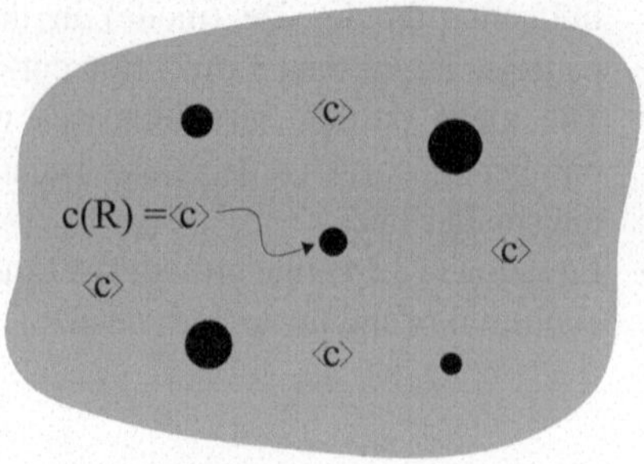

Fig. 13.6 Solute concentration field for source-limited kinetics [5]

$$\sum_{part} R^2(c^{eq}(R) - \langle R\rangle) = 0. \tag{13.21}$$

From Eq. (13.4),

$$\langle c\rangle = c^{eq}(\infty)\left(1 + \frac{2\gamma\Omega R}{kT\langle R^2\rangle}\right). \tag{13.22}$$

Equations (13.20) and (13.22) give rise to an expression for the growth rate of any particle of a particular size distribution,

$$\frac{dR}{dt} = \frac{2Kc^{eq}(\infty)\gamma\Omega^2}{kT}\left(\frac{R}{\langle R^2\rangle} - \frac{1}{R}\right). \tag{13.23}$$

On integration, one gets

$$R^2(t) - R^2(0) = \frac{4Kc^{eq}(\infty)\gamma\Omega^2}{kT}\left(\int_0^t \frac{\langle R\rangle R}{\langle R^2\rangle}dt - t\right). \tag{13.24}$$

Equation (13.24) can be averaged and the solution is

$$\langle R^2(t)\rangle - \langle R^2(0)\rangle = \frac{64Kc^{eq}(\infty)\gamma\Omega^2}{81kT}t = K_S t. \tag{13.25}$$

This is a case of coarsening that is source-limited, and K_S is the rate constant.

13.1.2 Exceeding the Classical Mean-Field Theory of Coarsening

The mean-field theory has some limitations. (a) Assuming that the mean concentration around all particles and the existence of spherically symmetric diffusion fields around every particle in a large medium to be like what would appear around a single particle. (b) These approximations become cruder with larger total volume fraction of closely spaced particles. (c) In the classical model these volume-fraction effects are ignored. So, the classical theory is correct when the volume fraction of the coarsening phase $\rightarrow \infty$, and the kinetics and the size distributions are strongly precipitate volume fraction dependent. The diffusion-limited coarsening rate, given by Eq. (13.19), remains valid for all volume fractions, but the rate constant K_D increases monotonically with ϕ (Eq. 13.9).

Particle coarsening rates, influenced by the volume fraction, have been measured experimentally. Normally, liquid + solid system data are more reliable than solid + solid systems (no coherency stresses act on the former). Low-volume-fraction solid, liquid + solid Pb-Sn mixtures of different volume fractions were studied experimentally under microgravity conditions without significant buoyancy effects. In 0.1–0.2 volume-fraction solid specimens the approach to steady state was very slow. The initial stages, of coarsening involved significant solid - particle coarsening compared with a randomly arranged particles, abundant particle clustering was present. Elastic interactions of particles changed their shapes, and this affected coarsening, e.g., aligned precipitate structures in Ni-Al alloys on prolonged aging. The alignment of coherent misfitting precipitates along the cube axes are formed, with an attendant lowering of the elastic energy. Elastic energy also causes stress coarsening, as in nickel-based superalloys, where single crystals coarsen under applied, uniaxial stress and depending on whether the stress is tensile or compressive, form either platelets or rods. If no load is resent, cuboidal particles form—which is called rafting. A high creep resistance is associated with such alloys. The equilibrium precipitate shape is determined by elastic anisotropy, elastic inhomogeneity, applied stress state, and stress-free transformation strain of precipitates.

Numerical simulations of coarsening with appropriate boundary curvatures to estimate that such shape changes are possible.

Large particles in aged Ni-Si and Ni-A1 alloys split into pairs or several smaller particles with aging, which is an experimental result. Triumph of reduced elastic energy over an increase in interphase boundary area is responsible for this.

13.2 Grain Growth

This is defined as a kinetic process that increases the average grain size of a polycrystalline material. This is present prominently at elevated temperatures (Fig. 13.7).

The larger grains grow. The smaller grains gradually disappear, decreasing the total number of grains. Then, the average grain size is greater. The lowering of the total interfacial energy of the system resulting from grain growth is the reason. Realistic models should comprise grain size and shape distribution. 2D grain growth models are reasonably accurate. Physically meaningful models for 3D grain growth are absent. Computer simulation alone has helped a bit, not theoretical reasoning.

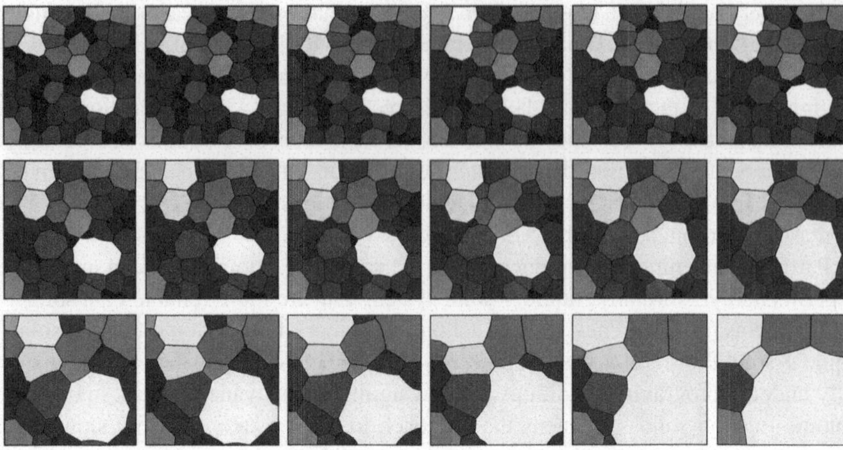

Fig. 13.7 Isotropic and uniform 2D grain growth simulations were performed using the software, Surface Evolver. The *(N-6)* rule in grains that intersect the edge of the domain [6]

13.2.1 Two-Dimensional Grain Growth

In 2D only three grains meet at any grain boundary vertex. (Vertices resulting from the convergence of more than three boundaries are unstable). They will decompose immediately into three-grain vertices connected by new boundary segments. The three boundaries at each vertex are locally equilibrated with respect to each other. For isotropic and equal boundary energies, this equilibration requires that the net capillary force on the vertex is zero. Therefore, the boundaries meet at 120° angles. Geometrical growth parameter, which is important, is the average number of sides per grain, (N), in the ensemble. Euler's theorem states that

$$N^G - N^B + N^V = 1 \tag{13.26}$$

with N^G the number of grains, N^B the number of boundary segments, and N^V the number of vertices in the system. The relationship

$$N^B = \left(\frac{3}{2}\right) N^V \tag{13.27}$$

is valid. Every boundary segment provides a side for two grains. Or $\langle N \rangle = \left(\frac{2N^B}{N^G}\right)$. N^G, N^B, N^V are all greater than 1. Therefore, $N^G + N^V \approx N^B$. From Eq. (13.27) $N^B = 3N^G$. Therefore,

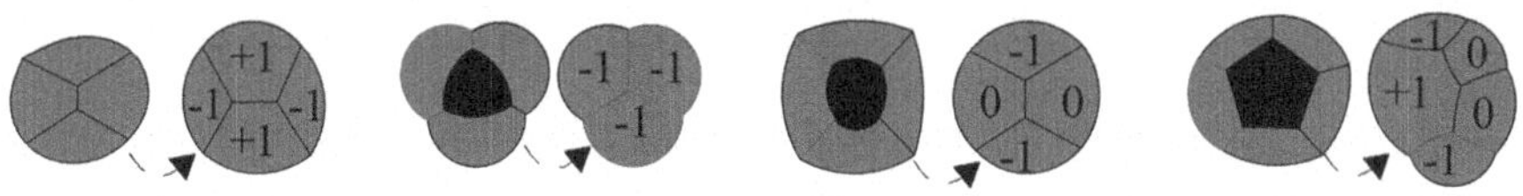

Fig. 13.8 Switching events during 2D grain growth [7]

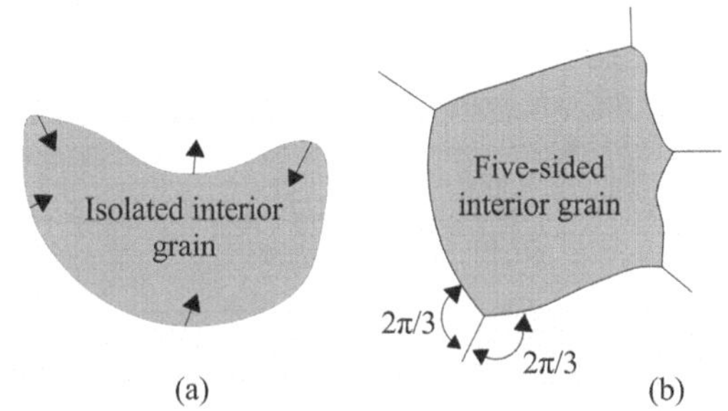

Fig. 13.9 (**a**) 2D isolated grain. Low pressures on the boundary are indicated by arrows. (**b**) Five-sided grain in a polycrystalline material. Isotropic grain-boundary energies [4]

$$\langle N \rangle = \left(\frac{2N^B}{N^G} \right) = 6. \tag{13.28}$$

2D Grain Growth Model: Here, the shapes and sizes of grains are a function of time – a distribution function $F(A, N, t)$. At a time t, it has N sides and an area A. It evolves when grains grow or shrink without a change in the number of their sides or discontinuously changing the number of their sides by grain switching (Fig. 13.8).

Two vertices meet on Fig. 13.8a (each lose a side). In Fig. 13.8b, a three-sided grain disappears. One side is lost by all three grains. In Fig. 13.8c a four-sided grain disappears. Two grains each lose a side. Two other grains retain their sides. In Fig. 13.8d a five-sided grain disappears. Two grains lose one side each. Two grains retain their sides. Grains with $N \geq 6$ are not likely to shrink. This way, all important switching events are present (Fig. 13.8). The distribution change rate, $F(A, N, t)$, is the sum of (i) the accumulation rate of grains of area A and N sides due to the growth or shrinkage of grains without switching, and (ii) the accumulation rate due to grain side switching. To determine (i), the expression for the rate of growth or shrinkage of a single grain without switching of grain boundary sides is needed. It is computed as follows.

Two assumptions (i) all boundaries are isotropic, and (ii) they all have same grain boundary energy, γ, are made.

See the single isolated interior grain (Fig. 13.9a).

Any portion of a boundary of a grain experiences a normal pressure directed toward its concave side, given by $P = \gamma\kappa$ (Fig. 13.9a). The net boundary velocity is taken to be proportional to pressure, i.e.,

$$v = M_B P = M_B \gamma \kappa \tag{13.29}$$

where M_B is a mobility that links velocity with pressure. If the local velocities of the boundaries are integrated along the perimeter of a grain, the area change rate of the grain interior results. That is,

$$\frac{dA}{dt} = -\int_{GB} v ds = \int_{GB} M_B \gamma \kappa ds. \tag{13.30}$$

If M_B and γ are constants that are not dependent on the boundary orientation and curvature, one gets,

$$\frac{dA}{dt} = -M_B \gamma \int_{GB} \kappa ds = -M_B \gamma \int_{GB} \frac{d\theta}{ds} ds = -2\pi M_B \gamma. \tag{13.31}$$

θ is the boundary normal angle, a fixed direction in the specimen plane. The curvature is the change rate of boundary normal as the line integral is completed, i.e., $\kappa = \frac{d\theta}{ds}$ (θ varies between 0 and 2π in the integration as the normal rotates by 2π). Then, regardless of grain shape,

$$\frac{dA}{dt} = -constant. \tag{13.32}$$

Even if this mass and some segments leave and enter it along other segments of the boundary this result is valid (Fig. 13.9a).

In Fig. 13.9b, Eq. (13.30) decides the area change rate. Then, discontinuities in $\left(\frac{d\theta}{ds}\right)$ at the vertices are present due to a change in the boundary orientation. If N sides are present in a grain, then there are N vertices. Integrating over all boundary segments,

$$\frac{dA(N)}{dt} = -M_B \gamma \left(\int_{seg1} d\theta + \int_{seg2} d\theta + \ldots + \int_{segN} d\theta \right) = M_B \gamma (2\pi - N\Delta\theta). \tag{13.33}$$

where $\Delta\theta$ is the external angle change at each vertex, i.e.,

$$\Delta\theta = \theta_{begin}(segment\ i+1) - \theta_{end}(segment\ i+1).$$

If the boundaries are locally equilibrated at each vertex, the angles between boundaries are $\left(\frac{2\pi}{3}\right)$. So $\Delta\theta = \left(\frac{4\pi}{3}\right) - \pi$. Therefore,

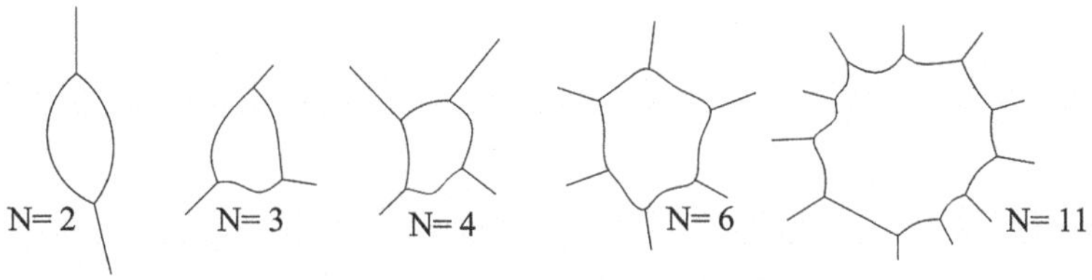

Fig. 13.10 Shapes of N-sided grains in polycrystals (2D)

$$\left(\frac{dA(N)}{dt}\right) = M_B\gamma\frac{\pi}{3}(N-6). \tag{13.34}$$

When the sides are more than 6, grains will grow. Those with fewer than 6 will shrink. This generalizes an earlier result for the *(N−6)*-rule. Then, topology decides the grain growth rate and not the grain size (Fig. 13.10). When the sides are more than 6, the fixed angles at each vertex will force the boundary segments to be concave away from the grain center and make them grow. This is curvature caused growth. When the number of sides are six, the sides are flat and static. Less than six sides causes shrinking.

The distribution change rate of the distribution, F (A, N, t) can be derived from Eq. (13.34) as,

$$\frac{\partial F(A,N,\ t)}{\partial t} = \left[-\frac{\partial F(A,N,t)}{\partial A}M_B\gamma\frac{\pi}{3}(N-6)\right] + I_{N-1}^{+} + I_{N+1}^{-} - I_N^{+} - I_N^{-}. \tag{13.35}$$

The rate of accumulation of the grain area A and sides due to the growth or shrinkage of grains is present inside the brackets—see Eq. (13.35). The second is the rate of accumulation due to switching. (I_{N-1}^{+}, for example, is the rate at which grains of area A with (N−1) sides gain a side by switching to become grains of area A with N sides). Equation 13.35 has not been fully derived, e.g., when a four-sided grain disappears, four grains are picked up randomly and a side is removed from the two that have the smallest areas. The grain that has the largest area gains a side when the grain with 5 sides disappears. With many such shortcuts, Fig. 13.11 and others like it are obtained. The experimental results of 2D bubble growth of soap froths are included for comparison.

Reasonably good predictions are found in Fig. 13.11a. After an initial transient, with time of the grain area increases linearly with time (both predicted and measured), i.e.,

$$\langle A\rangle \propto t. \tag{13.36}$$

Computer simulation was used to predict 2D grain growth. Monte Carlo methods, direct tracking methods, and vertex models (the evolution of 2D grain structure is

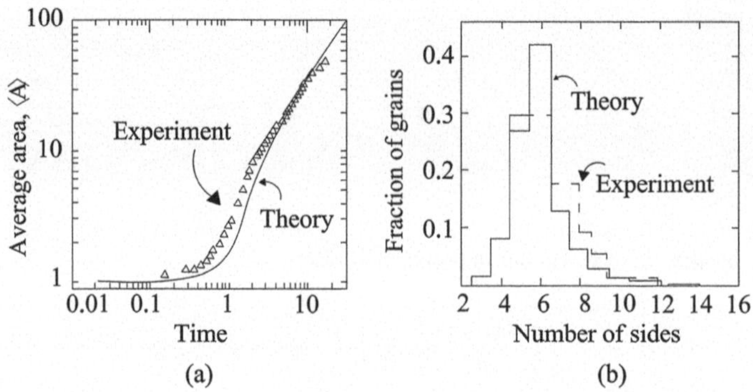

Fig. 13.11 (**a**) Average grain area vs. time. (**b**) Grain fraction having N sides. Figures comparing the "best" predictions with experimental findings of the growth soap bubbles [8, 9]

described in terms of the vertices motion). After initial transients, all these simulations result in statistical self-similarity during growth. Average grain area linear increase with time is according to Eq. (13.36) (the grain structure later looks statistically like itself, except for a uniform enlargement—structure scaling.)

If statistical self-similarity holds, individual grains obey Eq. (13.34). Let a large area, A, contain G grains. Then, $\langle A\rangle = \left(\frac{A}{G}\right)$ and $\langle A^2\rangle = \left(\frac{1}{G}\right)\sum_{i=1}^{i=G} a_i^2$, where a_i is the area of the *I th* grain. When summed over all grains, this can also be written as,

$$\langle A\rangle \frac{\langle A^2\rangle}{\langle A\rangle^2} = \frac{1}{A}\sum_{i=1}^{G} a_i^2. \tag{13.37}$$

Using self-similarity and differentiating Eq. (13.37) with $\left(\frac{\langle A^2\rangle}{\langle A\rangle^2}\right) = constant$, and employing self-similarity along with Eq. (13.33), one gets,

$$\frac{\langle A^2\rangle}{\langle A\rangle^2}\frac{d\langle A\rangle}{dt} = \frac{1}{A}\sum_{i=1}^{G} 2a_i\frac{d}{dt} = \frac{2\pi M_B\gamma}{3A}\sum_{i+1}^{G}(a_i(N_i-6) \tag{13.38}$$

where N_i is the number of sides of the *i*th grain. But

$$\sum_{n=1}^{G} a_i(N_i-6) = \sum_{N=2}^{\infty}(A_N N - 6A_N) \tag{13.39}$$

where A_N is the total area occupied by N-sided grains. The second sum begins with N = 2, because one-sided grains do not exist. Therefore, using Eq. (13.39) in Eq. (13.38) one can get

$$\left(\frac{d\langle A\rangle}{dt}\right) = \frac{2\pi M_B\gamma}{3}\frac{\langle A\rangle^2}{\langle A^2\rangle}\sum_{N=2}^{\infty}\frac{A_N}{A}(N-6) = C = Constant. \tag{13.40}$$

Here, $\frac{A_N}{A}$, the fraction of the total area occupied by N-sided grains, is constant because of self-similarity.

If R_{rms} is the effective root-mean-square radius of the approximately equiaxed grains,

$$\frac{d\langle A\rangle}{dt}\frac{d}{dt}\left(2\pi R_{rms}^2\right) = C. \tag{13.41}$$

On integrating,

$$R_{rms}(t) = R_{rms}^2(0) = \frac{C}{\pi}t. \tag{13.42}$$

The growth laws expressed by Eqs. (13.36) and (13.42) are consistent with the behavior of many alloys, models, and computer simulation. During thin polycrystalline film growth rate, 2D grain growth could be impeded by surface grooves formation at the boundary intersections. The boundaries can then get trapped at the grooves and stagnate growth.

13.2.2 Three-Dimensional Grain Growth

The geometry of 3D grain growth is more difficult to picturize and simulate. Figure 13.12 shows the relatively simple structure of an idealized polycrystal. (All the grains are tetrakaidekahedral—which possesses 14 faces, arranged on a BCC lattice.

The 3D grain structure is described through grains, boundaries, edges, and vertices. Two grains meet at a common boundary. Three grains meet along edges. Four edges and four grains meet at vertices.

When local equilibrium is ensured, the net capillary force on each junction is 0. if γ^B is isotropic and equal for all boundaries, a trihedral symmetry acts along each

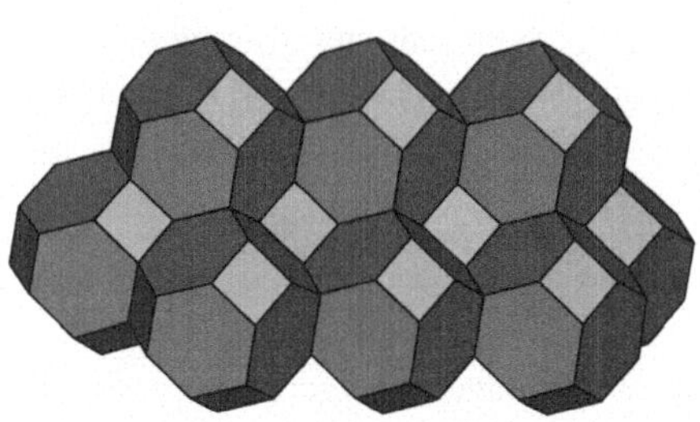

Fig. 13.12 A three-dimensional polycrystal with identical grains. Each grain is a tetrakaidecahedron—6 square and 8 hexagonal faces

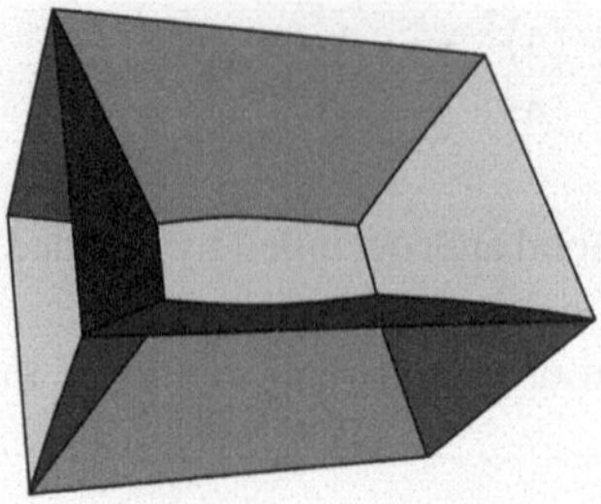

Fig. 13.13 A six grain polycrystal of orthorhombic shape meeting at four vertices. One face of the polycrystal belongs to each grain. Four grains share each vertex. At each edge three grains meet (triple line). Each vertex meets four edges. All edges and vertices are in local equilibrium. Plateau's laws are obeyed. All boundary energies are taken as equal and isotropic. The figure was computed using the software Surface Evolver [6]

edge. Along each vertex a tetrahedral symmetry is present. That is, at three-boundary junctions, two faces have a dihedral angle $cos^{-1}\left(\frac{-1}{2}\right) = 120^\circ$, $cos^{-1}\left(\frac{-1}{3}\right) = 109^\circ 28'$ and at four-edge junctions, each pair of edges makes an angle $cos^{-1}\left(\frac{-1}{3}\right) = 109^\circ 28'$. These local equilibrium symmetry conditions are referred to as Plateau's laws. The grain boundary displayed in Fig. 13.12 does not obey these laws. For obeying Plateau's laws, each grain face must be curved, e.g., Fig. 13.13. Here a polycrystal made of six grains meet at four vertices. One face of the polycrystal is occupied by a grain. Three grains meet at each edge (Fig. 13.12), four grains meet at each vertex, and four edges meet at each vertex.

Grain growth is accompanied by a grain size (volume) distribution with different numbers of faces and edges. At all junctions Plateau's laws are obeyed. According to Euler's theorem,

$$B\text{-}E + V = G. \tag{13.43}$$

(E = number of edges, V = number of vertices, G = number of grains, B = number of faces). Every boundary provides two faces for grains. Therefore, the average number of faces per grain,

$$\langle f\rangle_{syst}, \text{ is } \langle f\rangle_{syst} = \frac{2B}{G}. \tag{13.44}$$

Let $\langle\cdot\rangle_{syst}$ represent any average value. As three grains meet at an edge, each edge provides three edges for grains. Then, the average number of edges per grain, $\langle e\rangle_{syst}$, is

$$\langle e\rangle_{syst} = \frac{3E}{G}. \tag{13.45}$$

Since four edges emerge from a vertex and every edge connects two vertices,

$$E = \frac{4V}{2} = 2V. \tag{13.46}$$

By combining Eqs. (13.43)–(13.46),

$$\langle e \rangle_{syst} = 3\langle f \rangle_{syst} - 6. \tag{13.47}$$

As every edge on a grain provides two edges for faces on that grain, the average number of edges per face on grains is $\langle e_f \rangle_{syst} = \frac{2\langle e \rangle_{syst}}{\langle f \rangle_{syst}}$. Combining this result with Eq. (13.47) gives

$$\langle f \rangle_{syst} = \frac{12}{6 - \langle e_f \rangle_{syst}}. \tag{13.48}$$

By inspection, Eqs. (13.47) and (13.48) are obeyed for the grain structure presented in Fig. 13.12.

The (N−6) rule (this is equivalent to Gauss-Bonnet theorem in mathematics) for 2D grain growth in a system of isotropic and uniform boundary energies (Eq. 13.34) is an illustration of a principle that equivalent grains grow at identical rates. That is, grains that have at least one topological feature in common. In the 2D case, this means that grains with the same number of sides, N will grow at the same rate. No known simple equivalence class describes grain growth in three dimensions.

No exact growth law exists for the 3D case. But approximate laws have been found. In 2D growth, the growth law is identical to the statistical law. Here the number of grain sizes, *N*, is selected as the topological feature defining the class.

A statistical growth law is fine for 3D growth if *V* and *E* are chosen as the topological features to define the equivalence class. For symmetric grains of volume *V* and number of faces *B*, the average growth rate, $\left(\left\langle \frac{dv}{dt} \right\rangle\right)_{\{v,f\}}$, varies as $\sqrt{B}$ as *B* becomes large. In Eq. (13.47), where $\langle e \rangle_{syst}$ is ~ proportional to $\langle f \rangle_{syst}$ for large *B* and *E*. This indicates possibly similar behavior on average if *E* is chosen rather than *B*, and this suggests the expansion

$$\left\langle \frac{dV}{dt} \right\rangle_{\{E,V\}} = \left(\frac{dV}{dt}\right)_0 \left(aE^{1/2} + b + cE^{-1/2} + \ldots.\right). \tag{13.49}$$

where $\left\langle \frac{dV}{dt} \right\rangle_{\{e,v\}}$ is the average volumetric growth rate of grains belonging to the equivalence class $\{E, V\}$. $\left(\frac{dV}{dt}\right)_0$ is the shrinkage rate of a sphere of the same volume embedded in a matrix grain. The undetermined coefficients, *a, b, c, ...* have no explicit dependence on *E*. To obtain the behavior for $E >> 1$, the third- and higher-order terms are neglected, and in analogy with soap bubbles in 2D, b is set equal to 1. This represents an offset from the mean pressure of the bubble. By summing over all classes and using the conservation of total volume,

$$\sum\nolimits_{\{e,v\}}\left(\frac{dV}{dt}\right)=0. \tag{13.50}$$

A statistical growth law for grains with large edge number, E, results. This is given by,

$$\left\langle\frac{dV}{dt}\right\rangle_{\{E,V\}}=\left[\frac{dV}{dt}\right]_0\left(\sqrt{\frac{E}{E_{crit}}}-1\right). \tag{13.51}$$

where E_{crit} is given by

$$E_{crit}=\left(\frac{\sum_{\{E,V\}}V^{1/3}\sqrt{E}}{\sum_{\{E,V\}}}\right)^2. \tag{13.52}$$

A comparison with direct simulations of grain growth reveals that Eq. (13.51) is correct for small and large E. Unlike 2D growth, here e is the typical topological feature, and it is not constant but depends on the grain-boundary structure. No progress has been there in treating grain switching in 3D. But evidence for statistical self-similarity has been found in terms Eq. (13.42) both in experiments and by simulation.

The best applications of the above grain growth equations are in understanding sintering of polycrystalline materials and grain growth accompanied by dislocation motion. See, for example [10, 11].

Problems

1	In a material, particles with an initial radius of 0.5 μm are undergoing coarsening due to capillary forces. The radius of the particles grows over time, and after 100 h, the average radius is 1.5 μm. Assuming a classical coarsening model, calculate the growth rate of the particles **Solution:** $\left(\frac{r(t)}{r_0}\right)=\left(\frac{t}{t_0}\right)^{1/3}$ $\left(\frac{1.5}{0.5}\right)=\left(\frac{100}{t_0}\right)^{1/3}$ $3=\left(\frac{100}{t_0}\right)^{1/3}$ $27=\left[\frac{100}{t_0}\right]$ $t_0=\frac{100}{27}\approx 3.7 hours$
2	In a material undergoing coarsening, the number of particles decreases with time. If the initial number of particles is 1000 and the coarsening leads to a 50% reduction in particle number after 10 h, apply the classical mean-field theory to estimate the average particle growth rate **Solution:**

(continued)

	$N(t) = N_0\left(\frac{t_0}{t}\right)^{\alpha}$ $500 = 1000\left(\frac{t_0}{10}\right)^{\frac{1}{3}}$ $0.5 = \left(\frac{t_0}{10}\right)^{\frac{1}{3}}$ $0.125 = \frac{t_0}{10}$ $t_0 = 1.25$ hours
3	Beyond the classical theory, coarsening is affected by additional factors such as local interactions between particles. If the characteristic interaction time between particles is 5 hours and after this time, the average particle radius increases by 20%, calculate the effective coarsening rate beyond the classical mean-field theory **Solution:** Growth Factor = 1 + 0.2 = 1.2 Given the interaction time $t_{int} = 5$ h, the effective coarsening $\frac{r(t)}{r_0} = 1.2$

References

1. Greenwood, G. W. (1956). The growth of dispersed precipitates in solutions. *Acta Metallurgica, 4*(3), 243–248.
2. Ardell, A. J. (1969). *Experimental confirmation of the lifshitz-wagner theory of particle coarsening* (Vol. 116). Institute of Metals.
3. Ardell, A. J. (1988). Precipitate coarsening in solids. Modern theories, chronic disagreement with experiment. In G. W. Lorimer (Ed.), *Phase transformation* (Vol. 87, pp. 485–490). Institute of Metals.
4. Balluffi, R. W., Allen, S. M., & Craig Carter, W. (2005). *Kinetics of materials*. A John Wiley & Sons.
5. Ratke, L., & Voorhees, P. W. (2002). *Growth and coarseing*. Springer–Verlag.
6. Brakke, K. A. (1992). The surface evolver. *Experimental Mathematics, 1*(2), 141–165.
7. Thompson, C. V. (2001). Grain growth and evolution of other cellular structures. *Solid State Physics, 55*, 269–314.
8. Marder, M. (1987). Soap-pubble growth. *Physical Review A, 36*(1), 438–440.
9. Glazier, J. A., Gross, S. P., & Stavans, J. (1987). Dynamics of two–dimensional soap froths. *Physical Review A, 36*(1), 306–312.
10. Pease, L. F., & West, W. G. (2002). *Fundamentals of powder metallurgy*. Metal Powder Industries Federation.
11. Anderson, P. M., Hirth, J. P., & Lothe, J. (2017). *Theory of dislocations* (3rd ed.). Cambridge University Press.

Chapter 14
Phase Diagrams And Phase Transformations

Contents

K. A. Padmanabhan et al., *Thermodynamics and Kinetics of Materials*, Engineering Materials, https://doi.org/10.1007/978-3-032-14527-7_14

14.1 Fundamentals of Phase Equilibria: A Recap

14.1.1 The Phase Equilibrium Thermodynamic Condition

A phase is a section of the system that has homogeneous attributes and composition and differs physically from other sections of the system. The components of a system refer to the elements or chemical compounds that constitute it, and the composition of a system can be represented by the proportions of each component [1].

Phase transformations involve the change of one or two phases into a new phase or a combination of phases. Transformations occur between the unstable phase and a stable final state. The stability of a phase is explained by thermodynamics. The Gibbs free energy (G) of a system determines its relative stability for transformations that occur at constant temperature and pressure. G is defined as:

$$G = H - TS \tag{14.1}$$

where, H is enthalpy (is defined as the amount of heat content), T is the absolute temperature, and S is entropy, a degree of randomness in the system.

Enthalpy (H) is further defined as:

$$H = E + PV \tag{14.2}$$

where E is internal energy, which arises from atomic kinetic and potential energies, P is pressure, and V is volume.

14.1.1.1 Phase Equilibria

Phase equilibria are situations in which two or more phases coexist (are in equilibrium). Since the stable phase has the lowest chemical potential at a given temperature and pressure, the chemical potential can be used to predict phase stability. The phase diagram is an important tool that illustrates conditions (such as pressure and temperature) under which different thermodynamic phases form and coexist in equilibrium [1].

Thermodynamic properties are classified as:

- **Intensive properties**: Regardless of systems' magnitude (e.g., temperature, pressure, etc.).
- **Extensive properties**: Proportional to system size (e.g., Gibbs free energy, entropy, enthalpy, volume, energy, etc.).

Extensive properties are often expressed as molar quantities (per mole of material).

Since there is no effort and no mass or heat transfer across the system's boundary, the equilibrium state denotes the lack of a tendency for the system's properties to change on their own. In other words, the properties of the system are independent of time. It's critical to recognize the distinction between a steady state and an equilibrium state. To reach the equilibrium, the single-phase system must have uniform intensive properties. Uniformity (having the same value from point to point) is required for the intensive properties. These requirements are not necessary for steady state, meaning that a system may be in steady state without being in equilibrium [2].

14.1.2 Gibbs's Phase Rule

The Gibbs phase rule is a key concept in thermodynamics that determines how many independent variables or degrees of freedom can be varied without affecting the number or types of phases present in an equilibrium system. For a system to be in equilibrium, its various phases must have the same temperature and pressure. In addition, every component in every equilibrium phase has the same chemical potential. As a result, the Gibbs phase rule was developed, which states that

$$\mathrm{F} = \mathrm{C} - \mathrm{P} + 2 \tag{14.3}$$

where F stands for the minimum number of degrees of freedom needed to replicate the system C for the number of independent components, and P for the number of phases. Gibbs phase rule can be applied to calculate the number of phases in a multicomponent system at a specific temperature and pressure [1].

14.1.3 Stability and Critical Phenomena

The phase with the lowest Gibbs free energy will be the most stable when a pure material exists in multiple phases. Changes in temperature, pressure, or volume can affect stability. Changes in temperature or pressure can change the Gibbs energy of different phases, leading to phase transformations when the energies of two phases become equal [1]. The stable phase is determined by the conditions, and phase diagrams help to locate these stability regions based on the principle that the phase with the minimum Gibbs energy is achieved. Since G is a state function of T and P, such that

$$\mathrm{dG} = \mathrm{VdP} - \mathrm{SdT} \tag{14.4}$$

it follows that

$$(\mathrm{dG}/\mathrm{dT})\mathrm{P} = -\mathrm{S} \tag{14.5}$$

Thus, for a certain stage, G falls with increasing temperature. If the G(T) curve for the two phases α and β crosses (see Fig. 14.1), when phases α and β are in

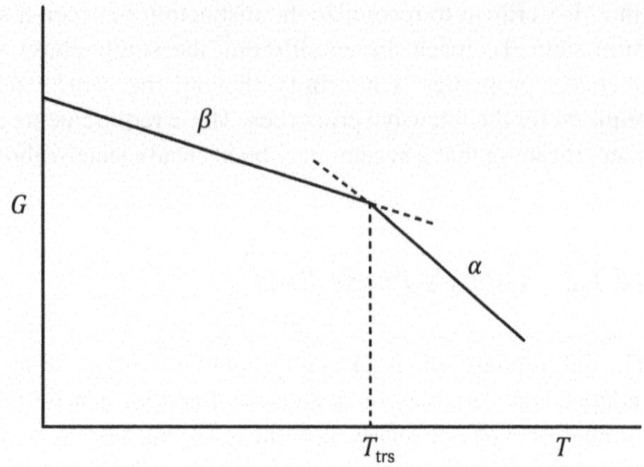

Fig. 14.1 Gibbs energy vs temperature, displaying a phase change at T_{trs}. A stable phase with the lowest G value is represented by the solid line. The α phase would be metastable relative to the β phase at T [1]

equilibrium, $G_\alpha = G_\beta$, then the phase with the lower G will be stable. When G_α is greater than G_β, it means that phase β is more stable than phase α. However, if the kinetic barrier for $\alpha \rightarrow \beta$ is too large in these conditions, α may exist in place of β. In that scenario, since α is at a local energy minimum rather than the global energy minimum, it would be metastable with respect to β. For instance, liquid water is metastable with respect to ice for $T < 0$ °C and can be supercooled below 0 °C. The kinetic barrier is responsible for the presence of supercooled water.

14.1.4 Phase Transitions types

A phase transition is a critical change in the macroscopic properties of a material as a function of variables, such as temperature or pressure. One of the most well-known examples occurs in ferromagnets, where abrupt changes in magnetization can be observed when the system approaches a critical point.

Phase transitions can be broadly classified into two types [1]:

A. **First-Order Phase Transitions**

1. Categorized by a discontinuous change in thermodynamic properties.
2. Involves the coexistence of distinct phases at the transition point.
3. Example: Melting of a solid, where solid and liquid phases coexist.
4. In ferromagnets, spin-up and spin-down states coexist at the transition.
5. Correlation length remains finite.

B. **Second-Order Phase Transitions**

1. A continuous transition where thermodynamic properties change smoothly.
2. Correlation length diverges, leading to large-scale fluctuations.

3. Example: The transition from paramagnetic to ferromagnetic states.
4. Magnetization smoothly approaches zero at the critical point.
5. The system exhibits long-range collective fluctuations.

A defining characteristic of second-order transitions is that fluctuations become correlated across all length scales, driving the system into a distinct critical phase. As the correlation length approaches infinity, magnetization gradually decreases, resulting in a clearly identifiable universality class

14.1.4.1 Universality and Critical Phenomena

The concept of universality states that phase transitions exhibit similar behavior regardless of microscopic details, being determined primarily by fundamental symmetries.

- **Ginzburg-Landau Theory:** A phenomenological approach describing critical phenomena using a free energy functional constrained by symmetry considerations.
- **Universality Classes:** Systems with similar symmetry properties and dimensionality fall into the same class, exhibiting identical critical behavior.
- Example: The Ising model, describing one-component ferromagnets, shares critical behavior with the liquid-gas transition.
- Similarly, superconductors belong to the same universality class as the two-component (XY) Heisenberg ferromagnet.

14.1.4.2 Critical Behavior

The unique behavior near the critical point is defined through a set of critical exponents. These exponents delineate the non-analyticity of several thermodynamic functions. Remarkably, transitions as disparate as liquid-gas and ferromagnetic transitions can be characterized by the same set of critical exponents and are classified within the same universality class.

14.1.4.3 Order Parameter

According to the definition, multiple equilibrium phases may exist along a coexistence line. The order parameter is a thermodynamic variable that varies between these phases and can be utilized to distinguish them.

- In a uniaxial magnet, the order parameter is often the total magnetization or magnetization density.

- Without an external magnetic field, magnetization is zero in the paramagnetic phase and non-zero in the ferromagnetic phase. This behavior is contingent upon the reduced temperature.
- The unique behavior of the order parameter along the coexistence line is characterized by a critical exponent.
- Along the critical isotherm, the correlation between the order parameter and the external field is determined by an additional critical exponent.

In the liquid-gas transition, the density difference from the critical density serves as the order parameter.

14.1.4.4 Response Functions

Critical systems exhibit significant sensitivity to external disturbances. At the liquid-gas critical point, compressibility becomes infinite. The system's response functions display distinctive behavior dictated by critical exponents [3].

- **Magnetic Susceptibility**: In a ferromagnet, the magnetic susceptibility without an external field diverges around the critical point, exhibiting power-law behavior defined by a critical exponent.
- **Heat Capacity:** The thermal response function, given by the heat capacity, characterizes the singularity in energy fluctuations.

In most cases, the exponents remain the same on both sides of the transition, i.e., Plots illustrating the relationship among phases in equilibrium relative to composition, pressure, and temperature are termed phase diagrams, also referred to as equilibrium diagrams. The unary diagrams are drawn between Temperature (in the y-axis) and pressure on the x-axis). The binary phase diagrams are drawn between Temperature (in the y-axis) and composition on the x-axis. The composition is expressed in weight percent. These diagrams show how phases change in response to variations in composition or temperature. Since a system always strives for stability, phase diagrams are always created at equilibrium. The alloy systems comprise multiple components. The phase diagrams are categorized as follows.

1. **Unary phase diagram** (single component system)
2. **Binary phase diagram** (two-component system)
3. **Ternary phase diagram** (three-components system)
4. **Quaternary phase diagram** (four-component system), and so on.

Unary systems can be categorized into two-dimensional phase diagrams as follows:

1. Field–field phase diagrams
2. Field–density phase diagrams
3. Density–density phase diagrams

Field variables are intrinsically intensive qualities, including pressure (p), temperature (T), and chemical potential (μi). Density variables are intensive qualities

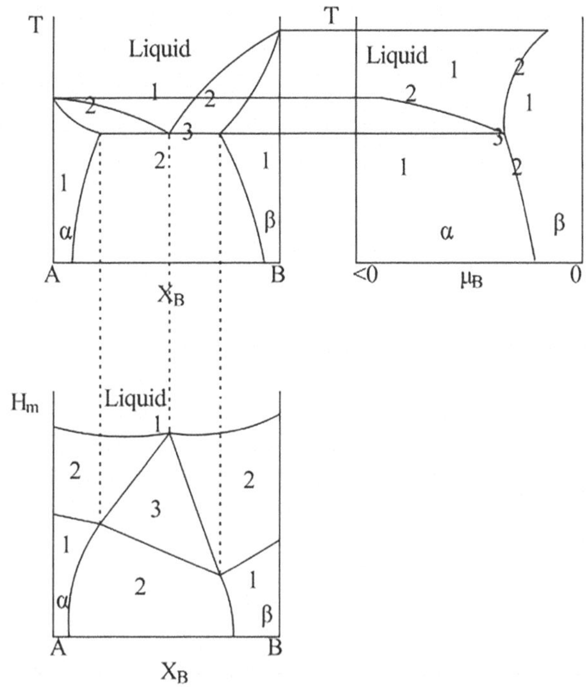

Fig. 14.2 Illustrations of field–density, field–field, and density–density binary phase diagrams. The numerals denote the quantity of phases existing in that section of the diagram [4]

obtained from extensive quantities by dividing by mass or amount, generally referred to as particular quantities.

The identical classification pertains to phase diagrams for binary systems; however, the inclusion of an additional component introduces an extra independent variable. Thus, three independent variables are necessary to characterize the features of a single-phase binary system. To depict these features or phase diagrams in two dimensions, one variable must remain constant. In Materials Science and Engineering, phase diagrams are generally depicted at a constant total pressure of 1 bar. Common examples of two-dimensional charts for binary systems include T–μ_i, T–x_i, and H_m–x_i diagrams.

Schematic representations of these three are shown in Fig. 14.2.

14.2 Unary Systems and Phase Diagrams

14.2.1 Unary Systems

A unary system denotes a material system that consists of only one chemical component. It is the simplest type of system studied in phase diagrams and thermodynamics. Examples of unary systems include pure metals like iron (Fe), copper (Cu), or pure substances like water (H_2O).

In unary systems, the phase behavior depends primarily on external variables, including temperature and pressure. By modifying these factors, the system can manifest in many phases, including solid, liquid, or gas.

Examples of phase changes in unary systems include:

- Ice (solid water) melting into liquid water.
- Liquid iron solidifying into solid iron during cooling

14.2.1.1 Unary Diagram of H_2O

To understand the phase behavior of water (H_2O) under varying temperature and pressure conditions, we apply the Gibbs phase rule, which is given in Eq. (14.3):

In a single-phase region (where only one phase exists, such as liquid water, ice, or steam), the system consists of one component (**C = 1**) and one phase (**P = 1**). Applying Gibbs phase rule:

$$F = 1 - 1 + 2 = 2$$

Since **F = 2**, Temperature and pressure can be independently adjusted. A single-phase equilibrium is defined by a region in the phase diagram.

In equilibrium between two phases (e.g., liquid and vapor, solid and liquid), the number of phases is $P = 2$, whereas the number of components remains $C = 1$. Utilizing Gibbs' phase rule:

$$F = 1 - 2 + 2 = 1$$

indicating that either pressure or temperature may be altered independently, but not concurrently. This indicates that two-phase equilibrium occurs only along the phase boundaries in the diagram.

At the triple point, all three phases (solid, liquid, and gas) coexist in equilibrium, indicating $P = 3$. Utilizing Gibbs' phase rule:

$$F = 1 - 3 + 2 = 0$$

since $F = 0$, the system possesses no degrees of freedom; thus, neither pressure nor temperature can be altered. The equilibrium of three phases occurs at a single point in the phase diagram, known as the triple point. The phase diagram of H_2O elucidates the equilibrium states of water under varying temperature and pressure circumstances, delineating stability zones for solid, liquid, and gas phases, along with phase borders and the distinctive triple point.

Consider the phase diagram of iron (Fe), which is given in Fig. 14.3. The graphic depicts several phases as a function of temperature and pressure.

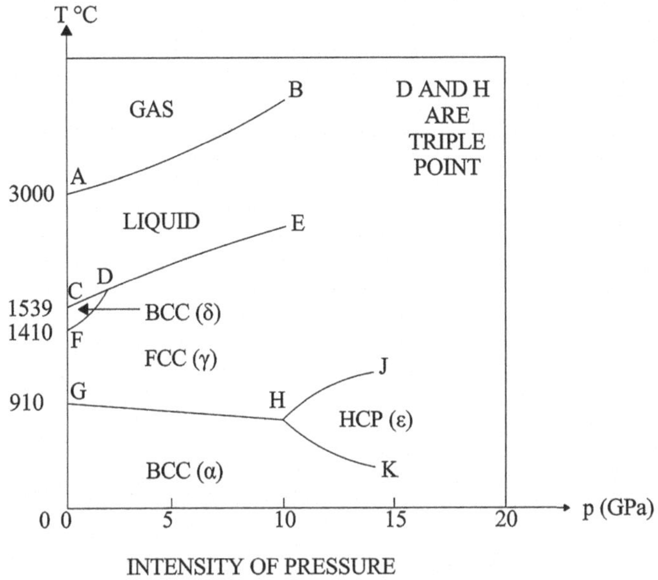

Fig. 14.3 Unary phase diagram of a single-component iron system illustrating vapor, liquid, and solid phases as a function of temperature and pressure [1]

Iron exists in liquid, solid, and gaseous (vapor) states. In a two-phase equilibrium, the borders AB, CDE, FDE, GHI, and GHK represent phase boundaries. In this instance, F = 1 indicates that either pressure or temperature can be adjusted. Modifications in temperature and pressure are essential to maintain two-phase equilibrium at these limits.

Three phase boundaries intersect at points D and H. We designate these points as triple points. F = 0, and three-phase equilibria are present. Given that the degree of freedom is 0, the system is restricted, and neither temperature nor pressure can be altered. Elevated temperatures lead to the production of several crystalline structures of iron, including body-centered cubic (BCC) (α), face-centered cubic (FCC) (γ), and body-centered cubic (BCC) (δ). A pressure of approximately 15 GPa induces the transformation of the BCC (α) structure into the HCP (ε) structure.

14.2.2 Phase Changes

The addition or removal of energy, usually as heat, can induce phase changes, which are the transitions among solid, liquid, and gaseous forms of matter. Sublimation (solid to gas), vaporization (liquid to gas), condensation (gas to liquid), melting (solid to liquid), freezing (liquid to solid), and deposition (gas to solid) exemplify frequent phase transitions. Processes such as melting, vaporization, and sublimation

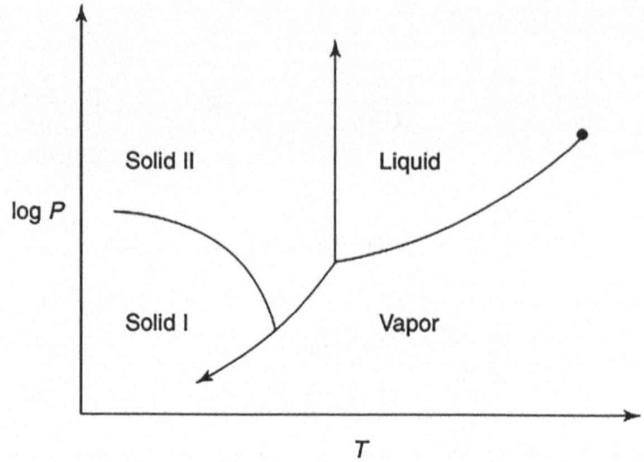

Fig. 14.4 A schematic equilibrium P–T phase diagram for a pure substance exhibiting two stable solid phases [1]

are endothermic, necessitating energy absorption, whereas freezing, condensation, and deposition are exothermic, resulting in energy release. During a phase change, the temperature of a substance remains constant even as energy continues to be transferred; this energy is used to alter the internal structure of the material rather than raising its temperature. Phase changes are fundamental to everyday phenomena like boiling water, freezing ice, and forming frost, and are also critical in industrial processes such as casting, distillation, and freeze-drying. Figure 14.4 illustrates the equilibrium pressure-temperature phase diagram for a pure substance exhibiting two stable solid phases.

14.2.3 Polymorphism

Polymorphism in pure materials refers to the formation of different stable phases under varying conditions such as pressure, temperature, magnetic field, electric field, and others. Even when considering temperature, complex materials can exhibit intermediate phases termed mesophases that display partial disorder. Generally, when the temperature rises, materials exhibit increased disorder [5].

Two major types of mesophases can occur in molecular materials with particular molecular shapes. At intermediate temperatures, the material can form a liquid crystalline phase if its molecules are rod- or disc- shaped, allowing them to easily glide past one another. Due to the molecules' ability to pass one another in liquid crystalline mesophases, the substance can conform to the shape of its container and exhibit liquid like behavior. However, the material also displays crystal-like characteristics like light diffraction because the molecules stay somewhat aligned. Therefore, liquid crystal is a very appropriate term. These substances usually form an isotropic liquid at higher temperatures and a normal ordered solid at lower temperatures [5].

An orientationally disordered mesophase may arise if the molecules are nearly spherical and the barrier to reorientation is sufficiently low. C_{60} molecules, for instance, form a face-centered cubic lattice but rotate quite freely, resulting in orientational disorder. The molecules of C_{60} become more ordered as the temperature drops [5].

Both orientationally disordered solids and liquid crystals are examples of polymorphism, i.e., the existence of various solid phases for a specific composition. Polymorphism is prevalent in molecular materials, where the Gibbs energy landscape features multiple local minima that can be accessible as external parameters such as temperature and pressure vary.

14.3 Binary Phase Diagrams

14.3.1 Introduction to Binary Phase Diagrams

Binary phase diagrams are graphical representations that illustrate the relationship between temperature, composition, and phase equilibrium in an alloy. These diagrams are essential for understanding microstructural evolution, as various microstructures develop due to phase transformations that occur with temperature changes, usually during cooling. Such transformations may involve the formation, dissolution, or phase transitions. By analysing binary phase diagrams, one can predict phase transformations and the resulting microstructure, and determine whether they are in equilibrium or not [5].

Binary phase diagrams are categorized according to component solubility, phase transitions, and equilibrium processes. These diagrams provide critical insights into phase stability, solidification behavior, and microstructural evolution.

14.3.1.1 Isomorphous Phase Diagram

An isomorphous system is characterized by the total mutual solubility of its two components in both liquid and solid states, resulting in a continuous single-phase solid solution throughout the full composition range. This indicates that, irrespective of composition, just a single solid phase exists. The boundaries of phase transformation are defined by the liquidus and solidus lines, which indicate the temperature limits of melting and solidification. Unlike other binary systems, isomorphous systems do not exhibit any other invariant reactions, simplifying their phase behavior and making them ideal for applications requiring homogeneous single-phase microstructures [5].

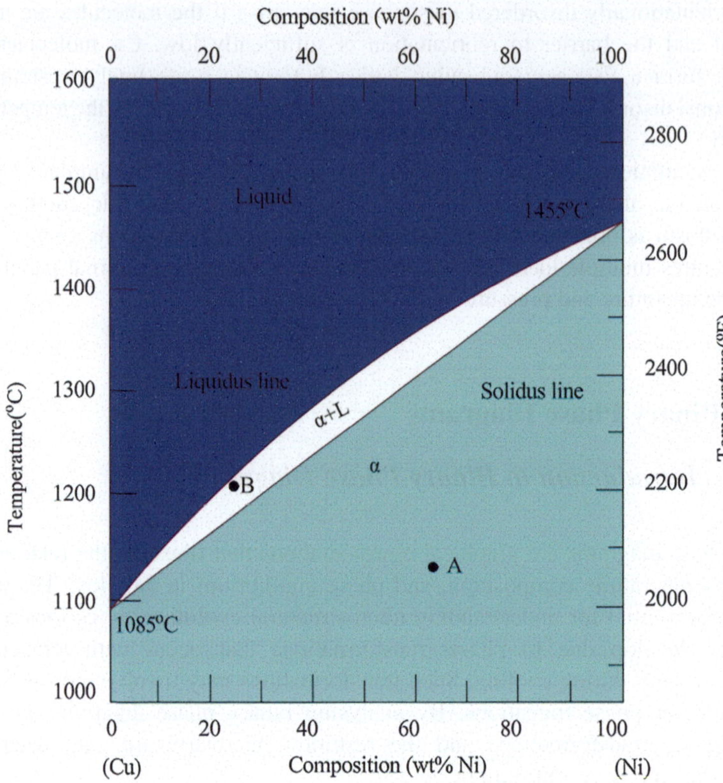

Fig. 14.5 Isomorphous system [5]

Example
The **copper-nickel (Cu-Ni) system** is a classic example of an isomorphous binary phase diagram (Fig. 14.5). Due to their similar atomic sizes and crystal structures (both FCC), Cu and Ni exhibit complete solid solubility.

14.3.2 The Lever Rule

Phase diagrams are essential instruments for comprehending the equilibrium states of alloys, specifically in ascertaining the relative quantities of phases at any given composition and temperature. The method for determining phase fractions is contingent upon whether the system is in a single-phase or two-phase region.

14.3.2.1 Phase Fractions in Single-Phase Regions

If the composition and temperature of an alloy correspond to a point lying within a single-phase region on a phase diagram, then the interpretation is relatively simple. Since only one phase is stable in that region, the entire alloy consists of that phase alone. In other words, the fraction of that phase is equal to 1.0 (or 100%).

For instance, examine a 35 wt.% Ni – 65 wt.% Cu alloy at 1100 °C (Fig. 14.6). This composition and temperature reside wholly within the α phase area, indicating that the alloy comprises solely the α phase. There is no presence of a second phase, and therefore, the alloy is 100% α phase.

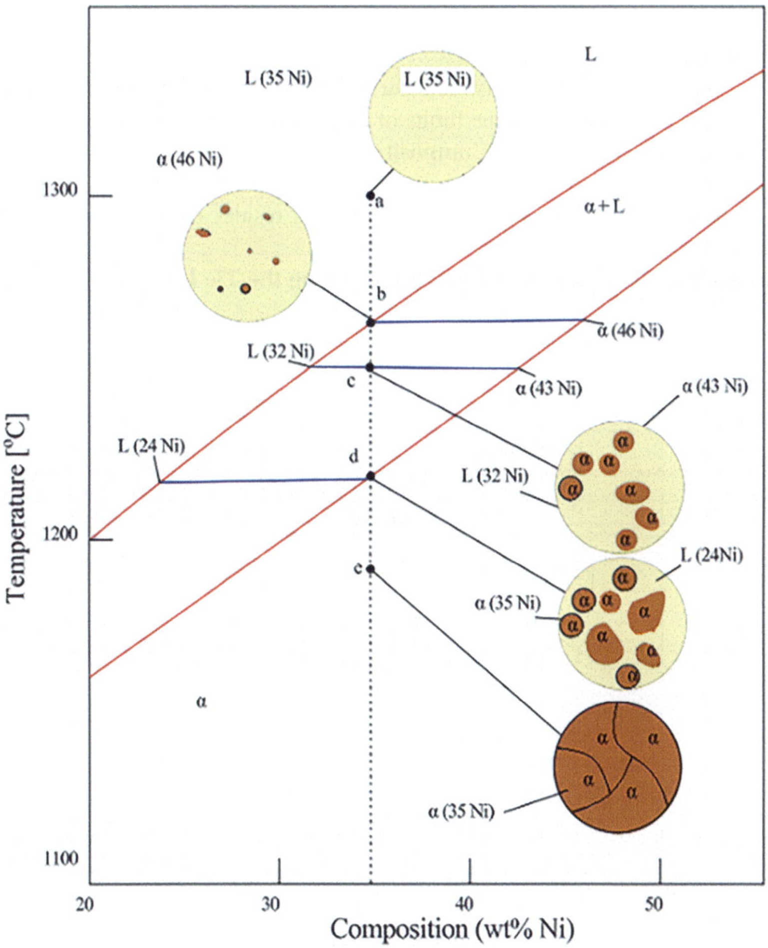

Fig. 14.6 Ni-Cu isomorphous phase diagram [5]

14.3.2.2 Phase Fractions in Two-Phase Regions

When an alloy's composition and temperature reside inside a two-phase region of a binary phase diagram, ascertaining phase fractions necessitates a more meticulous methodology. Unlike single-phase regions, where only one phase is present, two-phase regions contain a mixture of distinct phases coexisting in equilibrium. In such cases, the lever rule, often referred to as the inverse lever rule, must be applied in conjunction with the tie line method [6].

The tie line is an isothermal line drawn within the two-phase region at the specified temperature. It delineates the phase boundaries of the two coexisting phases and supplies the requisite information to compute the fraction of each phase at equilibrium.

The process for determining phase fractions using the lever rule is as follows (see Fig. 14.7):

1. **Constructing the Tie Line**
   A horizontal tie line is constructed at the specified temperature, traversing the two-phase zone to intersect the limits of the two coexisting phases.
2. **Locating the Overall Alloy Composition**
   The alloy's total composition is indicated along the tie line. This composition denotes the overall bulk alloy composition and resides between the two-phase boundaries.
3. **Calculating the Fraction of Each Phase Using the Tie Line**

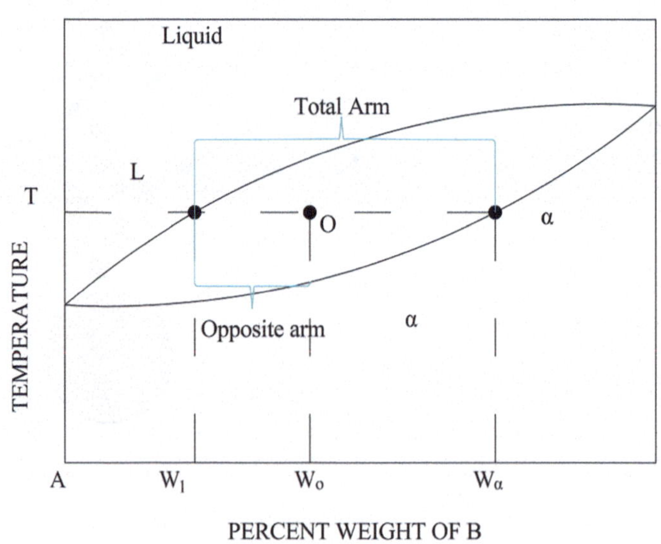

Fig. 14.7 Lever rule is applied on the phase diagram

The proportion of each phase is ascertained by measuring the segment of the tie line that extends from the alloy composition to the boundary of the opposing phase.

$$f_{\alpha} = \frac{opposite\ arm}{total\ arm} = \frac{W_o - W_l}{W_{\alpha} - W_l} \tag{14.6}$$

where:

- W_0 = Overall composition of the alloy
- W_{α} = Composition of the α phase at equilibrium
- W_l = Composition of the liquid phase at equilibrium

The proportion of one phase is calculated by measuring the length of the tie line from the overall composition to the phase boundary of the opposing phase, and then dividing this length by the total tie line length.

4. **Calculating the Mass of Each Phase**

   Since phase diagrams typically use weight percent as the composition scale, the fractions obtained from the lever rule represent mass fractions. The fraction of each phase is proportional to its mass in relation to the overall mass of the alloy.

   If phase percentages are required, each fraction is simply multiplied by 100.

14.3.2.3 Example of Applying the Lever Rule in a Binary Phase Diagram

Examine a Cu-Ni alloy comprising 35 wt.% Ni and 65 wt.% Cu at a temperature of 1250 °C (Refer to Fig. 14.6). Utilizing a phase diagram for the Cu-Ni system, we ascertain the following:

- The liquid phase (L) at 1250 °C comprises 32 wt.% Ni.
- The solid phase (α) at 1250 °C comprises 43 wt.% Ni.

Step 1: Construct the Tie Line
Draw a horizontal tie line at 1250 °C that extends across the two-phase (L + α) region and intersects the phase boundaries at 32 wt.% Ni (L) and 43 wt.% Ni (α).

Step 2: Locate the Overall Composition
The alloy composition is 35 wt.% Ni, situated between the liquid and solid phase compositions.

Step 3: Apply the Lever Rule
Using the formula:

$$\text{Fraction of } \alpha = \frac{C_L - C_0}{C_L - C_{\alpha}}$$

$$\text{Fraction of } L = \frac{C_0 - C_\alpha}{C_L - C_\alpha}$$

where:

- $C_0 = 35\%$ (Overall composition of the alloy)
- $C_L = 32\%$ (Composition of the liquid phase)
- $C_\alpha = 43\%$ (Composition of the solid phase)

Now, calculating the phase fractions, we get, at 1250 °C, that the alloy consists of 27.27% solid (α-phase) and 72.72% liquid (L-phase). This means the alloy is in a partially solidified state, with liquid and solid coexisting in equilibrium.

Upon further cooling of the alloy, an increased formation of the solid α-phase will occur, resulting in a reduction of the liquid phase fraction, ultimately culminating in total solidification.

14.3.2.4 Practical Considerations in Applying the Lever Rule

To accurately apply the lever rule in real-world applications, the lengths of tie-line segments must be measured correctly. There are two primary ways to determine these segment lengths:

1. **Direct Measurement**
   - A straightforward approach is to use a linear scale, preferably one that is finely graduated (e.g., in millimetres), to measure the physical distance between points on the phase diagram. These measurements can subsequently be employed to directly calculate phase fractions.
2. **Compositional Subtraction**
   - Instead of measuring physical distances, one can read the composition values directly from the phase diagram and use numerical subtraction to determine tie line segment lengths. This method is particularly useful for high-precision calculations and computational modelling.

14.3.3 Partial Molar Properties and the Common Tangent Method

Partial molar properties describe how a certain thermodynamic property varies with the addition of a small quantity of a component, while keeping temperature, pressure, and composition constant. These properties are crucial for understanding phase equilibria and chemical potential variations in binary phase diagrams.

The partial molar amount $\bar{M}_i$ for component i in a solution is defined for an extended thermodynamic quantity M (such as Gibbs free energy G, enthalpy H, or volume V) as:

$$\overline{M}_i = \left(\frac{\partial M}{\partial n_i}\right)_{T,P,n_j} \tag{14.7}$$

where n_i represents the number of moles of component I, and the derivative is computed while maintaining the number of moles of other components, temperature, and pressure constant.

14.3.3.1 The Chemical Potential and Gibbs Free Energy

The Gibbs free energy (G) of a mixture is an essential thermodynamic function in phase diagrams. The increase in Gibbs free energy upon the addition of one mole of a material to a system is referred to as the chemical potential of that system.

Therefore, the chemical potential (μ_i) represents the partial molar Gibbs free energy of the system and is expressed as:

$$\mu_i = \left(\frac{\partial G}{\partial n_i}\right)_{T,P,n_j} \tag{14.8}$$

In a binary solution (A-B), the total Gibbs free energy is expressed as:

$$G_{mix} = n_A\mu_A + n_B\mu_B \tag{14.9}$$

μ_A and μ_B represent the chemical potentials of components A and B, respectively.

Phase stability is greatly affected by the curvature of the Gibbs free energy vs. composition curve (Fig. 14.8).

14.3.3.2 Common Tangent Construction

The common tangent rule is used to determine phase equilibrium in binary phase diagrams. The underlying premise is that each component's chemical potential must be equal in both phases at equilibrium.

Graphical Interpretation

Examine the Gibbs free energy G as a function of composition, X_B (mole fraction of component B). A single-phase system exhibits a continuous Gibbs free energy

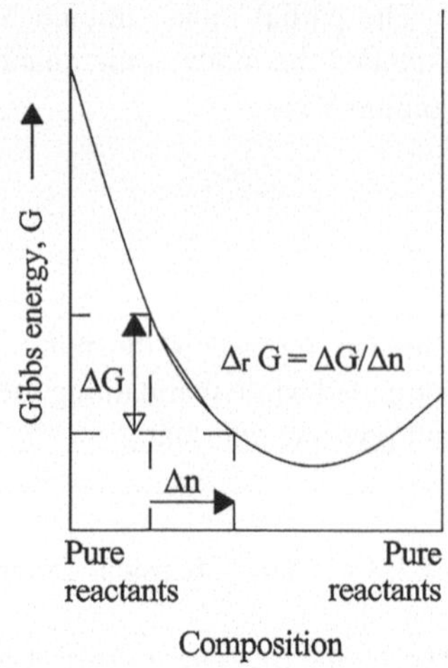

Fig. 14.8 Composition curve against Gibbs free energy [2]

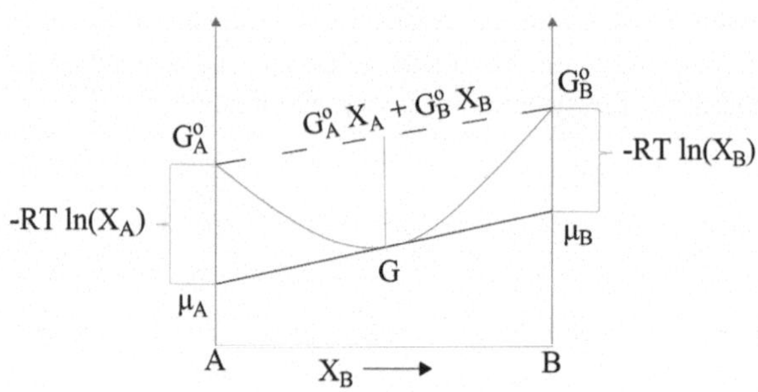

Fig. 14.9 Gibbs free energy of mixing (G_{mix}) for a binary solution [2]

profile. Two points on the G Vs. X curve can be joined by a common tangent if the system promotes phase separation.

The diagram (Fig. 14.9) illustrates the Gibbs free energy of mixing (G_{mix}) for a binary solution as a function of composition (X_B), highlighting both ideal and real mixing behaviours. The dotted line represents the ideal Gibbs energy ($G°_A X_A + G°_B X_B$), while the green curve depicts the actual G_{mix}, accounting for enthalpy and entropy contributions. The downward curvature indicates stability and miscibility, with chemical potentials (μ_A, μ_B) derived from the tangent slopes. If the curve becomes concave, phase separation occurs, requiring the common tangent rule to determine equilibrium compositions.

The common tangent construction illustrated in Fig. 14.10a serves to establish phase equilibrium between the α (solid) and L (liquid) phases at a specified temperature (T = 1000). The common tangent touches the Gibbs free energy curve at two distinct compositions (X_2 values marked in red and green), specifying the equilibrium compositions of the coexisting phases. At equilibrium, the chemical potentials of both components in the α and L phases are equivalent. The area delineated by

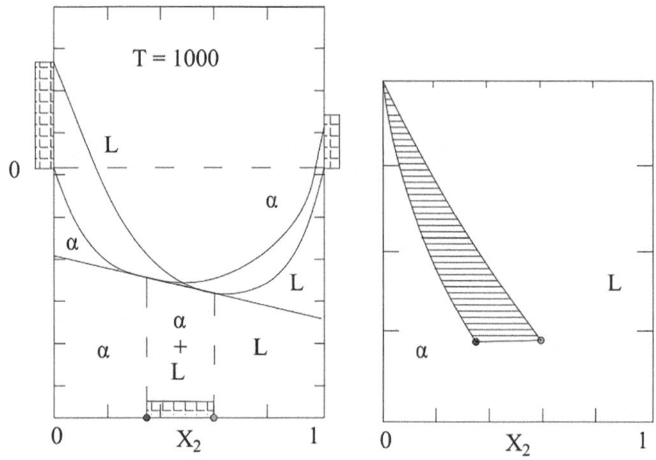

Fig. 14.10 Gibbs free energy curve at two distinct compositions [7]

these two points signifies the two-phase (α + L) region, characterized by phase separation instead of the formation of a homogeneous mixture.

The right diagram (Fig. 14.10) shows the corresponding phase diagram, with the shaded region representing the miscibility gap, reinforcing the role of the common tangent in predicting phase stability and equilibrium compositions.

Mathematical Condition

For two coexisting phases, alpha(α) and beta (β), the tangent condition ensures:

$$\mu_A^\alpha = \mu_A^\beta \text{ and } \mu_B^\alpha = \mu_B^\beta \tag{14.10}$$

which ensures thermodynamic equilibrium between phases.

14.3.3.3 Applications of Common Tangent Construction

- Used to determine phase separation in non-ideal solutions.
- Helps identify eutectic, Peritectic, and monotectic phase boundaries in phase diagrams.
- Important in understanding miscibility gaps in alloys and ceramic systems.
- Utilized in electrochemical energy storage systems to analyse phase stability.

14.3.4 Heterogeneous Binary Systems and Ideal Solutions

Binary systems, consisting of two components, exhibit different thermodynamic behavior depending on their miscibility and interactions. They can exist as either homogeneous (single-phase) or heterogeneous (multi-phase) systems. Here, we understand heterogeneous binary systems, which are characterized by phase separation, and perfect solutions, which are characterized by the absence of phase separation and compliance with Raoult's law.

An ideal solution is a homogeneous mixture of two components that obeys to Raoult's Law and assumes that intermolecular interactions between dissimilar molecules are equivalent to those between similar molecules. The behavior of perfect solutions is dictated by thermodynamic laws, namely the Gibbs free energy of mixing and the entropy of mixing.

A binary solution of components A and B is considered ideal if:

1. $\Delta H_{mix} = 0 \rightarrow$ No enthalpy changes on mixing.
2. $\Delta V_{mix} = 0 \rightarrow$ No volume changes on mixing.
3. Adheres to Raoult's Law: The partial vapor pressures of components are directly proportional to their mole fractions.

$$P_A = P_A^o X_A \tag{14.11}$$

$$P_B = P_B^o X_B \tag{14.12}$$

where P_A^0, P_B^0 are the vapour pressures of the pure components, and X_A and X_B are their respective mole fractions.

Figure 14.11 depicts Raoult's Law for an ideal solution, demonstrating the linear correlation between vapor pressure and mole fraction. The x-axis denotes composition (x_A or x_B), and the y-axis signifies vapor pressure. The dashed lines represent the partial pressures of components A and B. Following $P_A = x_A P_A{}^0$ and $P_B = x_B P_B{}^0$, with the total vapor pressure given by $P_T = P_A + P_B = x_A P_A{}^0 + x_B P_B{}^0$. The linearity of the curves validates ideal behavior, indicating that intermolecular forces

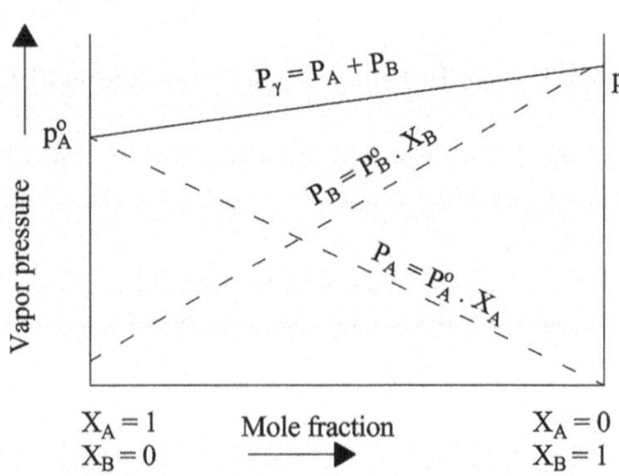

Fig. 14.11 Vapour pressure of liquid solutions [7]

in the solution resemble those in pure components, ensuring adherence to Raoult's Law without deviation.

The Gibbs free energy of a solution is expressed as:

$$G = \Delta H - T\Delta S \tag{14.13}$$

Given that ΔHmix = 0 for an ideal solution, the alteration in Gibbs free energy is exclusively dictated by entropy.

$$G_{mix} = -T\Delta S_{mix} \tag{14.14}$$

Entropy is a quantitative measure of randomness or disorder in a system. The mixing of two pure substances results in an increase in disorder. The entropy of an ideal solution can be derived through a statistical approach utilizing Boltzmann's equation:

$$S = k\ln\Omega \tag{14.15}$$

where Ω is the number of possible microscopic arrangements.

The change in configurational entropy upon mixing for a binary solution is expressed as follows:

$$\Delta S_{mix} = -R(X_A ln\ X_A + X_B\ ln X_B) \tag{14.16}$$

where R is the universal gas constant.

Substituting the value of ΔS_{mix} in Eq.(14.14)

$$\Delta G_{mix} = RT(X_A ln X_A + X_B ln X_B) \tag{14.17}$$

The equation indicates that the Gibbs free energy of mixing is consistently negative, signifying that mixing is inherently spontaneous for ideal solutions.

14.3.4.1 Significance of the Derivation

Spontaneous Mixing: Since $\Delta G_{mix} < 0$, ideal solutions mix completely at all compositions.

Minimum Free Energy: The lowest Gibbs free energy corresponds to complete miscibility.

Raoult's Law Validity: This derivation is consistent with ideal solution behavior, confirming that vapor pressure follows Raoult's law.

14.3.5 *The Gibbs Phase Rule, Eutectic and Peritectic Reactions in Heterogeneous Binary Systems*

For a **binary system (C = 2)**, this equation simplifies to:

$$F = 4 - P$$

- When **P = 1** (single-phase region), **F = 3**, meaning three variables (temperature, pressure, and composition) can be independently varied.
- When **P = 2** (two coexisting phases), **F = 2**, meaning only two variables can be changed independently.
- When **P = 3** (invariant reaction, such as eutectic or Peritectic), **F = 1**, meaning the system has a fixed temperature and composition at equilibrium.

14.3.5.1 Eutectic Reactions

A eutectic reaction takes place in a binary system when a liquid phase converts into two solid phases at a defined temperature and composition.

The reaction is represented as:

$$L \rightarrow \alpha + \beta \tag{14.18}$$

where:

L is the liquid phase,

α and β are two solid phases.

The eutectic point represents the minimum temperature at which the liquid phase can coexist, characterized by a distinct melting point akin to that of a pure substance. The phase diagram features a horizontal tie line, signifying an isothermal transformation in which the liquid phase undergoes an instantaneous transition into two distinct solid phases. In a eutectic phase diagram, the liquidus curve delineates the temperature threshold above which solely the liquid phase is present, whereas the solidus curve indicates the boundary below which only solid phases exist. At the eutectic composition, three phases are present in equilibrium: liquid (L) and two solid phases (α and β).

A well-known example of a eutectic system is the Lead-Tin (Pb-Sn) alloy (Fig. 14.12), which exhibits a eutectic reaction at 183 °C and 61.9% Sn. This system is widely utilized in soldering applications due to its low melting point and excellent wettability.

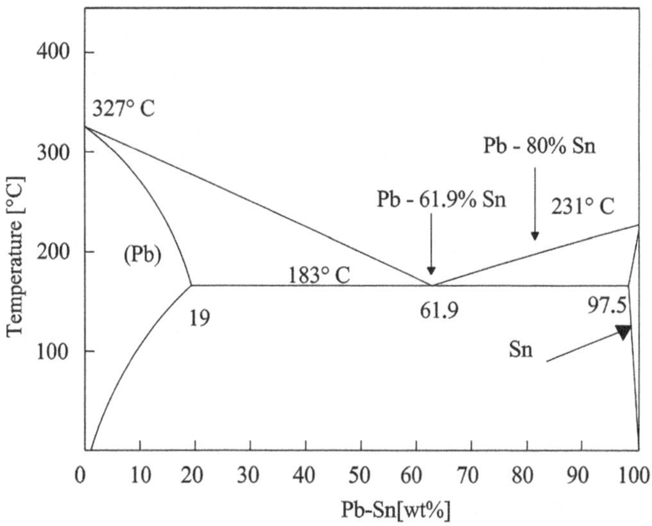

Fig. 14.12 Eutectic phase diagram [5]

14.3.5.2 Peritectic Reactions

A Peritectic reaction is defined as the process in which a solid phase and a liquid phase interact to form a new solid phase during the cooling process. The reaction is expressed as follows:

$$L + \alpha \rightarrow \beta \tag{14.19}$$

A Peritectic phase diagram is categorized by a distinct Peritectic reaction occurring at a specific temperature, where a liquid phase and a solid phase react together to form a new solid phase. This transformation is often sluggish due to the necessity of diffusion through an existing solid phase, which can lead to incomplete transformation and the formation of non-equilibrium structures. As a result, Peritectic reactions are generally slower compared to eutectic reactions.

An established example is the Iron-Nickel (Fe-Ni) system (Fig. 14.13), in which δ-ferrite and liquid phase transform γ-austenite at the Peritectic temperature. The transformation is essential in steel processing and alloy development, affecting the final microstructure and mechanical properties.

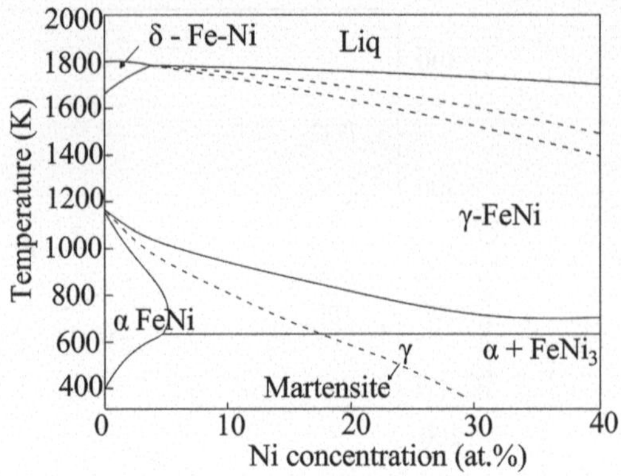

Fig. 14.13 Peritectic reaction [8]

14.3.6 *Binary Phase Diagrams: Temperature–Chemical Potential Diagrams*

A μ-T diagram illustrates how the chemical potential of each phase varies with temperature at a specified composition. The diagrams present a thermodynamic foundation for the phase transformations identified in binary phase diagrams.

14.3.6.1 Single-Phase Regions

In homogeneous single-phase regions, whether liquid, solid, or gas, the chemical potential of each component decreases as temperature rises, due to changes in entropy.

- In the liquid phase, chemical potential decreases very quickly with temperature than in solids due to the greater entropy of liquids.
- In the solid phase, the chemical potential exhibits a gradual decrease attributed to the ordered crystalline structure.
- At a fixed composition, the most stable phase at a given temperature is the one with the lowest chemical potential.

14.3.6.2 Two-Phase Equilibrium

At equilibrium, the chemical potentials of two coexisting phases are equal. This is illustrated by a common tangent construction in the Gibbs free energy versus composition diagram and is depicted in μ-T diagrams as two curves intersecting at the equilibrium temperature.

For example, in a liquid-solid equilibrium (L ↔ S): $\mu_L = \mu_S$.

where μ_L and μ_S represent the chemical potentials associated with the liquid and solid phases, respectively. This equality guarantees that there is no net transfer of atoms between phases, thereby preserving equilibrium.

Several types of phase transformations can be analysed using μ-T diagrams.

(a) **Eutectic Reaction**

At the eutectic temperature, the chemical potential of the liquid equal that of both solid phases (α and β), resulting in a horizontal tie-line in the phase diagram.

$$\mu_L = \mu_\alpha = \mu_\beta$$

(b) **Peritectic Reaction**

At the Peritectic temperature, the chemical potentials of the three phases (L, α, β) are in equilibrium.

$$\mu_L = \mu_\alpha = \mu_\beta$$

Owing to the reduced rate of diffusion within the solid phase, Peritectic transformations are often incomplete, leading to non-equilibrium microstructures.

(c) **Solid-Solid Transformations**

In some systems, solid phases transform into one another at specific temperatures (allotropic or polymorphic transformations). The discontinuity in the μ- T curve at transformation temperatures marks the phase boundary.

For example, in the **Fe-C system**, α-iron (BCC) transforms into γ-iron (FCC) at 912 °C:

$$\alpha \rightarrow \gamma$$

At this temperature:

$$\mu_\alpha = \mu_\gamma$$

14.3.6.3 Construction of Common Tangents and Analysis of Phase Equilibrium

The common tangent rule is essential for analysing phase equilibrium through μ-T diagrams. At equilibrium, the chemical potential of each component is required to be identical in both phases.

In a binary system consisting of phases A and B, the equilibrium compositions can be ascertained by constructing a common tangent to the Gibbs free energy curve. For two coexisting phases, the mathematical condition requires that the slopes of the tangent lines be equal.

$$\frac{dG}{dX_A} = \mu_A^\alpha = \mu_A^\beta \tag{14.20}$$

$$\frac{dG}{dX_B} = \mu_B^\alpha = \mu_B^\beta \tag{14.21}$$

In this context, X_A and X_B represent the mole fractions of components A and B, respectively, while α and β denote the coexisting phases.

Applications of μ-T Diagrams

- Predicting Phase Stability: Determines stable phases at given temperatures.
- Understanding Diffusion-Driven Transformations: Helps analyse kinetics in Peritectic and eutectic reactions.
- Material Processing and Alloy Design: Optimizing temperature-dependent phase stability for industrial applications.

14.3.7 Non-Ideal Solutions: Dilute and Regular Models

It is noted here that some of the ideas presented in this section can also be found in the earlier chapters. But for the convenience of the reader, they are collated here with a bias towards application.

In many practical cases, solutions deviate from ideality due to intermolecular interactions between different components. Non-ideal solutions demonstrate deviations from Raoult's Law, either positive or negative, as a result of variations in intermolecular forces. Understanding these deviations is critical in thermodynamics, phase equilibria, metallurgy, and chemical engineering.

Non-ideal solutions are modelled using different approaches based on the degree of deviation and the system's characteristics. Two significant models for such solutions are:

- Dilute solution models, which describe systems where one component is in very low concentration.
- Regular solution models, which incorporate enthalpic deviations but assume ideal entropy of mixing.

14.3.7.1 Non-Ideal Solutions and Their Characteristics

A solution is classified as non-ideal when it fails to adhere to Raoult's Law, indicating that the partial vapor pressure of each component diverges from its theoretical value.

$$P_i \neq X_i P_{ipure} \tag{14.22}$$

where:

- P_i represents the partial vapor pressure of component i
- X_i represents the mole fraction of component i.
- P_{ipure} denotes the vapor pressure of the pure component i.

Analysis of Positive and Negative Deviations from Raoult's Law

- **Positive Deviation:** The solution exhibits intermolecular forces that are less robust compared to the pure components, resulting in a vapor pressure that exceeds the predictions made by Raoult's Law. Example: Ethanol and Acetone.
- **Negative Deviation:** The solution has stronger interactions than the pure components, leading to lower vapor pressure. Example: Acetic Acid-Water.

Modifications in the expressions for Gibbs free energy, enthalpy, and chemical potential are required due to these deviations, which are addressed in non-ideal solution models (Fig. 14.14).

14.3.7.2 Dilute Solution Models

A dilute solution is characterized by one component being present in a very small concentration. These solutions follow Henry's Law rather than Raoult's Law and require a different thermodynamic treatment.

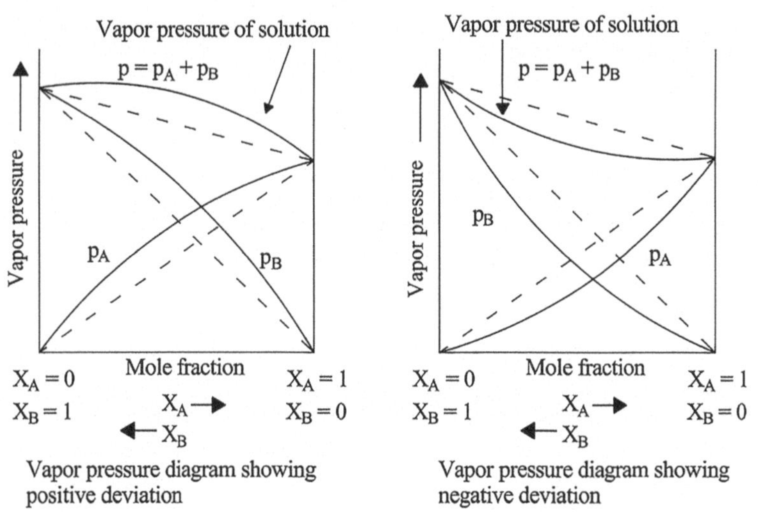

Fig. 14.14 Deviations from Raoult's Law: positive and negative aspects [9]

a) Henry's Law for Dilute Solutions

For a dilute solution, the solute obeys Henry's Law, which states:

$$P_{solute} = k_H X_{solute} \tag{14.23}$$

where k_H represents Henry's Law constant, which depends on temperature and solvent properties.

This law is particularly useful in describing gases dissolved in liquids and solute behavior in dilute metallic solutions.

b) Chemical Potential in Dilute Solutions

The expression for the chemical potential of a solute in a dilute solution is represented as follows:

$$\mu_i = \mu_i^0 + RTlnX_i \tag{14.24}$$

where:

μ_i^0 represents the standard chemical potential, Xi denotes the mole fraction of the solute, R signifies the universal gas constant, and T indicates the absolute temperature.

In the context of non-ideal dilute solutions, the activity coefficient (γ) is utilized to account for deviations from ideal behavior.

$$\mu_i = \mu_i^0 + RTlna_i \tag{14.25}$$

where activity a_i is defined as:

$$a_i = \gamma X_i$$

The activity coefficient, γ, measures the extent of deviation from ideal behavior.

- If $\gamma > 1$, the solute experiences repulsive interactions.
- If $\gamma < 1$, the solute experiences attractive interactions.

14.3.7.3 Regular Solution Model

The Regular Solution Model is applicable for systems where both components are present in substantial quantities and exhibit deviations from ideal behavior. This model includes the enthalpy of mixing while presuming an ideal entropy of mixing.

Assumptions of the Regular Solution Model

1. Random Mixing: Components are randomly distributed.
2. No Volume Change: The total volume remains unchanged upon mixing.
3. Ideal Entropy of Mixing: The entropy contribution follows the ideal solution model.
4. Nonzero Enthalpy of Mixing: Unlike molecules interact differently, contributing to excess enthalpy.

Gibbs Free Energy of Mixing (ΔG_{mix})

For a binary solution of components, A and B:

$$\Delta G_{mix} = RT(X_A ln\, X_A + X_B\, ln\, X_B + \Omega X_A X_B) \tag{14.26}$$

where:

- The first term represents ideal entropy of mixing.
- The third term ($\Omega X_A X_B$) accounts for the enthalpy contribution, where Ω is the interaction parameter:

Chemical Potential in a Regular Solution

The chemical potential of component A:

$$\mu_A = \mu A_0 + RTlnX_A + \Omega X_{B2} \tag{14.27}$$

Similarly, for component B:

$$\mu_B = \mu B_0 + RTlnX_B + \Omega X_{A2} \tag{14.28}$$

This shows that chemical potential depends on both composition and interaction energy.

Phase Separation and Critical Temperature

If $\Omega > 2RT$, the system undergoes phase separation at lower temperatures, forming a miscibility gap. The critical temperature (T_c) for miscibility is:

$$T_c = \Omega 2\mathrm{R} \tag{14.29}$$

At temperatures below T_c, the system phase-separates due to repulsive interactions.

14.3.7.4 Graphical Representation and Diagrams

(a) Excess Gibbs Free Energy (G_{excess} vs. Composition)

- Ideal solutions: $G_{excess}=0$
- Regular solutions: $G_{excess}=\Omega X_A X_B$
- Strongly non-ideal solutions: Large deviations create miscibility gaps.

(b) Phase Diagrams

- For $\Omega < 0$: The system favor mixing, forming intermetallic compounds.
- For $\Omega > 0$: The system shows immiscibility gaps, with phase separation at low temperatures (Fig. 14.15).

14.3.8 T–X_i Phase Diagrams for Solutions That Are Strictly Regular

The temperature-composition (T-X_i) phase diagrams provide insights into phase equilibria for binary solutions. In the case of strictly regular solutions, these diagrams are derived based on the excess Gibbs free energy, which accounts for non-ideal solution behavior. Unlike ideal solutions that follow Raoult's Law perfectly, strictly regular solutions exhibit deviations due to enthalpy interactions between different species.

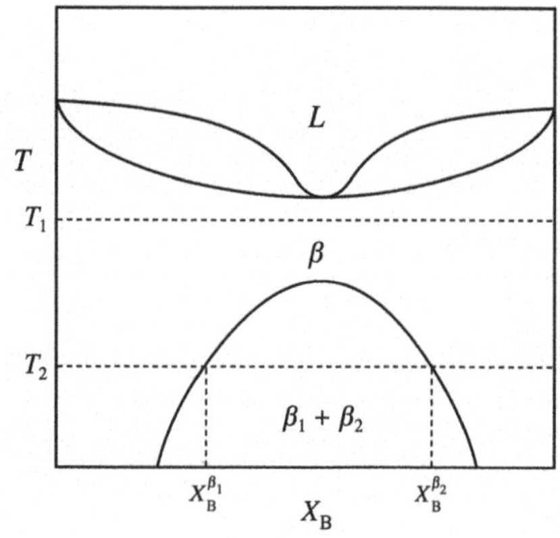

Fig. 14.15 A temperature-composition diagram for a theoretical binary system exhibiting a minimum in the solidus and liquidus curves, along with a miscibility gap in the solid state [4]

14.3.8.1 Regular Solution Model

A regular solution assumes:

- No volume changes upon mixing.
- No preference for like or unlike interactions, except for a symmetric enthalpy term.
- The activity coefficients are governed by the excess Gibbs free energy (**G**^{e}).
- It is possible to express the Gibbs free energy of mixing (ΔG_{mix}) for a binary system in the following manner:

$$G_{mix} = RT(X_1 lnf_0X_1 + X_2 lnf_0X_2) + \Omega X_1X_2 \tag{14.30}$$

where Ω is the regular solution parameter representing interaction energy.

If $\Omega > 0$, the solution exhibits immiscibility tendencies; if $\Omega < 0$, ordering tendencies dominate.

14.3.8.2 T-x_i Phase Diagram Construction

Through the process of minimizing the Gibbs free energy, the equilibrium between the liquid (L) and solid (S) phases may be calculated. It is necessary for the chemical potentials of the components in both phases to be precisely the same:

$$\mu_{iL} = \mu_{iS}$$

It is possible to calculate the chemical potential of a component in the liquid phase of a normal solution as follows:

$$\mu_{iL} = \mu_i^* + RTlnf_0X_i + \Omega(1 - X_i)^2 \tag{14.31}$$

where the standard state chemical potential is denoted by the symbol μi*

Using the condition for phase equilibrium and incorporating the entropy of fusion (ΔS_f), we obtain an expression for the liquidus curve:

$$T_L = Tf_l - R\Delta S_f\, lnf_0X_i - \Omega\Delta S_f(1 - X_i)^2 \tag{14.32}$$

In this equation, T_f represents the temperature at which the pure component melts.

Similar to the solidus curve, the solidus curve is derived using a related expression, which guarantees that the solubility limits of components in the solid phase are taken into consideration.

14.3.8.3 Miscibility Gaps and Spinodal Decomposition

For positive Ω, there may exist a miscibility gap, where the system phase-separates into two different compositions. The critical temperature for phase separation can be determined from the condition:

$$\partial^2 G/\partial X^2 = 0$$

which helps in identifying the Spinodal—the boundary within which decomposition occurs spontaneously.

Applications

- Predicting phase stability in metallic and ceramic systems.
- Designing alloy compositions for specific temperature-dependent properties.
- Understanding solubility limits in non-ideal binary systems.

14.4 Advanced Phase Diagram Representation

14.4.1 True Phase Diagrams

From the corresponding Gibbs equation one can determine the unique set of natural variables associated with each thermodynamic potential. For example, it is evident that S and V are the natural variables of U from dU = TdS − pdV. If a thermodynamic potential is stated as a function of the system's natural variables, then it is possible to determine each and every thermodynamic attribute that characterizes the system. Therefore, in the case of a multicomponent system, the knowledge that G = G (p, T, n_i) enables the determination of all other thermodynamic properties. If, on the other hand, G is stated as G (V, T, ni), then this information is insufficient to completely characterize the system [4].

Each extensive thermodynamic function has a corresponding conjugate variable that serves as an intensive thermodynamic potential. The conjugate pairs are derived from the comprehensive thermodynamic definition of work. The relationship is clearly demonstrated by the fundamental equation of thermodynamics.

$$dU = Tds - pdV + \sum_i \mu_i dn_i \quad (14.33)$$

T and S represent conjugate variables, similar to −p and V, as well as μ_i and n_i.

Plotting any pair of conjugate variables does not yield a valid phase diagram, defined as a diagram where the phase constitution at each condition is uniquely determined by each point. Instead, plots of conjugate variables result in property diagrams, which depict thermodynamic properties but do not uniquely define phase boundaries.

14.4.2 Alternative Representations

Numerous equations similar to Eq. (14.33) that are valid from thermodynamics can be derived. For illustration purposes, examine the constant-pressure Gibbs–Duhem equation applicable to a binary system:

$$0 = SdT + n_A d\mu_A + n_B d\mu_B \tag{14.34}$$

This equation can be rewritten in two different forms: first, by dividing by n_B, and second, by dividing by $n_A + n_B$.

$$-d\mu_B = \frac{S}{n_B} dT + \frac{n_A}{n_B} d\mu_A \tag{14.35}$$

$$-d\mu_B = S_m dT + x_A d(\mu_A - \mu_B) \tag{14.36}$$

At a constant total pressure, the following types of true phase diagrams can be constructed.

Field-field: T versus μ_A
Field-density: μ_A *versus* $\frac{S}{n_B}$ *and T versus* $\frac{n_A}{n_B}$
Density – density: $\frac{S}{n_B}$ versus $\frac{n_A}{n_B}$
Plots of $\frac{S}{n_B}$ versus T or of $\frac{n_A}{n_B}$ versus μ_A do not give true phase diagrams

Utilizing the molar quantities from eq. (14.36), it is possible to construct the following types of true phase diagrams under constant pressure conditions:

Field-field: T versus $\mu_A - \mu_B$
Field-density: $\mu_A - \mu_B$ *versus* S_m *and T versus* x_B
Density- density S_m *versus* x_B

Various sets of conjugate variables may be employed, leading to distinct pairs of variables that can either constitute a valid phase diagram. Due to the non-conjugate nature of field variables, a field–field diagram consistently represents an accurate phase diagram. Density–density diagrams exhibit similar properties provided that both functions utilize consistent mass or quantity units [4].

On the other hand, field-density diagrams need to be handled with greater care. Figure 14.16 illustrates this idea by presenting various variable selections for the eutectic system of (Pb- Sn) alloys based on Eqs. (14.35) and (14. 36). Three of these diagrams, as seen in the figure, are considered true phase diagrams while Fig. 14.16c

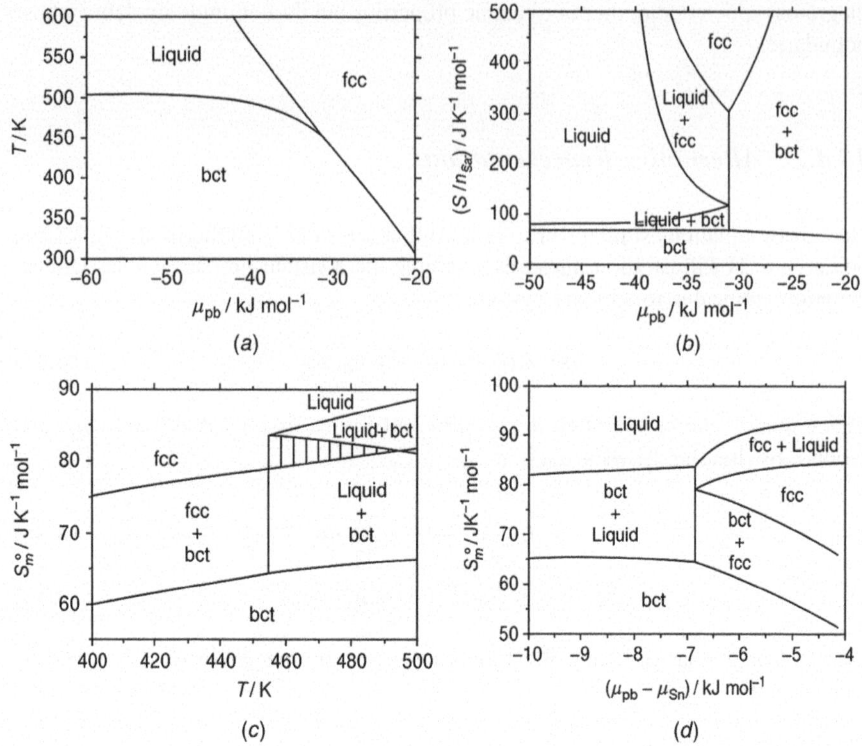

Fig. 14.16 Presents plots of various thermodynamic functions for the Pb–Sn system, indicating which plots represent true phase diagrams and which do not. (**a**) true field–field phase diagrams and (**b**) true density–field phase diagrams (**c**) Conjugate variable density-field diagrams and (**d**) True density-field phase diagram. States referenced from the Standard Element Reference (SER) have been implemented [4]

is not. In this field-density diagram, the shaded area indicates a region where the phase constitution at a given state is not uniquely determined by any one point [4].

Different categories of phase diagrams possess practical importance. H_m–x_i plots serve a significant purpose in conducting heat balance calculations. Nevertheless, the T–x_i diagram represents the most critical type. This section will provide a detailed examination of the key aspects of these diagrams.

14.4.3 Alternative Representations: T vs. X

When both solution phases are assumed to be strictly regular, the liquid and solid (α) phases yield results at constant pressure and temperature.

$$\mu_A^1 = \mu_A^\alpha \tag{14.37}$$

$$\mu_A^{o,\alpha} + \Delta\mu_A^{id,\alpha} + \mu_A^{E,\alpha} = \mu_A^{o,1} + \Delta\mu_A^{id,1} + \mu_A^{E,1} - \left(\mu_A^{o,1} - \mu_A^{o,\alpha}\right) \tag{14.38}$$

The final term in Eq. (14.38) can be represented as

$$\left(\mu_A^{o,1} - \mu_A^{o,\alpha}\right) = \Delta_{fus}H_m^o(A) - T\Delta_{fus}S_m^o(A) \tag{14.39}$$

$$\approx \left[T_{fus}^o(A) - T\right]\Delta_{fus}S_m^o(A) \tag{14.40}$$

The working equation for component A can be derived as follows:

$$RT\,log_e\left(1 - x_B^1\right) + L_0^1\left(x_B^1\right)^2 = RT\,log_e\left(1 - x_B^\alpha\right) + L_0^\alpha\left(x_B^\alpha\right)^2 - \left[\left(T_{fus}^o(A) - T\right)\Delta_{fus}S_m^o(A)\right. \tag{14.41}$$

$$RT\,log_e\left(\frac{x_A^1}{x_A^\alpha}\right) = \left[L_0^\alpha\left(x_B^\alpha\right)^2 - L_0^1\left(x_B^1\right)^2\right] - \left[\left(T_{fus}^o(A) - T\right)\Delta_{fus}S^o(A)\right. \tag{14.42}$$

By applying a similar formula for component B, it is possible to compute binary phase diagrams for regular solutions through the resolution of the two equations.

Based on Eq. (14.42) and the corresponding equation for B, it is evident that the positioning of phase boundaries is contingent upon:

I. $T_{fus}^o(A)\ and\ T_{fus}^o(B)$
II. $\Delta_{fus}S_m^o(A) and\ \Delta_{fus}S^o(B)$
III. $L_0^1\ and\ L_0^\alpha$

Changes in these factors can significantly influence the topology of phase diagrams for strictly regular solutions, as illustrated below. The entropy of fusion ($\Delta fusS_m^{\circ}$) in most metals is generally approximately 10 $J{\cdot}K^{-1}{\cdot}mol^{-1}$, whereas in semiconductors, it is usually considerably elevated. In order to analyze the impact of pure component properties, two sets of parameters have been established: one set features components with a minimal difference in fusion temperatures (T_{fus}°), while the other set includes components with a significant difference.

$T_{fus}^o(A) = 1000K$	$T_{fus}^o(A) = 500K$
$T_{fus}^o(B) = 1200K$	$T_{fus}^o(B) = 3000K$
$\Delta_{fus}S_m^o(A) = 10\,Jmol^{-1}K^{-1}$	$\Delta_{fus}S_m^o(A) = 10Jmol^{-1}K^{-1}$
$\Delta_{fus}S_m^o(B) = 10\ Jmol^{-1}K^{-1}$	$\Delta_{fus}S_m^o(B) = 10\,Jmol^{-1}K^{-1}$

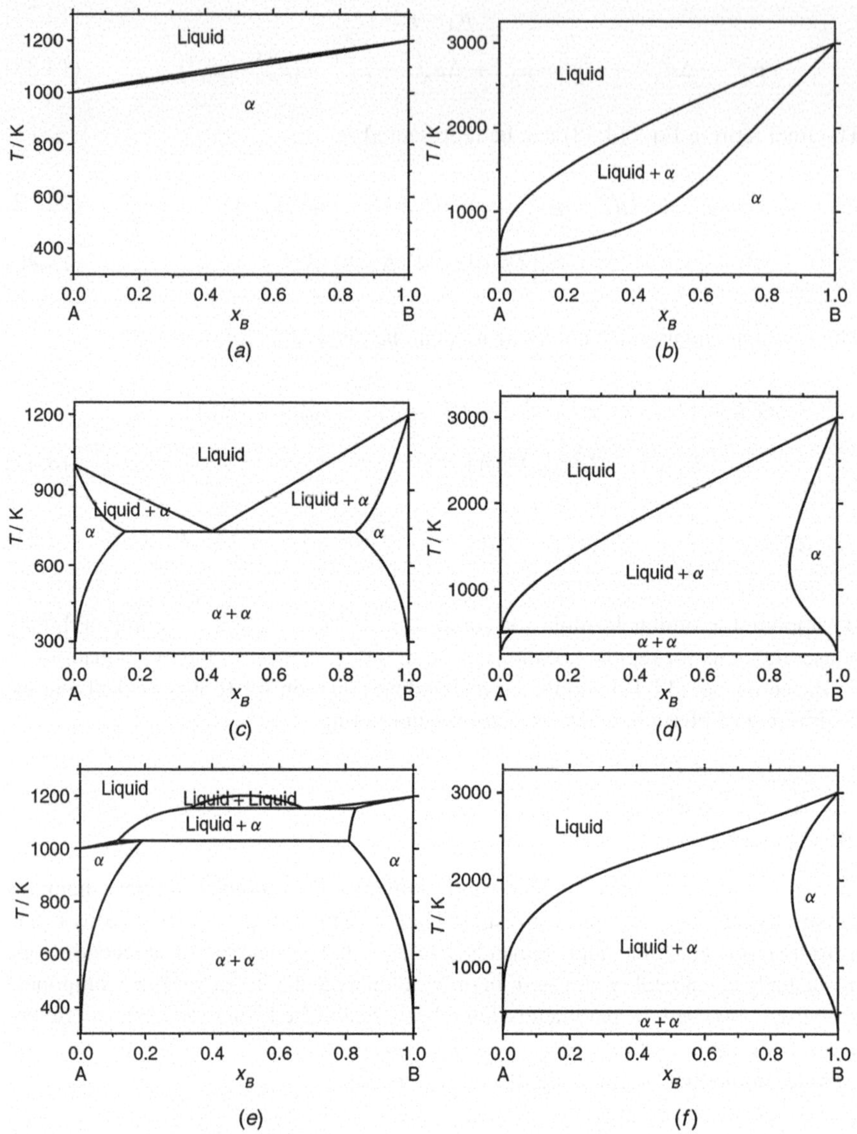

Fig. 14.17 Illustrates phase diagrams that encompass two phases, each characterized as strictly regular solutions. The melting points exhibit a difference of 200 K for the entries in the left column, while a difference of 2500 K is observed for the entries in the right column. (**a, b**) $L_0^1 = 0, L_0^\alpha = 0$; (**c, d**) $L_0^1 = 0; L_0^\alpha = +15\ KJmol^{-1}$; (**e, f**)$L_0^1 = +20\ kJmol^{-1}$ [45]

Figure 14.17 illustrates the phase diagrams on the left, produced with the initial set of parameters that exhibit a minimal variation in the fusion (melting) points of the components. The diagrams on the right are derived from an alternative set of parameters, characterized by a significant disparity in the fusion points of the components.

The impact of altering the solution parameters of the two phases is evident in the figure.

The phase diagrams for strictly regular solutions illustrate that a diverse array of phase behaviours can emerge by adjusting only a limited number of model parameters. Analysis of the diagrams presented in Figs. 14.17 and 14.18 reveals general trends and conditions that promote the formation of various types of simple phase diagrams.

14.4.3.1 Complete Solubility Phase Diagrams

Complete solubility is achieved when both the liquid and solid phases exhibit ideal behavior i.e., when $L_0^1 = L_0^\alpha = 0$. However, complete solubility can still be observed even in non-ideal systems, particularly when $L_0^1 \approx L_0^\alpha$. Examples of such alloy systems include Ag–Au, Ag–Pd, Cd–Mg, and Co–Ni.

When both L_0^1 and are negative, the phase diagram exhibits characteristics similar to those of an ideal solution. In contrast, when $L_0^1 \approx L_0^\alpha \gg 0$, miscibility gaps may develop at reduced temperatures. This behavior is observed in systems such as Au–Ni, where $L_0^\alpha \gg 0$ is significantly greater than zero.

Additional observations:

i. If $L_0^1 \approx L_0^\alpha \gg 0$, the melting region is wider than in the ideal case. If $L_0^1 \approx L_0^\alpha < 0$, the melting region becomes narrower.
ii. The melting points and entropies of fusion of the constituent elements are critical components that define the width of the two-phase region. For instance, germanium (Ge) and silicon (Si) display high fusion entropies, resulting in a wide freezing range for the Ge–Si system.

14.4.3.2 Peritectic Phase Diagrams

Peritectic phase diagrams are typically observed under specific conditions:

(a) The fusion temperatures exhibit a notable disparity $T_{fus}^0(A)$ and $T_{fus}^0(B)$.
(b) The interaction parameters satisfy ≫0, indicating strong positive deviations from ideality.
(c) The critical temperature Tc for the metastable miscibility gap in the solid phase intersects the metastable solidus curve.

Examples of real alloys demonstrating Peritectic behavior include the Ag–Pt and Co–Cu systems.

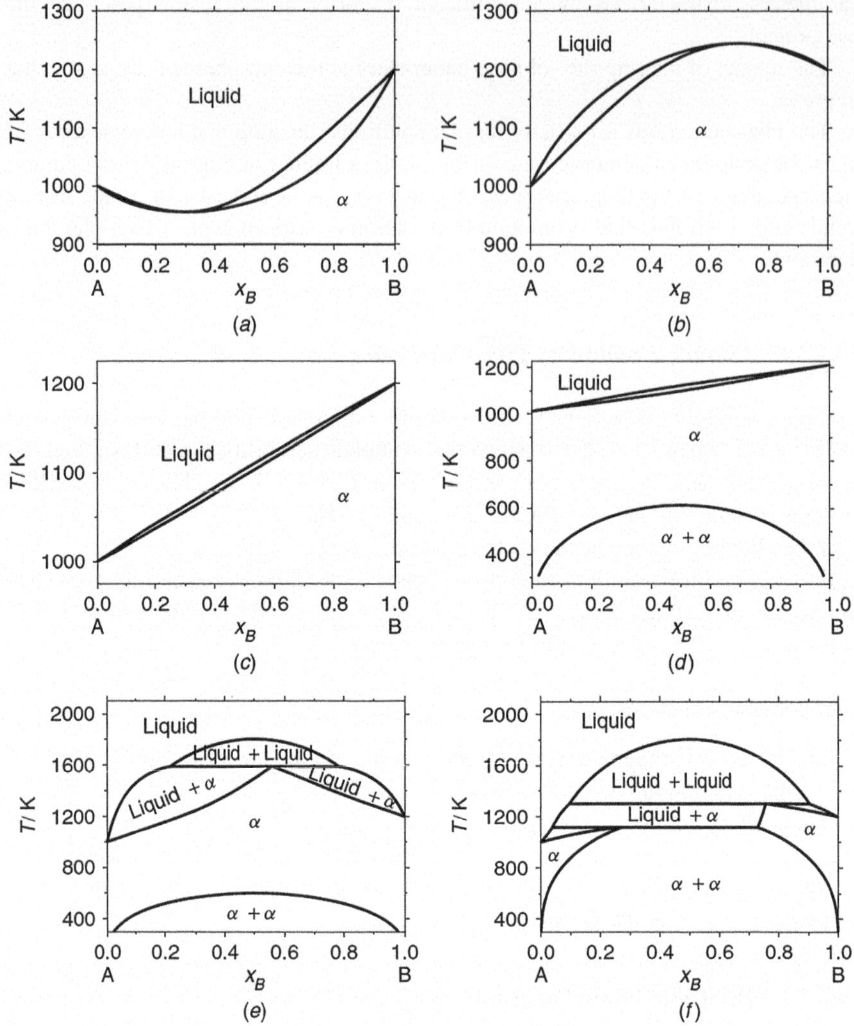

Fig. 14.18 Illustrates phase diagrams that encompass two phases, with each phase being represented as a strictly regular solution. All diagrams represent systems in which the two components display a 200 K variance in melting points. (**a**) $L_0^1 = -5\,kJmol^{-1}, L_0^{\alpha} = 0$; (b) $L_0^1 = 0$, $L_0^{\alpha} = -5\,kJmol^{-1}$; (c) $L_0^1 = -10\,kJmol^{-1}$; $L_0^{\alpha} = -10\,kJmol^{-1}$; (d) $L_0^1 = +10\,kJmol^{-1}$, $L_0^{\alpha} = +10\,kJmol^{-1}$; (e) $L_0^1 = +30\,kJmol^{-1}, L_0^{\alpha} = +10\,kJmol^{-1}$ (f) $L_0^1 = +30\,kJmol^{-1}, L_0^{\alpha} = +20\,kJmol^{-1}$ [4]

14.4.3.3 Phase Diagrams with Maximum and Minimum Congruent Melting Points

A maximum in the phase diagram is observed when the solid phase exhibits greater stability compared to the liquid phase, occurring under conditions where both interaction parameters L_0^1 and L_0^α are small. If L values are large and positive, miscibility gaps form instead. When $L_0^1 < L_0^\alpha$ is not too large, a congruent maximum appears in the liquid/solid region—see systems like Pb–Ti. Conversely, when $L_0^1 > L_0^\alpha$ is not too large, a congruent minimum forms, as observed in Co–Pd and Cr–Mo alloy systems.

14.4.3.4 Eutectic Phase Diagrams

When $L_0^\alpha$is greater than or equal to zero, and the melting point difference between components is moderate, a eutectic phase diagram is established. An illustrative instance of this behavior can be observed in the Cu–Ag alloy system.

14.4.3.5 Monotectic, Syntectic, and Monotectic + Syntectic Phase Diagrams

Miscibility gaps occur in the liquid phase when $L_0^1$is significantly greater than zero. A genuine alloy system that demonstrates this behavior is Pb–Zn.

14.4.3.6 More on Miscibility Gaps

By examining the compositional dependency of the Gibbs free energy (G) and its derivatives, the conditions at the critical temperature can be ascertained. In the Al–Zn system, the stable phase diagram shows a miscibility gap over a limited temperature range, with the metastable extension shown in Fig. 14.19a. The critical point is situated near 622.3 K.

The conditions at the critical temperature can be determined by analysing the composition dependency of the Gibbs free energy (G) and its derivatives. In the Al–Zn system, the metastable extension is shown in Fig. 14.19a, whereas the stable phase diagram exhibits a miscibility gap over a limited temperature range. The critical point is around 622.3 K.

At this point, the second and third derivatives of the Gibbs energy of mixing regarding x_{Zn} vanish:

$$\frac{d^2G}{dx_{Zn}^2} = \frac{d^3G}{dx_{Zn}^3} = 0$$

but the first derivative,

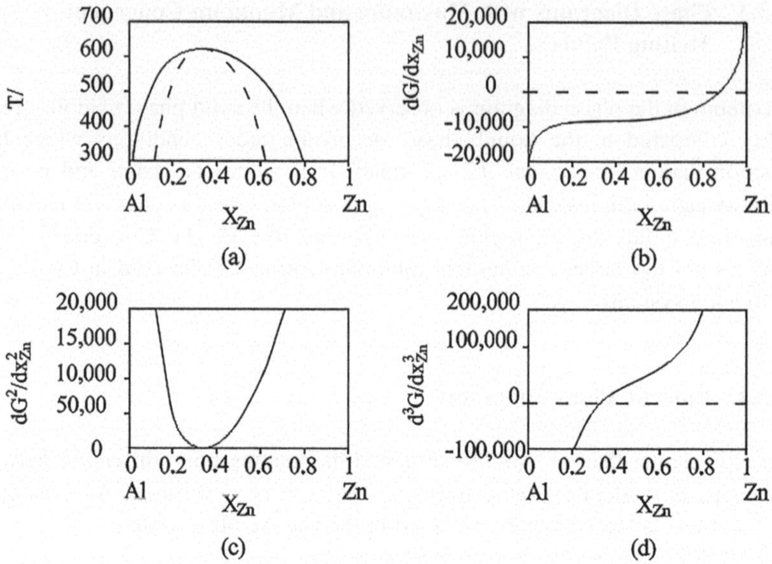

Fig. 14.19 Illustrates the first, second, and third derivatives of G concerning x_{zn} for fcc Al–Zn at a critical temperature of 622.29 K: (**a**) Miscibility gap in FCC phase; (**b**) First derivative of G; (**c**) Second derivative of G; (**d**) Third derivative of G. The critical composition is observed at $x_{zn} = 0.347$ [4]

$$\frac{dG}{dx_{Zn}} \neq 0$$

indicating that it is not a regular solution (where all three derivatives vanish at $x = 0.5$).

This behavior influences the application of the phase rule at the critical point of a miscibility gap. Innocently implementing the principle,

$$F = C – P + 1 = 2,$$

seems incorrect because both temperature and composition are fixed, implying $F = 0$.

However, the correct form includes the Gibbs–Duhem relation and additional constraints, resulting in:

- Number of independent chemical potentials R_{μ}= 3(due to vanishing derivatives),
- Degrees of freedom:

$$F = rC + 1 - R_{\mu} = 1 \times 2 + 1 - 3 = 0$$

which aligns with the observed behavior.

14.4.4 The Chemical Spinodal

The dashed curve in Fig. 14.20a illustrates the chemical spinodal, which is defined as the collection of compositions at which the second derivative of the Gibbs energy with respect to composition $^{d^2G}/_{dx_B^2}$ is 0. This curve marks the limit of metastability and indicates the onset of spontaneous phase decomposition without the need for nucleation. The bimodal curve delineates the phase boundary for stable two-phase equilibrium, established through the application of the common tangent construction.

Figure 14.20 illustrates the Spinodal limits represented by points S1 and S2, further emphasizing the importance of the Spinodal curve. Thermodynamic instability and two-phase breakdown arise when a single-phase alloy, with a composition situated between specific points, is subjected to quenching from a high temperature ($T > T_c$) to a lower temperature that falls within the miscibility gap. Two mechanisms for phase decomposition can be identified.

(a) **Spinodal Decomposition:**

For compositions located between the Spinodal lines, the G versus x_B curve exhibits a concave shape. This indicates that the homogeneous phase has a higher Gibbs energy compared to any combination of compositions, such as c and d, that could arise from its decomposition. The system experiences spontaneous decomposition due to ongoing composition fluctuations, a phenomenon referred to as Spinodal decomposition.

(b) **Nucleation and Growth:**

Between the binodal and spinodal, the G vs. x_B curve is convex, meaning local decomposition raises Gibbs energy, so fluctuations are disfavoured. Consequently, decomposition proceeds only via nucleation—overcoming an energy barrier—followed by growth. Both spinodal decomposition and nucleation are vital for phase transformations and alloy microstructural evolution.

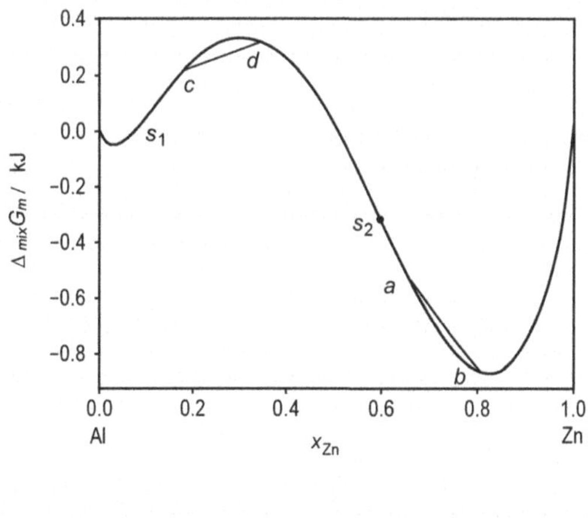

Fig. 14.20 The $\Delta_{mix}G_m^0$ for fcc (Al, Zn) at 300 K. Within the spinodal region, any fluctuations in composition exhibit instability. Conversely, in the metastable region, which lies between the binodal and the spinodal, fluctuations are stable [4]

14.4.5 Temperature– Chemical Potential Diagrams

So far, we've focused on temperature–composition phase diagrams, which are constant-pressure slices of a 3D pressure–temperature–composition diagram. These are the most commonly used 2D diagrams for binary systems due to two key reasons:

1. Phase equilibrium is typically measured in temperature–composition space.
2. Alloys are generally produced and used within this space.

Thermodynamically, this implies binary systems—especially alloys—are treated as closed systems.

Still, the $T–x_i$ diagram isn't always the most practical or only way to present phase equilibrium data.

It was noted earlier that a true phase diagram uniquely defines the phase at each point. A field–field diagram is always true, as field variables aren't conjugate. For open systems (where composition varies), the most useful field–field plot is temperature (T) versus chemical potential (μ_i) at constant pressure. If components or their compounds are volatile at relevant temperatures, composition becomes dependent, and μ_i becomes the control variable.

First the general aspects of $T–\mu_i$ diagrams are discussed, followed by examples. While the Standard Element Reference (SER) is important for equilibrium calculations, it is rarely used in $T–\mu_i$ phase diagrams. Instead, setting the pure component at temperature T and pressure 1 bar as the reference state is more practical. This approach makes the $T–\mu_i$ diagram similar to a $T–x_i$ diagram, as μ_i and x_i are monotonically related in single-phase regions.

For gases, this reference poses no issues. But for condensed phases, we must choose between the pure solid or liquid. In this chapter, we'll use the pure solid at standard conditions, meaning $\mu^i = \mu_i(T, 1\ \text{bar}, x_i) - \mu_i°(T, 1\ \text{bar}, xi = 1)$. For focus on $T–\mu_B$ diagrams of an A–B system, B(s) at T is the reference.

Example 1: System with Two Ideal Solution Phases

To introduce $T–\mu_i$ diagrams, we begin with the simple case: both the solid and liquid phases are ideal solutions. Figure 14.21 illustrates the $T–x_B$ phase diagram, derived using $\Delta fusS° = 10\ J{\cdot}K^{-1}{\cdot}mol^{-1}$ for each component. The melting points are indicated, and at 1432 K, a liquid alloy with a $x_B = 0.1$ is in equilibrium with an α-phase alloy with $x_B = 0.374$.

With B(s) as the reference state, we must now plot the line for pure B(l) on the $T–\mu_B$ diagram. This is calculated as follows:

$$\Delta_{fus}\mu_B^0 = \mu_B^{0,1}(T) - \mu_B^{0,S}(T) = \Delta_{fus}G_m^0(B,T) \approx \prec \Delta_{fus}H_m^0(B) \succ -T \prec \Delta_{fus}S_m^0(B) \succ \quad (14.43)$$

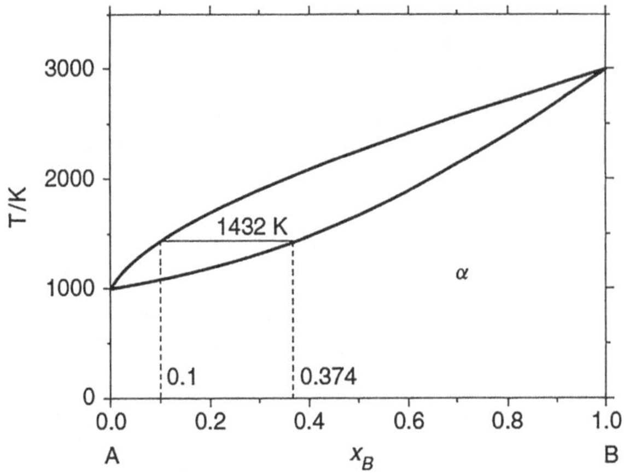

Fig. 14.21 The T–x_B phase diagram illustrates the conditions under which both solid and liquid phases form ideal solutions [4]

In the last equation, a linear approximation for $\Delta \text{fus} G_m^0$ was used, with bracketed terms as averages. This method is reliable for high temperatures, though the averages, based on high-T data, are applied down to 0 K. The construction lines are

(i) At T = 0 K; $\Delta_{fus}\mu_B^o(0\ K) = \prec \Delta_{fus}H_m^o(B) \succ$
(ii) At $T = T_{fus}^o$; $\Delta_{fus}\mu_B = 0$

This enables the construction of a line connecting the two points on the T–μ_B diagram, as illustrated in Fig. 14.22. This can be extended to alloys by incorporating lines for different compositions in both solid and liquid phases. Under the assumption of ideal solutions, the chemical potentials in each phase, referenced to pure solid B, are expressed as follows:

$$\mu_B^S(x_B) - \mu_B^{o,S} = RT \log_e x_B^S \tag{14.44}$$

$$\boldsymbol{\mu_B^S(x_B) - \mu_B^{o,S} = RT\ log_e x_B^1 + \Delta_{fus} G_m^0(B)} \tag{14.45}$$

These two equations result in the formation of a composition span on the $T-\mu_B$ diagram.

Figure 14.22 illustrates the T-μ_B diagram, which is similar to the T-x_B diagram presented in Fig. 14.20; however, the two-phase region has been condensed into a singular curve. The line corresponding to $x_B = 0.1$ in the liquid phase initiates at the 0 K point designated as $\Delta \text{fus} H°_m$ and intersects the solid-phase line for $x_B = 0.374$, which originates at the coordinate (0,0), at a temperature of 1432 K.

Both diagrams convey the same information. While μ_B could be added to the T-x_B plot, it would clutter the diagram. Thus, the T-μ_B diagram is more practical when μ_B is externally controlled.

Gas-phase properties may be incorporated into both T–x_B and T–μ_B diagrams. In this example, with the assumption that $\Delta_{vap}\ G_m{}^0 = 340{,}000 + 85\ T$ J/mol (boiling

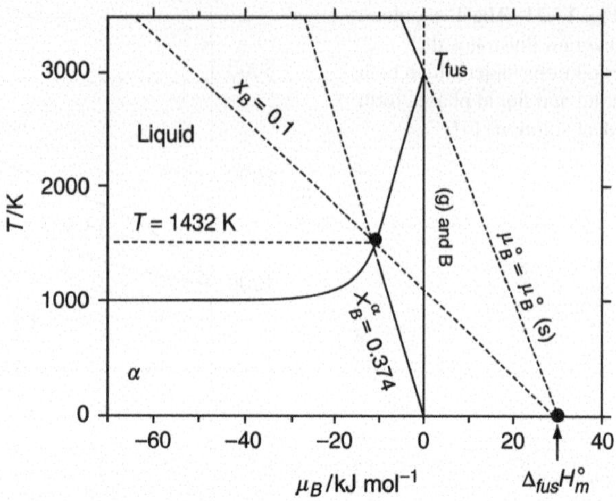

Fig. 14.22 The T-μ_B phase diagram illustrates the conditions under which both solid and liquid phases behave as ideal solutions [4]

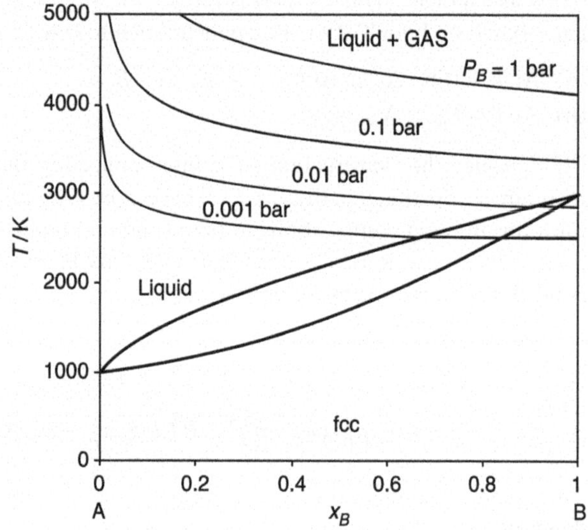

Fig. 14.23 Inclusion of lines of constant p_B on the temperature-composition diagram [4]

point: 4000 K), curves for specific values of p_B can be incorporated into the T–x_B diagram, as illustrated in Fig. 14.23. Gas-phase partial pressure fans can be represented on the T-μ_B diagram using the same methodology applied for liquid and solid composition fans, as illustrated in Fig. 14.22. Phase equilibrium necessitates that the chemical potential of component B in the gas phase, $\mu_B{}^g$, is equal to the chemical potential of component B in the solid phase, $\mu_B{}^S$. Additionally, gas-phase data should be referenced to pure solid B.

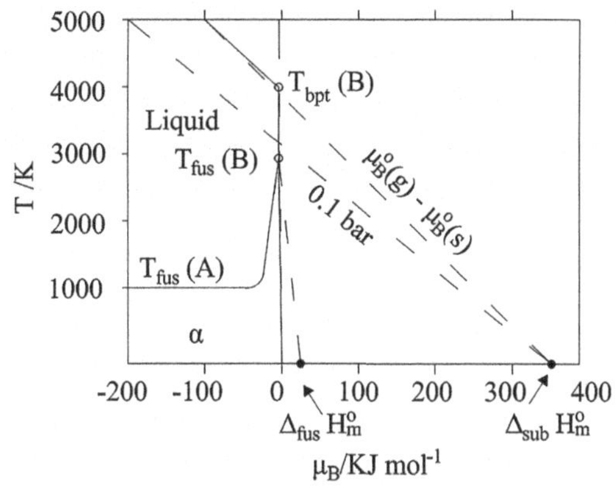

Fig. 14.24 Inclusion of lines of constant p_B on T–μ_B phase diagram [4]

The standard property difference between gas and solid during sublimation can be expressed as

$$\mu_B^{o,g}(x_B) - \mu_B^{o,s} = \Delta_{sub}G_m^o(B,T) \tag{14.46}$$

$$\simeq \prec \Delta_{sub}H_m^o(B) \succ -T \prec \Delta_{sub}S_m^o(B) \succ \tag{14.47}$$

Using the linear approximation again, Eq. (14.46) gives the construction lines as

(i) At T = 0 K; $\mu_B^{o,g}(0\,K) - \mu_B^{o,g}(0\,K) = \prec \Delta_{sub}H_m^o(B) \succ$
(ii) At T=T_{sub}^o; $\mu_B^{o,g}\left(T_{sub}^o\right) - \mu_B^{o,s}\left(T_{sub}^o\right) = 0$

Assuming ideal gas behavior, we can use

$$\mu_B^g(P_B) - \mu_B^{o,g} = RT\,log_e\left(\frac{P_B}{P_B^o}\right) \tag{14.48}$$

The chemical potential in relation to pure solid B is determined by integrating the information from Eqs. (11.47) and

$$\mu_B^g(P_B) - \mu_B^{o,g} = RT\,log_e\left(\frac{P_B}{P_B^o}\right) + \Delta_{sub}G_m^o(B,T) \tag{14.49}$$

The inclusion of the gas-phase data in Fig. 14.23 results in the T–μ_B phase diagram presented in Fig. 14.24. This diagram illustrates the solid or liquid compositions that are in equilibrium with a gas phase at a constant p_B across various temperatures T.

Example 2: Simple Eutectic System

Practicing the conversion between T–x_B and T–μ_B diagrams is essential. Figure 14.25 illustrates a more complex case. The three-phase line represented in the T–x_B diagram is reduced to a singular point in the T–μ_B diagram. The melting point

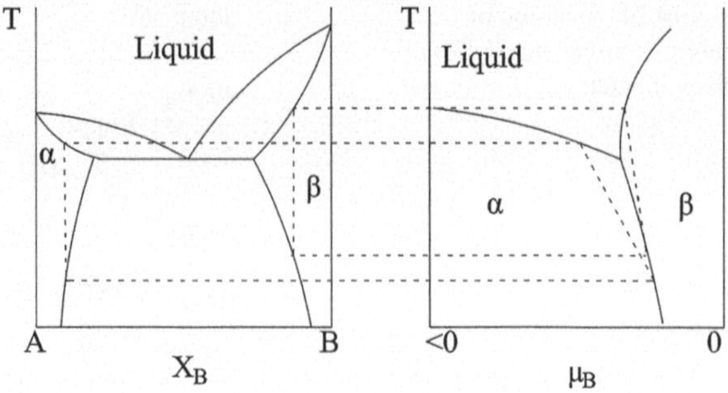

Fig. 14.25 Illustrates the interrelation between the $T - X_B$ and $T - \mu_B$ phase diagrams. The representation of concentration lines on the $T - \mu_B$ phase diagram is demonstrated [4]

of pure B is consistent with $\mu_B = 0$, whereas the melting point of pure A corresponds to $\mu_B = -\infty$. The locations of the two iso-composition curves are distinctly illustrated in both diagrams.

14.5 Ternary and Complex Phase Diagrams

The Complex realm of ternary systems is a natural extension of binary alloys. While most foundational concepts apply, they expand into a higher-dimensional space. In a ternary system (A–B–C), the Gibbs energy of a solution phase is a function of the composition of all three components, commonly represented as:

$$G = x_A G_A + x_B G_B + X_C G_C + RT(X_A lnx_A + x_B lnx_B + x_C lnx_C) + G_{excess} \quad (14.50)$$

Just like binaries, we use partial molar quantities and activities to describe the behavior of each component.

The Gibbs triangle (equilateral triangle plot) becomes the main geometric tool for visualizing composition space.

To estimate ternary alloy properties using binary data, one selects a binary composition that reflects the ternary point of interest. Geometric methods like the symmetric Kohler and Muggianu approaches use binary interaction parameters to extrapolate ternary properties, assuming regular solution behavior. While symmetric methods are effective for substitutional systems, asymmetric models may be required for systems with interstitial elements or strong ordering effects.

In a ternary system (A–B–C), the sum of the mole fractions is equal to one ($X_A + X_B + X_C = 1$), which results in two independent composition variables. The compositions can be represented symmetrically using an equilateral triangle, as illustrated for the Bi–Sn–Cd system in Fig. 14.26. The vertices of the triangle denote pure components, while the edges are indicative of binary systems: Bi–Sn, Sn–Cd, and Cd–Bi. Lines of constant X_{Bi} are oriented parallel to the Sn–Cd edge; likewise, X_{Sn} and X_{Cd} lines are oriented parallel to the Cd–Bi and Bi–Sn edges, respectively. At point a, the values are as follows: $X_{Bi} = 0.05$, $X_{Sn} = 0.45$, and $X_{Cd} = 0.50$.

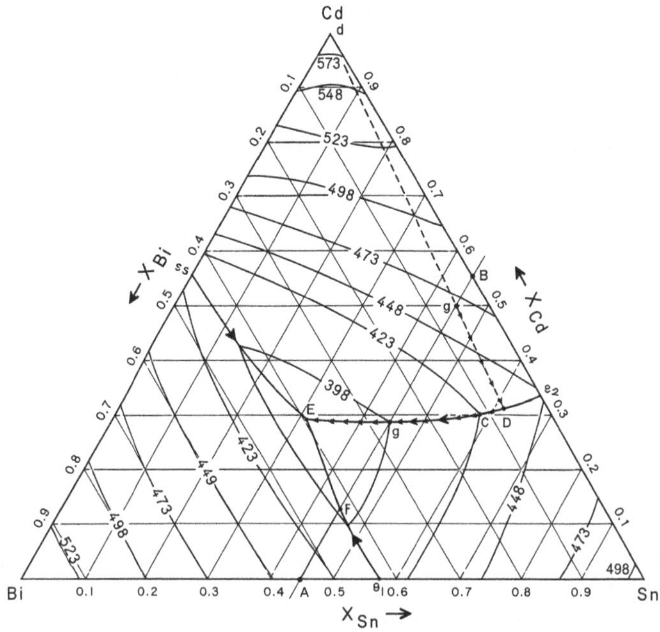

Fig. 14.26 Projection of the liquidus surface for the Bi–Sn–Cd system onto the ternary composition triangle (as detailed by Bray et al., 1961–1962). Small arrows indicate the crystallization path of an alloy with an overall composition represented at point a [10]

Weight percent can also be utilized in the context of similar triangles, as an alternative to mole fraction [10].

A ternary temperature–composition phase diagram at constant pressure is depicted as a three-dimensional model structured within a triangular prism. The base of the prism represents the composition triangle, while the vertical axis indicates temperature. Figure 14.27 demonstrates this concept within a basic eutectic A–B–C system. The vertical faces illustrate the binary phase diagrams (A–B, B–C, and C–A), each featuring eutectic points e1, e2, and e3. Within the prism, liquidus surfaces decrease from the melting points of pure A, B, and C, indicating liquid compositions in equilibrium with A-, B-, or C-rich solids [10].

In a ternary system maintained at constant pressure, the Gibbs phase rule can be expressed as $F = 4 - P$. In a system with one liquid and one solid phase are present ($P = 2$), the degrees of freedom are calculated as $F = 2$, resulting in a bivariant system. The liquidus represents a two-dimensional surface on which the temperature and one liquid composition variable influence the corresponding liquid component and solid compositions [10].

Figure 14.27 illustrates the intersection of the A- and B-liquidus surfaces along line e1, where liquid is in equilibrium with A-rich and B-rich solids at a pressure of 3. This condition is classified as univariant, with a degree of freedom (F) equal to 1. The term "valleys" refers to univariant lines. All three lines converge at the ternary

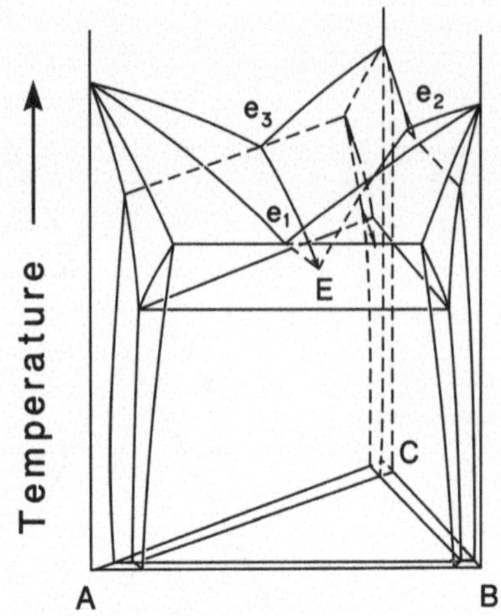

Fig. 14.27 Perspective view of a ternary space model representing a simple eutectic ternary system. e1, e2, and e3 represent the binary eutectics, while E denotes the ternary eutectic. The prism's base is an equilateral triangle configuration [10]

eutectic point E, characterized by P = 4 and F = 0, establishing it as an invariant point with defined temperature and phase compositions [10].

Within the set of 70 metallic elements, a total of 54,740 distinct ternary systems and 916,895 unique quaternary systems can be identified. The experimental determination of a single isothermal section of a ternary phase diagram necessitates significant effort; therefore, the use of predictive methods is crucial. The optimal method involves the use of thermodynamic modelling. The process begins with the analysis of phase diagrams and thermodynamic data for the binary subsystems to derive expressions for Gibbs energy.

These are then used with solution models to estimate ternary phase energies. Finally, the ternary diagram is computed using methods like tangent plane construction or Gibbs energy minimization [10].

14.6 Mathematical and Analytical Methods in Phase Equilibria Studies

This section examines the elements that influence the configuration or topology of phase diagrams. Although specific constraints delineate thermodynamically acceptable conditions, numerous older diagrams failed to meet these criteria.

A full understanding of phase diagram topology involves:

1. The phase rule defines the number of phases that can exist at each feature.
2. Combinatorial analysis establishes constraints on the quantity of features.
3. Schreinemaker's rules delineate the possible and impossible topographical arrangements.

4. The Gibbs–Konovalov equation delineates the variables that influence the slopes of phase boundaries.

14.6.1 Combinatorial Analysis

The phase rule does not specify the quantity of topographical features present in a system; however, combinatorial analysis determines the maximum potential number. It is important to note that not all potential features are represented in a specific phase diagram.

In a system characterized by C components and N phases, the maximum number of invariant points, defined by the conditions $F = 0$ and $\phi = C + 1$, can be determined using the combinatorial expression $\binom{N}{C+1}$:

Similar estimates apply to univariate and bivariate features. When a miscibility gap exists, both separated phases must be counted. Note that this method excludes invariant single-phase points such as congruent melting or critical points.

The combinatorial method facilitates estimating the maximum number of topographical features in a binary system with four phases. Under conditions of constant pressure, as dictated by the phase rule ($F = 0$), the maximum number of coexisting phases is restricted to three.

Thus:

- **Invariant features** (three-phase equilibria)
- **Univariate features** (two-phase equilibria)
- **Bivariant features** (single-phase regions)

  If the phases are labelled α, β, γ, and δ, all possible topographical features can be systematically enumerated.

Each invariant feature involving three phases can generate univariant lines by considering all possible two-phase combinations. For instance, the invariant αβγ leads to univariant equilibria αβ, αγ, and βγ (see Fig. 14.28 for illustrations on a T–μ_i diagram).

Combinatorial analysis identifies the maximum number of potential topographical features in a phase diagram; however, it does not differentiate between thermodynamically stable and metastable phases. The distinction is apparent in a binary system characterized by one stoichiometric compound and two eutectics positioned on either side (Fig. 14.29a). While four phases indicate the potential for four invariant points, the diagram may only display two, with the remaining points being metastable. The T–x_i and T–μ_i diagrams presented in Fig. 14.29 illustrate these points.

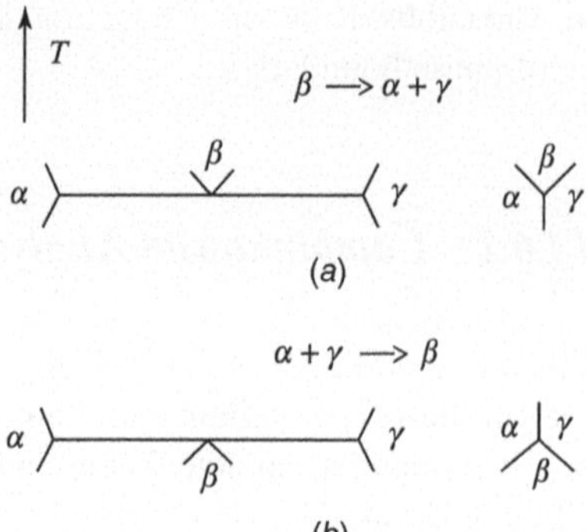

Fig. 14.28 Illustrates two types of invariant equilibria represented in both the (**a**) T –x_i and (**b**) T –μ_i planes [4]

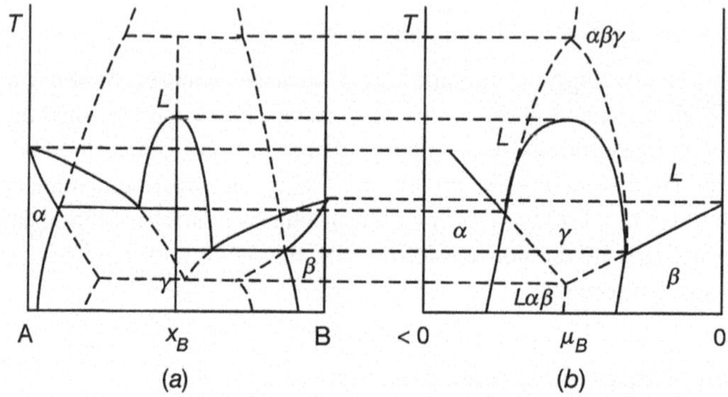

Fig. 14.29 (**a**) T–x_i and (**b**) T–μ_i phase diagrams for a system containing a stoichiometric compound. The diagrams indicate the locations of the metastable invariant equilibria [4]

14.6.2 Schreinemaker's Rules

Schreinemaker's rules are founded on the principle that the stability field of any phase is required to be limited by a more stable assemblage. These rules are most effectively applied to field–field phase diagrams due to their clearer representation of phase boundaries.

In a field–field diagram:

1. Adjacent univariant curves that intersect at an invariant point are required to create angles that are less than 180°.
2. The metastable extension of a stable univariant line is required to transition into an alternative single-phase field.

Figure 14.30 visually summarizes acceptable and unacceptable topologies. Extrapolation validity is easier to assess in field–field diagrams. The correctness of these rules in T–x_B phase diagrams is also evident in G–x_B curves, where metastable α/β

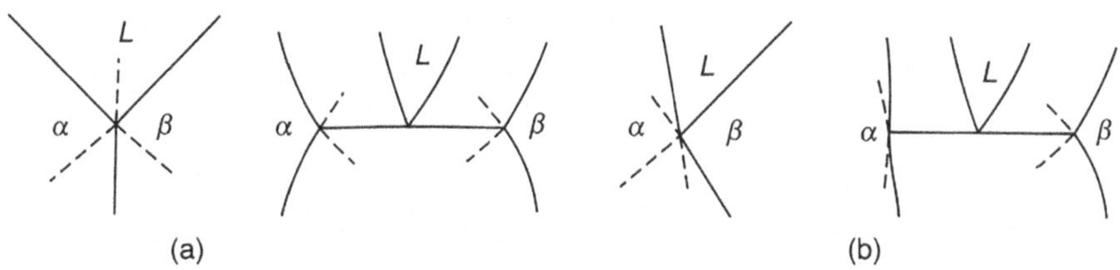

Fig. 14.30 Illustrates stable phase boundaries and their extrapolations into metastable regions in the T–μ_i and T–x_i phase diagrams. (**a**) acceptable; (**b**) unacceptable [4]

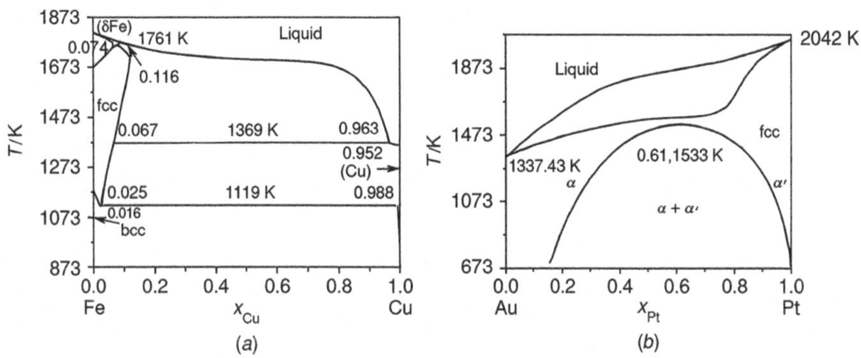

Fig. 14.31 Uncommon topological characteristics identified in phase diagrams: (**a**) Cu–Fe; (**b**) Au–Pt [4]

phase boundaries lie within stable liq/α and liq/β boundaries, confirming their applicability.

14.6.3 The Gibbs–Konovalov Equations

The phase rule, combinatorial analysis, and Schreinemaker's rules provide fundamental constraints on phase diagram topology, like the maximum number of invariant points and allowed phase connectivity. However, they don't reveal whether features such as miscibility gaps, congruently melting compounds, or the specific spatial arrangement of phases are present. To uncover these features, one must rely on the thermodynamic properties of individual phases.

These thermodynamic properties also determine the slopes of phase boundaries in both T–x_i and T–μ_i diagrams. Figures 14.31 and 14.32 illustrate phase diagrams with unusual features:

- Figure 14.38a shows retrograde solubility, where solid solubility increases with decreasing temperature, then decreases.

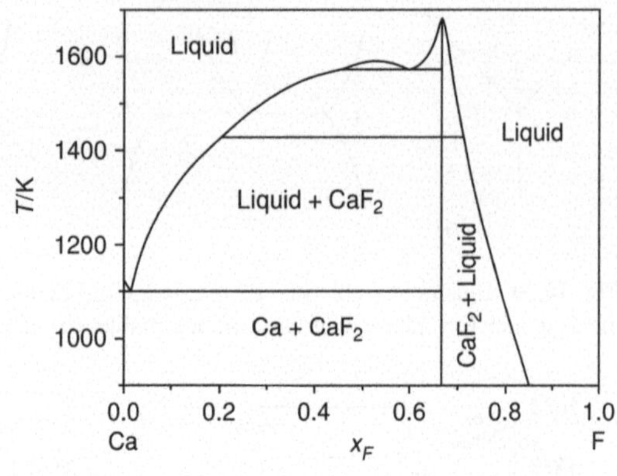

Fig. 14.32 Observation of a distinct sharp congruent point in the Ca–F system [4]

- Figure 14.38b presents a nearly flat solidus, which is rare.
- Figure 14.39 shows a sharp congruent melting point, highlighting distinct thermodynamic behavior.

These features can only be properly understood and experimentally validated through thermodynamic analysis of phase boundary slopes, crucial for avoiding errors in data interpolation or extrapolation. In calculated phase diagrams, such analysis is inherently included, ensuring thermodynamic consistency.

The aim is to determine the thermodynamic factors that influence the slopes of phase boundaries on isobaric phase diagrams, specifically for the T–μ_i and T-x_i types. Our objective is to formulate expressions for:

$$\left(\frac{\partial T}{\partial \mu_i}\right)_p \text{ and } \left(\frac{\partial T}{\partial x_i^\alpha}\right)_p$$

These expressions help us understand how temperature changes in response to chemical potential or composition along a phase boundary at constant pressure. The analysis shows how phase stability and boundary curvature are influenced by the partial molar properties and the Gibbs free energy differences between coexisting phases.

Both partial derivatives pertain to univariant equilibria, characterized by the presence of a single degree of freedom (i.e., F = 1 in accordance with the phase rule). At these invariant phase boundaries, the chemical potentials (μ_i) of each component are equivalent in both coexisting phases. The mathematical representation of this condition is as follows:

$$\mathbf{d}\boldsymbol{\mu}_\mathbf{A}^\alpha = \mathbf{d}\boldsymbol{\mu}_\mathbf{A}^\beta = \mathbf{d}\boldsymbol{\mu}_\mathbf{A}$$

$$\mathbf{d}\boldsymbol{\mu}_\mathbf{B}^\alpha = \mathbf{d}\boldsymbol{\mu}_\mathbf{B}^\beta = \mathbf{d}\boldsymbol{\mu}_\mathbf{B}$$

14.7 High Entropy Alloys, Nanostructured Materials, Metallic Glasses

14.7.1 High Entropy Alloys

High-entropy alloys (HEAs) consist of five or more principal elements combined in equal or near-equal proportions. These materials provide several benefits, including enhanced strength, resistance to corrosion, thermal stability, and improved irradiation resistance. Phase structure is key to their performance. Yeh et al. (2004) [11] popularized the term "high-entropy alloys" (HEAs), highlighting that configurational entropy in a binary alloy is maximized at equiatomic compositions, and the maximum entropy $\Delta S_{conf} = -R(X_A lnX_A + X_B lnX_B)$ increases with the number of elements. The significant influence of high mixing (configurational) entropy on phase stability, formation kinetics, lattice strain, and overall properties is notable. In high-entropy alloys, it enhances mutual solubility and results in simpler, unexpected phases and microstructures.

Recent studies have concentrated on the prediction of phase formation through the utilization of parameters such as configurational entropy, enthalpy of mixing, and atomic size difference. Computational methods such as CALPHAD are utilized to model high-entropy alloy phase diagrams with increased directness.

Cantor et al. [12] proposed a substitution strategy for equiatomic multicomponent alloys, consisting of 5 to 20 elements, which involves replacing elements in binary alloys with analogous elements in equal proportions. For example, in $Zr_{50}Cu_{50}$, Zr was partially replaced by Ti, Hf, and Nb, and Cu by Ag and Ni, forming a septenary alloy with good glass-forming ability, even in pseudo-ternary systems with added elements like Al. These substitutions also apply to intermetallic (e.g., NiAl with Fe or Co) and oxides (e.g., $BaTiO_3$ with Sr or Zr), enabling the creation of alloys with enhanced properties.

The Gibbs phase rule (F = C − P + 1) defines the degrees of freedom and maximum phases in a system, which for binary, ternary, quaternary, and quinary systems are 3, 4, 5, and 6, respectively, under invariant conditions. However, fewer phases are typically observed in HEAs due to their high configurational entropy, which promotes solid solution formation and limits phase separation. Additionally, low atomic diffusivities, due to multiple principal elements, kinetically control phase formation, enhancing material properties.

High-entropy alloys (HEAs) are composed of five or more elements in equiatomic or near-equiatomic proportions, with the presence of minor elements constituting less than 5 atomic percent. The elevated mixing entropy in high-entropy alloys (HEAs) promotes the stabilization of solid solution phases in preference to intermetallic phases, especially at elevated temperatures, thereby facilitating their synthesis and application. High-entropy alloys (HEAs) are generally characterized by each constituent element making up 5–35% atomic percent.

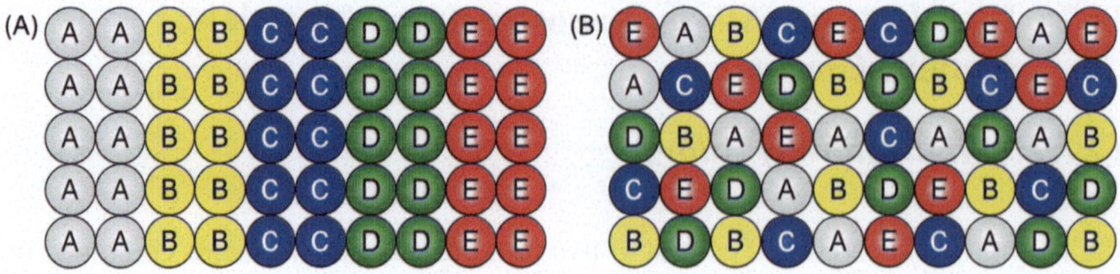

Fig. 14.33 (**a**) illustrates five components in an equiatomic ratio before mixing, and (**b**) depicts the process of mixing to create a random solid solution. The atomic sizes are considered to be equivalent [13]

High-entropy alloys are defined as multi-principal-element alloys due to their elevated configurational entropy. In the framework of statistical thermodynamics, configurational entropy is expressed through Boltzmann's equation:

$$\Delta S_{conf} = k \ln w \tag{14.51}$$

In this context, k represents Boltzmann's constant, while w denotes the quantity of potential atomic configurations. The configurational entropy per mole for a solid solution comprising n elements with mole fractions x_i is expressed as follows:

$$\Delta S_{conf} = -R\sum_{i=1}^{n} X_i \, ln \; X_i \tag{14.52}$$

In the case of an equiatomic alloy ($x_i = 1/n$), this simplifies to:

$$\Delta S_{conf} = R \ln n \tag{14.53}$$

This indicates that entropy rises with the number of components, which is the reason alloys containing five or more principal elements are classified as HEAs. Figure 14.33 depicts the process of forming a quinary equiatomic alloy composed of five distinct elements. The configurational entropy, calculated using the formula $\Delta S_{conf} = R \ln 5$, is determined to be 1.61 R. In contrast, a non-equiatomic alloy such as $Al_{1.5}CoCr_{0.5}FeNi_{0.5}$ ($Al_{33.3}Co_{22.2}Cr_{11.1}Fe_{22.2}Ni_{11.1}$ in atomic percent) exhibits a marginally reduced entropy value of 1.523 R. This indicates that the mixing entropy diminishes as alloy compositions diverge from equiatomic ratios.

Total mixing entropy comprises configurational, vibrational, magnetic, and electronic contributions, with configurational entropy being the most significant factor. Vibrational entropy, determined from Debye temperatures, can be negative, particularly in the presence of strong atomic pair attractions.

According to Richards' rule ($\Delta H_f/T_m > \approx \Delta S_f \approx R$), the entropy change on melting is ~R, and the enthalpy (ΔH_f) is about RT_m. Since this energy disrupts about 1/12 of the bonds in a solid, a configurational entropy of *R* in a solid solution significantly lowers the free energy ($\Delta G_{mix} = \Delta H_{mix} - T\Delta S_{mix}$) by RT—about

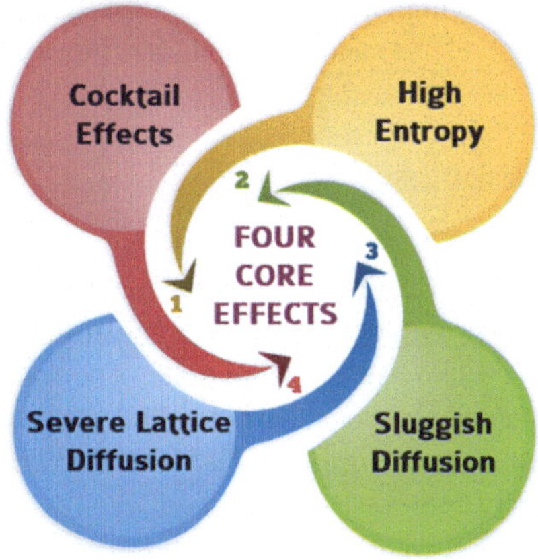

Fig. 14.34 Four fundamental effects of high entropy alloys [17]

8.3 kJ/mol at 1000 K. This reduction in $_{\Delta Gmix}$ increases the stability of disordered solid solutions over intermetallics, which have lower entropy due to their ordered structure, especially at high temperatures.

A ternary equiatomic alloy already exceeds 1 R in configurational entropy, and a quinary alloy exceeds it by 61%. For comparison, the normalized formation enthalpies of strong intermetallics like NiAl and TiAl are 1.38 R and 2.06 R, respectively [11]. Therefore, a configurational entropy of 1.5 R is considered sufficient to counteract strong atomic ordering, making it a practical threshold between high- and medium-entropy alloys. Similarly, 1 R is suggested as the cut-off between medium- and low-entropy alloys [14].

This entropy-based classification also extends to ceramics and polymers, which can be grouped as low-, medium-, or high-entropy materials depending on the number of components. Earlier definitions set the thresholds at 0.69 R and 1.61 R, based on the presence of 2 or 5 principal elements, whereas the updated approach uses entropy values directly [14–16].

The four core effects of high-entropy alloys, illustrated in Fig. 14.34, are interrelated and significantly influence their distinctive properties. The effects described are essential for the progression of high-entropy alloys, expanding the limits of materials science, and facilitating new opportunities in advanced materials research.

14.7.1.1 High Entropy Effect

The high-entropy effect is essential to high-entropy alloys, as it facilitates the formation of solid solutions and streamlines the microstructure. According to established principles of metallurgy, interactions among multiple elements can lead to the formation of binary, ternary, and higher-order compounds, which may result in complex and brittle structures. This perspective fails to consider the stabilizing function of high mixing entropy, which mitigates this complexity.

Mixing enthalpy (ΔH_{mix}) can be calculated using models developed by de Boer et al. (1988) and Takeuchi and Inoue (2000, 2005).

$$\Delta H_{mix} = 4 \sum_{i=1, J \neq i}^{n} \Delta H_{ij}^{min} X_i X_j + \sum_k \Delta H_k^{trans} X_k \tag{14.54}$$

In this equation, x_k denotes the mole fraction of the non-metallic element k within the system, while ΔH_{ij}^{mix} indicates the mixing enthalpy per mole for an equiatomic $i - j$ alloy in the solid state. The transformation enthalpies for the conversion of k-elements from their standard state to a metallic state are as follows: 30 kJ/mol for B, 180 kJ/mol for C, 310 kJ/mol for N, 37 kJ/mol for Si, 25 kJ/mol for Ge, and 17 kJ/mol for P.

Elemental phases exhibit minor negative values for ΔH_{mix} and ΔS_{mix}, primarily influenced by a single predominant element. Compound phases exhibit significant negative ΔH_{mix} values and minimal ΔS_{mix} values as a result of their ordered structures. Multicomponent solid solutions have medium negative ΔH_{mix} and the highest ΔS_{mix}. The medium ΔH_{mix} in these solutions results from the presence of unlike atomic pairs. For instance, in a random NiAl solution, half of the bonds are Ni–Al, whereas in the ordered NiAl compound, these bonds are more evenly distributed.

In a random solution state, the mixing enthalpy is approximately 50% of that observed in the fully ordered state. The mixing enthalpy of the random solution state for a quinary equiatomic alloy is 0.8 times that of the ordered state, while for a septenary alloy, it is 0.857. The inclusion of additional elements results in the mixing enthalpy of the random solution approaching that of the ordered state. The random solution state exhibits greater thermal stability compared to the ordered state, attributable to its elevated mixing entropy. While the mixing heats for dissimilar atoms may vary in practice, the overall trend towards greater disorder persists. This tendency renders partially ordered solid solutions with multicomponent compositions more favourable at equilibrium, particularly at elevated temperatures, as a result of the $2T\Delta S_{mix}$ effect. Miracle et al. (2014) [18] demonstrated that the ΔS_{conf} of HEA can destabilize 5–10% of intermetallic compounds at room temperature and up to 30–55% at 1500 K. This effect offers a novel method for regulating microstructure and properties in particulate-strengthened high-entropy alloys (HEAs).

14.7.1.2 Sluggish Diffusion

The sluggish diffusion of High-Entropy Alloys arises from their intricate atomic structure, characterized by a unique arrangement of elements at each lattice site. This diversity generates energy barriers that impede atomic movement, resulting in reduced diffusion and phase transformation kinetics relative to typical alloys. Atoms in high-entropy alloys experience diverse energy landscapes, which confine them to low-energy sites and diminish their mobility.

Studies show that HEAs, such as CoCrFeMnNi alloys, have higher activation energy for diffusion than pure metals, resulting in slower kinetics.

This sluggish diffusion enables nano-sized precipitate formation, enhancing strength and stability. HEA coatings serve as effective diffusion barriers, improving high-temperature performance. The gradual dispersion enhances creep resistance, rendering HEAs appropriate for applications necessitating long-term stability under stress and temperature. The intricate atomic interactions in HEAs eventually influence diffusion behavior, phase change, and material characteristics.

Severe Lattice Distortion Effect

In high-entropy alloys, each atom is encircled by diverse neighboring atoms owing to the multicomponent characteristics of the solid solution matrix. This leads to considerable lattice strain and stress, primarily due to atomic size mismatches and fluctuations in bonding and electronic configuration. Such non-uniform atomic surroundings cause severe lattice distortion, as illustrated in a five-component BCC lattice (Fig. 14.35). This distortion is much greater than in conventional alloys where neighbouring atoms are mostly the same.

The disparities in bonding energy and crystal structure among elements exacerbate the distortion. This significantly influences material properties, as hardness and strength are markedly improved by solution hardening. The equiatomic FCC CoCrFeMnNi alloy exhibits a Vickers hardness of 1192 MPa, markedly surpassing the 864 MPa anticipated under the mixture rule. The refractory BCC MoNbTaVW alloy attains a strength of 5250 MPa, nearly thrice the prediction based on the mixing rule.

FCC alloys tend to show less solution hardening than BCC ones, likely because FCC has 12 nearest neighbour compared to BCC's 8, leading to fewer unlike atomic pairs and smaller strain. Additionally, atomic size differences are smaller in FCC alloys like CoCrFeMnNi.

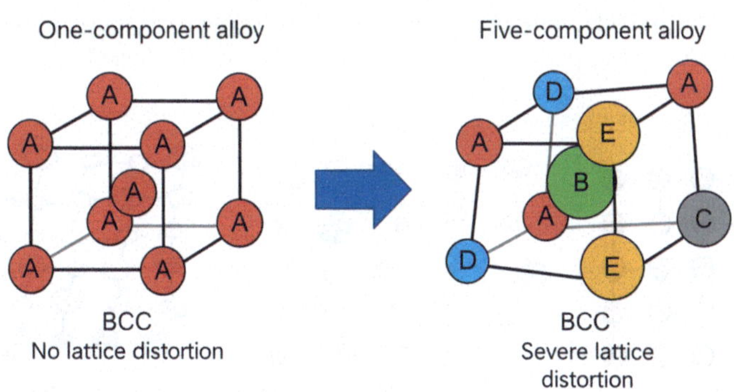

Fig. 14.35 presents a schematic picture illustrating significant lattice deformation inside the five-component BCC lattice [19]

Lattice distortion also impacts X-ray diffraction, causing diffuse scattering and reducing peak intensity. It increases electron and phonon scattering, lowering both electrical and thermal conductivities. However, these properties remain relatively stable with temperature, as thermal and vibration-induced distortion are minor compared to the intrinsic lattice distortion in HEAs.

14.7.1.3 Cocktail Effect

In high-entropy alloys, the "cocktail effect" denotes the improved characteristics resulting from the interaction of five or more components. High-entropy alloys can exhibit single or many phases, with each phase influencing properties according to its size, shape, distribution, and boundaries. Each phase functions as a multicomponent solid solution, akin to an atomic-scale composite. Properties are influenced not only by the combination rule but also by elemental interactions and lattice distortion, resulting in excess effects.

The cocktail effect spans atomic to microscale levels. For example, in $Al_xCoCrCuFeAl$ alloys, increasing Al content shifts phases from FCC to BCC (Yeh, 2006). Refractory HEAs suchas MoNbTaW and MoNbTaVW, developed by the U.S. Air Force Research Lab, have melting points above 2600 °C and high yield strength (400 MPa at 1600 °C), outperforming traditional superalloys [20, 21]. In $CoNiFe(AlSi)_{0.2}$, the combination of magnetic, mechanical, and electrical properties results from equiatomic ferromagnetic elements forming a ductile FCC phase and nonmagnetic elements enhancing lattice distortion (Fig. 14.36). These examples show how alloy design optimizes the cocktail effect, with multi-principal elements synergistically contributing to HEAs.

14.7.2 Alloy Design Utilizing CALPHAD Methodologies

Semi-empirical models combine theory and data to improve prediction accuracy, with CALPHAD being particularly useful for HEAs, where phase formation studies

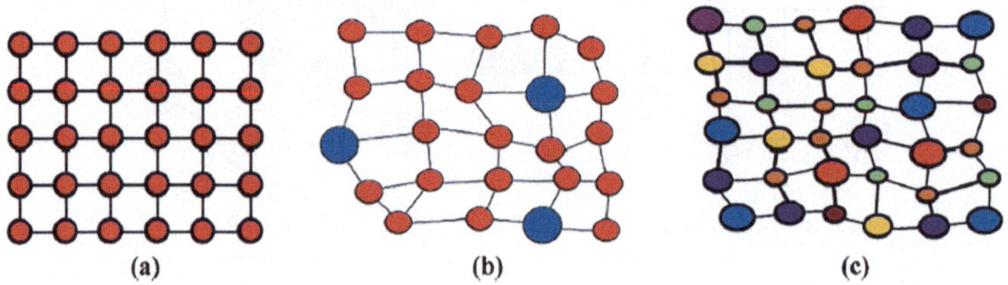

Fig. 14.36 (**a–c**). Lattice distortion in (**a**) pure metal, (**b**) typical alloy, and (**c**) high-entropy alloy crystal structures [22]

are complex and time-consuming. CALPHAD uses thermodynamic data from binary and ternary systems to predict phase stability in multicomponent systems. Introduced by Kaufman and Bernstein (1970), it integrates Gibbs energy, enthalpy of mixing, and heat capacity to determine phase equilibria, with software like Thermo-Calc, Pandat™, and FactSage using specialized databases for simulations. CALPHAD helps explain phase transitions in HEAs, such as the FCC to BCC shift in $Al_xCoCrCuFeNi$, and tracks phase evolution as a function of composition and temperature [22]. While the valence electron concentration (VEC) rule is not always consistent, databases like TCHEA1 improve predictions. CALPHAD significantly reduces experimental work, as shown by Gurao and Biswas [23, 24] narrowing 1287 alloys to 10 promising ones, and Senkov et al. [20, 21] screening 130,000 compositions, revealing that more elements reduce solid solution formation. It enables high-throughput screening, helping focus on stable compositions, especially in HEAs with sluggish diffusion and complex phases.

Example 1: The Co–Cr–Fe–Mn–Ni System [25]
The CoCrFeMnNi alloy was investigated by Cantor et al. [12, 26] and Otto et al. [27], with both studies identifying a single-phase FCC structure—Cantor in an as-cast alloy and Otto in an annealed variant. This observation is unexpected, given that the five elements exhibit distinct structures at room temperature. Nickel exhibits a face-centered cubic (FCC) structure, while chromium and iron display a body-centered cubic (BCC) arrangement. Cobalt is characterized by a hexagonal close-packed (HCP) structure, and manganese possesses an A12 structure. Analysis of the 10 binary systems in Co–Cr–Fe–Mn–Ni elucidates the formation of a face-centered cubic (FCC) phase. Calculations conducted with Pandat™ software indicate that within the Co–Fe–Ni ternary system, all three elements exhibit stability in an FCC phase across extensive temperature ranges. The binaries Co–Fe, Co–Ni, and Fe–Ni establish continuous FCC solid solutions, as illustrated in Fig. 14.37a–c.

Example 2: The Co–Cr–Fe–Mn–Cu System
Otto et al. [27] prepared five alloys by substituting Co, Cr, Fe, or Ni with similar elements to illustrate that entropy alone does not dictate phase stability; none of the alloys resulted in a single-phase FCC structure. One specific alloy, CoCrFeMnCu (where Ni is substituted by Cu), exemplifies this effectively. Calculations of phase diagrams indicate the distinctions between CoCrFeMnCu and CoCrFeMnNi. Figure 14.38 illustrates that Cu–X (where X represents Co, Cr, Fe, Mn) binary systems exhibit notable differences compared with the corresponding Ni–X systems. Ni–X binaries exhibit a broad range of FCC stability. In contrast, Cu is prone to forming a distinct FCC phase characterized by a miscibility gap, especially in relation to Fe and Cr, which demonstrate limited solubility in FCC Cu. In these systems, the entropy of mixing fails to compensate for repulsive atomic interactions and positive excess enthalpies, which inhibit the formation of a stable single phase.

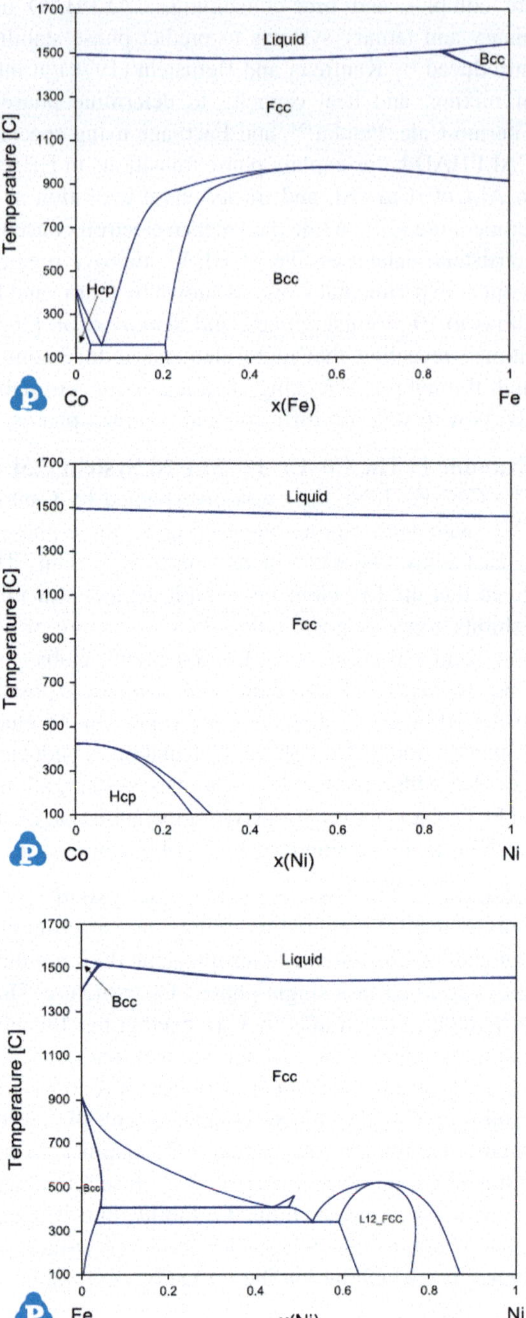

Fig. 14.37 The three constituent binaries in the Co–Fe–Ni ternary system are Co–Fe, Co–Ni, and Fe–Ni [25]

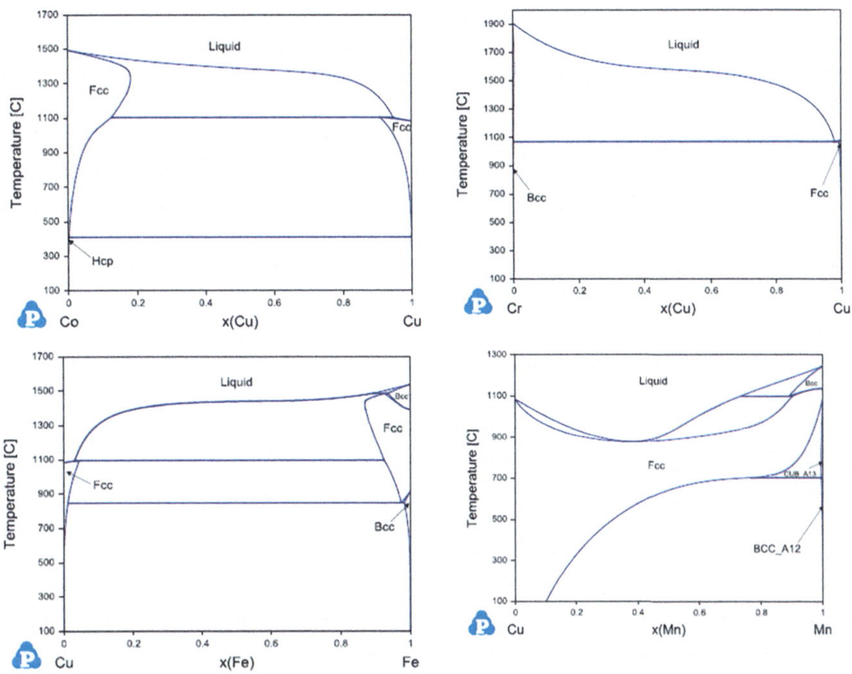

Fig. 14.38 Binary phase diagrams for Cu–X, where X denotes Co, Cr, Fe, and Mn [25]

14.7.3 Thermodynamics and Kinetics of Nanostructured Materials

In the past two or three decades, nanoscience has rapidly emerged as a major field, focusing on understanding matter on the nanoscale (1–100 nm). Nanomaterials lie between atoms and bulk matter, exhibiting unique size effects due to their high surface-to-volume ratios. These effects lead to discrete electronic structures and phenomena like "magic numbers" and quantum behavior, making nanomaterials fundamentally different from both their atomic and bulk forms.

One notable property is size-dependent melting point, first observed by Takagi [28] using TEM. For example, gold nanoparticles can melt up to 300 K below their bulk melting point. Such behavior has key implications for thin film technologies and nanoscale thermal stability.

To quantify these size-dependent changes, classical thermodynamics must be adapted, since it traditionally applies to systems larger than the nanoscale. This leads to thermodynamics of nanostructured state, which considers size as an additional degree of freedom. It is useful in interpreting nanoscale experiments, where results often vary due to sensitivity to measurement conditions.

Three main approaches have been proposed: (1) Hill's size-dependent thermodynamic model, (2) Tsallis' non-extensive entropy framework, and (3) pressure-based models using the Laplace-Young equation. Each offers insights into Nano-system-behavior, but a unified, quantitative Nano thermodynamic model is still missing. A deeper look at size effects on well-known processes like melting can help advance this understanding.

An understanding of thermodynamic phase stability is essential for controlling the nucleation, grain growth, and phase transformation in nanocrystalline materials, making it fundamental to their applications [29]. Nanocrystalline materials display a different thermodynamic behavior due to their high volume fraction of grain boundaries and large surface-to-volume ratio. Such behavior significantly influences phase stability, making thermodynamic analysis essential for understanding and predicting their structural evolution.

In nanocrystalline systems, the Gibbs free energy includes additional contributions from surface and interface energies. As a result, metastable phases that are not favoured in bulk form may become stable at the nanoscale. The total free energy G of a nanocrystalline phase can be expressed as:

$$G_{nano} = G_{bulk} + \Delta G_{surface} + \Delta G_{interface} \tag{14.55}$$

At a nonmetric scale, numerous meta-stable phases and polymorphs may be preferred over the thermodynamically stable phase as indicated in the phase diagram. The phase diagrams derived from bulk phase thermodynamics offer restricted insights for the synthesis of nanomaterials at the nanometre scale. The observed deviation from conventional phase diagrams primarily arises from the substantial impact of surface energy on the total energy of the nonmetric crystal. The impact of surface modifications on total energy and their subsequent influence on the phase diagram at a nonmetric scale are critical inquiries essential for understanding and guiding the synthesis of nanomaterials [30].

CALPHAD is a reliable technique for predicting bulk phase equilibria in multicomponent systems. It can be modified for nanoscale systems by incorporating size effects. This approach facilitates the modeling of nanostructure growth by incorporating surface effects associated with finite size, thereby minimizing the necessity for experimental trials through enhanced process simulation accuracy [31, 32].

CALPHAD is capable of predicting the thermodynamic properties of alloy nanoparticles that exceed approximately 5 nm in size. However, its reliability diminishes for nanoparticles smaller than this threshold, primarily due to reduced contributions from surface energy at edges and vortices. Tanaka et al. [33] extended CALPHAD to nanoscale transitions, while Park and Lee [34] applied tools such as Thermo-Calc to nanoscale phase diagrams. Lee and Sim [35] further refined the approach for different shapes, and others adapted it for intermetallics. Experimental thermal studies generally align well with CALPHAD predictions.

In CALPHAD, phase diagrams are constructed by specifying the Gibbs energy of all phases, determining the stable phase(s) that minimize Gibbs energy under constant temperature and pressure conditions. In nanoscale systems, surface energy has a substantial impact on Gibbs energy, necessitating the acquisition of supplementary data. This study employs methodologies from references [36, 37] to generate size- and shape-dependent phase diagrams for the In–Sb system. Surface energies were estimated utilizing the Butler equation for both liquids and solids, alongside density functional theory (DFT) for the In-Sb intermetallic. Additionally, phase diagrams for In–Sb nanoparticles and nanowires were calculated as a function of size [32].

The total Gibbs energy of a bulk phase comprising components A and B is expressed as

$$G^{Total,Bulk} = x_A^0 G_A + x_B^0 G_B + RT[x_A \ln x_A + x_B \ln x_B] + G^{Ex,Bulk} \tag{14.56}$$

In this context, x_i represents the molar fraction of component i, where i can be either A or B. Gi denotes the standard Gibbs energy of component i at a temperature of 298 K and a pressure of 1 atm. R and T denote the gas constant and absolute temperature, respectively. The excess Gibbs energy $G^{(Ex,Bulk)}$ is conventionally expressed using a Redlich–Kister polynomial [38] as follows:

$$G^{Ex,Bulk} = x_A x_B \sum L^{\nu}(x_A - x_B)^{\nu} \qquad (\nu = 0, 1, 2, \ldots.) \tag{14.57}$$

Here, L^{ν} is the interaction parameter, expressed as a function of temperature:

$$L^{\nu} = a + bT + cTlnT + \ldots. \tag{14.58}$$

The total Gibbs energy of a nanoparticle comprises a surface energy component, attributable to its elevated surface-to-volume ratio.

$$G^{total,nano} = G^{Bulk} + G^{surface} = x_A^0 G_A^{nano} + x_B^0 G_B^{nano} + RT[x_A ln\, x_A + x_B\, ln\, x_B] + G^{Ex,nano} \tag{14.59}$$

For a nanowire and an isotropic spherical particle, the surface contribution to the Gibbs energy, $G^{surface}$, is as follows:

$$G^{surface} = \frac{2C\sigma_i V_i}{r_{sphere}} \tag{14.59}$$

$$G^{surface} = \frac{C\sigma_i V_i}{r_{cylinder}} \tag{14.60}$$

In this case, C is a correction factor that takes into account uncertainties in surface tension, shape changes with size, and surface strain in particles below a certain size. σ_i is the surface energy, and Vi is the molar volume of component i. For liquids, C is typically taken as 1. For solids, C is estimated by comparing measured and calculated melting points of small particles.

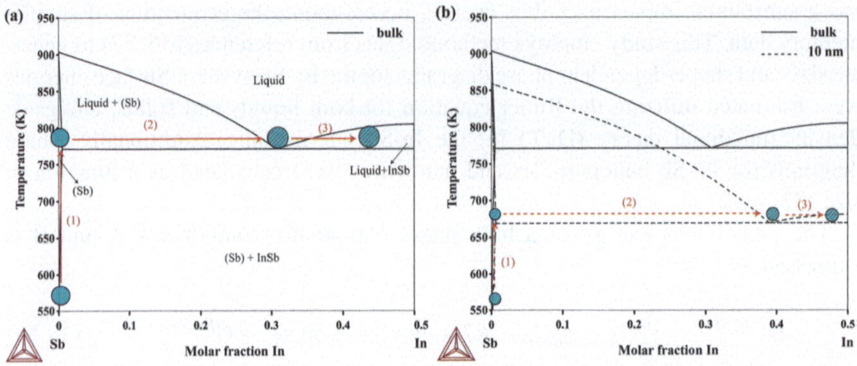

Fig. 14.39 (**a**) Illustrates the Sb-rich region of the bulk In–Sb phase diagram, along with the process of Sb-seed nucleation. Heating solid Sb near the eutectic point results in the formation of a liquid alloy. Upon super saturation, In-Sb initiates the crystallization process. Pure Sb and In–Sb alloy are represented in solid and patterned colours. (**b**) Compares the phase diagrams of bulk In–Sb (solid line) and 10 nm In–Sb (dashed line). Smaller particles necessitate approximately 8.0 atomic percent additional indium and exhibit lower melting temperatures as a result of capillary effects [32]

It is assumed that an alloy nanoparticle's excess Gibbs energy, $G^{(Ex,nano)}$ takes the same form as Eq. (14.61):

$$G^{Ex,nano} = x_A x_B \sum L^{\nu}_{nano}(x_A - x_B)^{\nu} \qquad (\nu = 0, 1, 2, \ldots.) \tag{14.61}$$

The term L^{ν}_{nano} *can be written as*

$$L^{\nu}_{nano} = f_1\left(\frac{1}{r}\right) + f_2\left(\frac{1}{r}\right)T + f_3\left(\frac{1}{r}\right)T\ln T + \ldots \tag{14.62}$$

where f_i is a size-dependent function assuming a simple linear form.

Case Study: Size-Dependent Phase Diagram of the In–Sb System

In In–Sb phase diagrams(Fig 14.39), it is found that nanoparticle melting points and eutectic temperatures drop significantly as size decreases, with a more pronounced effect, using In-Sb surface energy, and for particles below 5 nm, where In-Sb decomposes peritectically, calculations have been done. Nanowires show a smaller size effect due to lower surface energy contribution. Recent work, such as by Pendyala et al., [39] explains In- or Sb-seeded nanowire growth near their respective eutectics, with size-dependent diagrams clarifying Vapor–Liquid–Solid (VLS) nucleation behavior. To construct these diagrams, CALPHAD with surface energy corrections (using C = 0.975 for In, literature values for In and Sb, and DFT for In-Sb intermetallic), reveals that decreasing size lowers liquidus and eutectic temperatures while increasing solubility, especially in nanoparticles [32].

14.7.3.1 Effect of Pressure

The molar Gibbs energy of a system under high pressure, G_m^P is given by Eq. (14.63) [40],

$$G_m^P = G_m + \left(\frac{\partial G}{\partial P}\right)_{P\to 0}; P = G_m + VP \tag{14.63}$$

G_m represents the molar Gibbs energy of pure component i at atmospheric pressure, V denotes the molar volume of the solution, and P indicates the pressure. Conversely, the molar Gibbs energy of a system subjected to high pressure can be expressed as:

$$G_m^P = X_A G_A^P + X_B G_B^P + RT(X_A ln\, X_A + X_B\, ln\, X_B) + G^{Ex,P} \tag{14.64}$$

where G_i^P is the molar Gibbs energy of pure component i at pressure P, and $G^{Ex,\,P}$ is the excess Gibbs energy under pressure P. These are given by Eqs. (22) and (23), respectively [40].

$$G_i^P = G_i^o + \left(\frac{\partial G_i}{\partial P}\right)_{P\to 0} P = G_i^o + V_i P \tag{14.64}$$

where G_i^P is pure component I's standard molar Gibbs energy, and Vi is its molar volume at atmospheric pressure [40].

$$G^{Ex,P} = G^{Ex} + \left(\frac{\partial G^{Ex}}{\partial P}\right)_{P\to 0} P = G^{Ex} + V^{Ex} P \tag{14.65}$$

where V^{Ex} is the excess volume and G^{Ex} is the bulk excess Gibbs energy under atmospheric pressure. These can be written as

$$\begin{aligned} V^{Ex} &= V - V_{ideal} = V - (X_A V_A + X_B V_B) = \frac{M}{\rho} - \left(X_A \frac{M_A}{\rho_A} + X_B \frac{M_B}{\rho_B}\right) \\ &= \left(\frac{\partial G^{Ex}}{\partial P}\right)_{p\to 0} = X_A X_B \sum \left(\frac{\partial L_k^P}{\partial P}\right)_{P\to 0} (X_A - X_B)^K \end{aligned} \tag{14.66}$$

Here, M is the alloy's molar mass, ρ is its density, and V_{ideal} is the alloy's hypothetical volume under ideal mixing conditions. The molar mass and density of the pure component are shown by the symbols M_i and ρ_i, respectively. Under high pressures, up to several GPa, it is assumed that the extra volume stays constant. Equation (14.25) can therefore be used to simplify the system's excess Gibbs energy at high pressure [40].

$$G^{Ex,P} = G^{Ex} + V^{Ex}P = X_A X_B \sum L_K^P (X_A - X_B)^K \tag{14.67}$$

The effects of particle size and pressure can now be combined, and the excess Gibbs energy is represented by equation

$$G^{Ex,NP,P} = G^{Ex,NP} + V^{Ex}P = X_A X_B \sum L_K^{NP,P} (X_A - X_B)^K \tag{14.68}$$

14.7.3.2 Changes in the Phase Diagram

Figure 14.40 displays phase diagrams that were calculated using FactSage software and Gibbs energy minimization. Figure 14.40 compares diagrams for bulk, nanoparticles (5 nm), and nanoparticles at 2 GPa. Interaction parameters at atmospheric pressure were +15 kJ/mol for solid and −15 kJ/mol for liquid. These values were altered by size and pressure effects. The liquid excess volume parameter ranged from -10^{-6} to $+10^{-6}$ m^3/mol [40].

The melting points of pure A and B exhibited a decrease when in nanoparticle form, while an increase was observed under pressure conditions. The eutectic point also dropped for nanoparticles, further lowering under pressure, especially with

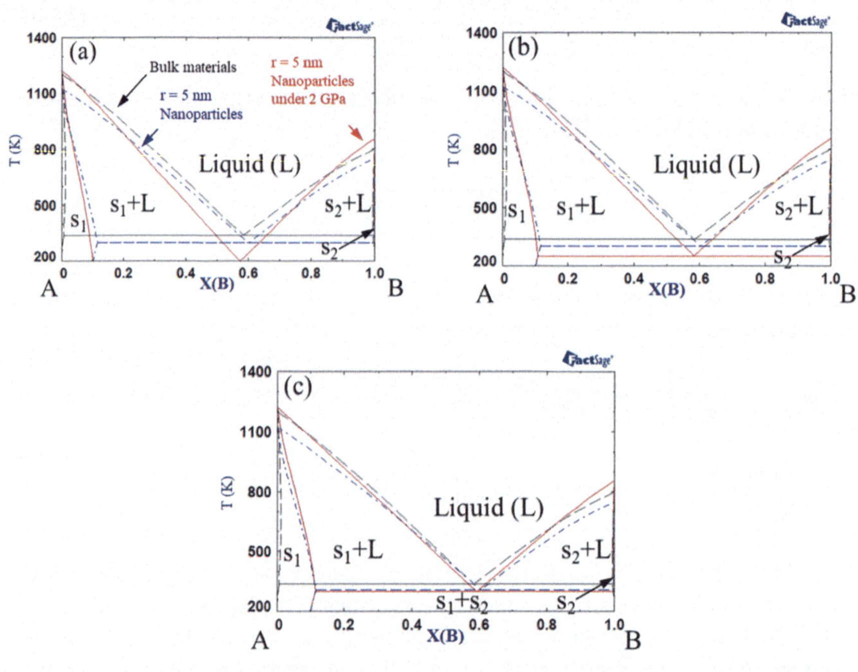

Fig. 14.40 A–B phase diagram for the respective cases: The black dashed line denotes the bulk material, the blue dash-dotted line indicates nanoparticles with a 5 nm radius, and the red solid line illustrates 5 nm radius nanoparticles subjected to a pressure of 2 GPa. (a) $V^{Ex} = -1\,cm^3/mol$ (**b**) $V^{Ex} = 0$ (**c**) $V^{Ex} = 1\,cm^3/mol$ [40]

negative excess volume. The eutectic temperature falls as solid interaction increases or liquid interaction decreases. At high pressure, positive excess volume raises the interaction parameters; negative volume change lowers them. Since solid excess volume is minimal, negative liquid excess volume here mimics a decrease in liquid interaction, reducing eutectic temperature with pressure [40].

14.7.4 Thermodynamics and Kinetics of Metallic Glasses

Metallic glasses, also known as amorphous metals, are characterized by their disordered atomic structures, in contrast to crystalline metals which exhibit long-range order. They form by rapidly cooling metallic melts (>10^5–10^6 K/s; there also systems in which metallic glasses can be produced at much lower cooling rates), preventing atoms from arranging into stable crystals and "freezing" the liquid structure into a solid glass. Following the discovery of Au-Si amorphous metal by Duwez et al. [41], advancements in the development of bulk metallic glasses (BMGs) with enhanced glass-forming ability (GFA) have been achieved through the alloying and optimization of compositions, particularly within Cu-, Fe-, Ni-, Pd-, Ti-, and Zr-based systems. Identifying optimal compositions experimentally presents challenges due to the vast number of possible combinations, thereby necessitating the use of computational prediction methods [42].

Although equilibrium phase diagrams do not fully describe metallic glass behavior, they still help understand thermodynamics and kinetics. Glass-forming ability (GFA) is linked to Gibbs free energy difference (ΔG); lower ΔG increases nucleation barriers and slows crystallization. Lele et al. [43] analysed temperature effects on ΔG and nucleation, noting that heating rate matters, and since measuring the specific heat difference (ΔCp) is experimentally challenging, various theoretical models are used to estimate ΔG [44].

Singh and Holz [45] presented the following expression relating ΔG_p to T:

$$\boldsymbol{\Delta G} = \frac{\boldsymbol{\Delta H_m \Delta T}}{\boldsymbol{T_m}}\left(\frac{\boldsymbol{7T}}{\boldsymbol{T_m + 6T}}\right) \tag{14.69}$$

Ji and Pan [46] related ΔG_p and T assuming hyperbolic variation between the variables:

$$\Delta G = \frac{2\Delta H_m \Delta T}{T_m}\left(\frac{T}{T_m + T} - \frac{\Delta T^2 T_m}{3(T_m + T)^3}\right) \tag{14.70}$$

Dubey and Ramchandra Rao [47] used the hole theory to derive $\boldsymbol{\Delta G}$

$$\Delta G = \frac{\Delta H_m \Delta T}{T_m} - \frac{\Delta C_p^m (\Delta T)^2}{2T}\left(1 - \frac{\Delta T}{6T}\right) \quad (14.71)$$

Lele et al. [43] derived ΔG considering temperature dependence in the undercooled region

$$\Delta G = \frac{\Delta H_m \Delta T}{T_m} - \frac{\Delta C_p^m (\Delta T)^2}{(T_m + T)} \quad (14.72)$$

14.7.4.1 Phase Diagrams Relevant to Metallic Glasses

Although metallic glasses are inherently non-equilibrium materials, equilibrium phase diagrams such as binary or ternary alloy diagrams remain highly useful to identify composition ranges favourable for glass formation, often located near deep eutectics characterized by low liquidus temperatures. Moreover, these diagrams help to understand the crystalline phases that may compete with glass formation during slow cooling or heat treatment and provide a supercooled liquid region, which is necessary for successfully producing metallic glasses.

Several methodologies have been proposed to calculate the glass-forming ability (GFA) and predict the glass-forming range (GFR) in alloy systems, based on thermodynamic considerations. One familiar approach involves the **T_0 curve**, which has shown effectiveness in certain binary and ternary systems. Another method employs **Miedema's semi-empirical model** [48] to estimate GFR, while a third approach uses **CALPHAD** techniques to calculate the thermodynamic driving force for crystallization in the supercooled melts, comparing these forces across the phases to assess the GFA. Particularly, the CALPHAD-based method has been successfully applied to various alloy systems, including Al-based alloys that exhibit distinctive glass-forming behavior. Additional thermodynamic and kinetic details, such as the suppression of intermetallic phase formation, also play critical roles in enhancing glass formation.

Combining thermodynamic predictions with kinetic models, researchers have evaluated GFA by applying CALPHAD-derived data to the Davies–Uhlmann model [49]. Assuming homogeneous nucleation and employing **Johnson-Mehl-Avrami isothermal kinetics** [50], time-temperature-transformation (TTT) curves were generated to estimate the time required to form a detectable crystalline phase at various temperatures. From the TTT curves, critical cooling rates were derived and used as a factor for assessing GFA in binary and ternary alloys [51].

Fundamentally, glass formation during cooling hinges on the suppression of crystallization until the material crosses the glass transition temperature. To predict which alloy compositions are most favourable for glass formation, one must closely analyse the nucleation behavior in the undercooled melt. The glass-forming range corresponds to those compositions where nucleation is least likely to occur. Over time, various models with differing levels of precision have been developed to assess the GFA, with the **T_0 curve method** remaining one of the most prominent [51].

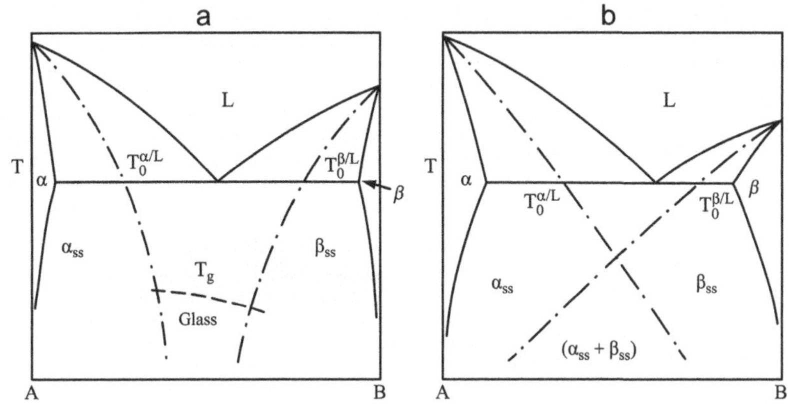

Fig. 14.41 Hypothetical T₀ curves for a binary eutectic A–B system. (**a**) T₀ curves drop steeply, indicating potential for glass formation. (**b**) T₀ curves intersect at lower temperatures, favouring partition-less crystallization [52]

The T_0 curve delineates the compositions and temperatures at which the Gibbs free energies of the liquid and solid phases are equal, offering essential insights into the minimum undercooling required for the formation of partition-less solids (Fig. 14.41). In eutectic systems, a deeply plunging T_0 curve typically indicates regions with a low tendency for crystallization and thus a higher potential for glass formation. Conversely, shallow T_0 curves suggest compositions where crystallization is more likely, reducing the chances of successfully forming a glass [51].

The T_0 curve presents a thermodynamic viewpoint on phase stability and glass formation. In addition, complementary models such as Miedema's serve as useful empirical instruments for measuring the energetic characteristics of alloy systems. Miedema's model provides an empirical method for estimating the heat of mixing in binary systems, applicable to both solid and liquid phases. The calculation of formation enthalpy for metallic glasses (ΔH^{amor}), solid solutions (ΔH^{SS}), and intermetallic compounds (ΔH^{inter}) is performed utilizing specified equations [53–55].

$$\Delta H^{amor} = \Delta H^{chem}(amor) + \Delta H^{topo} \quad (14.73)$$

$$\Delta H^{ss} = \Delta H^{chem}(SS) + \Delta H^{elastic} + \Delta H^{structure} \quad (14.74)$$

$$\Delta H^{inter} = \Delta H^{chem}(inter) \quad (14.75)$$

Here, ΔH^{chem} (amor) is the chemical mixing enthalpy of the amorphous state, and ΔH^{topo} represents the topological contribution. ΔH^{chem} (SS) and $\Delta H^{elastic}$ refer to the chemical and elastic enthalpies of a solid solution. ΔHstructure accounts for enthalpy changes due to structural rearrangements, while ΔH^{chem} (inter) denotes the chemical mixing enthalpy of intermetallic compounds.

The chemical enthalpy of mixing for a solid solution can be expressed as:

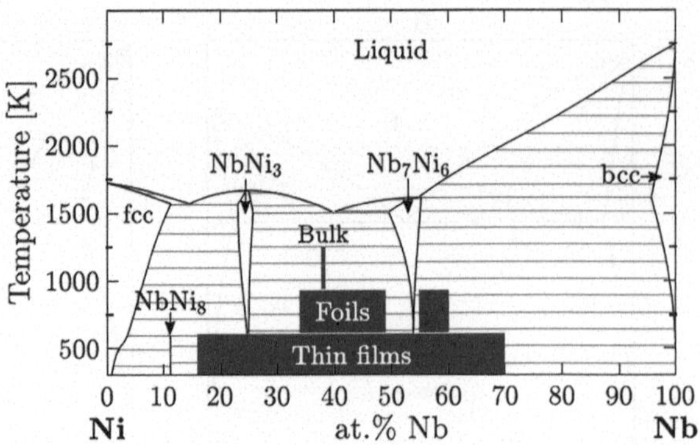

Fig. 14.42 Shows the glass-forming ranges (purple areas) for Nb–Ni alloys synthesized by sputtering, rapid solidification, and casting [57]

$$\Delta H^{chem} = x_A x_B \left[x_A \Delta H^{ss}_{B\ in\ A} + x_B \Delta H^{ss}_{A\ in\ B} \right] \tag{14.76}$$

Here, x_A and x_B are the mole fractions of elements A and B, respectively, and ΔH^{SS} refers to the enthalpy of solution of one element in the other at infinite dilution.

For metallic glasses, the enthalpy of mixing (ΔHmix) is key to preventing phase separation and has guided alloy design. ΔHmix, a specific case of the enthalpy of formation, is typically referenced for the liquid phase. Small positive ΔHmix values (e.g., Zr-Nb: 4 kJ/mol) improve glass-forming ability, while larger values (>15 kJ/mol) can cause liquid immiscibility, leading to two amorphous phases, as seen in Nb-Ni-Y. Though vapor quenching can produce glasses from immiscible elements, this is rare at slower cooling rates [56].

The extended regular solution model is commonly used to describe the enthalpy of mixing in alloys.

$$\Delta H_{mix} = \sum_{ij,\, i \neq j} x_i x_j \Omega_{ij} \tag{14.77}$$

In this model, x_i and x_j represent the atomic fractions of components i and j, while Ω_{ij} is the binary interaction parameter between them [56].

Materials with a large negative enthalpy of mixing fall into two groups: metal-metal systems and metal–metalloid/non-metal systems. Metal–metalloid bonds are often covalent, with localized electrons, and tend to form stable ceramic-like phases. For example, the Nb–Ni system (Fig. 14.42) has a negative mixing enthalpy of –30 kJ/mol and forms three intermetallic phases with lower melting points than the pure elements. In metal–metalloid systems, the enthalpy of mixing can exceed –100 kJ/mol, commonly leading to stable compound phases that strongly affect the glass-forming ability and thermal stability of metallic glasses.

Thermodynamic models, such as the extended regular solution framework, provide essential insights into the enthalpic contributions that influence alloy stability and glass formation. Additionally, it is crucial to consider kinetic factors, which are effectively represented by Time-Temperature-Transformation (TTT) and

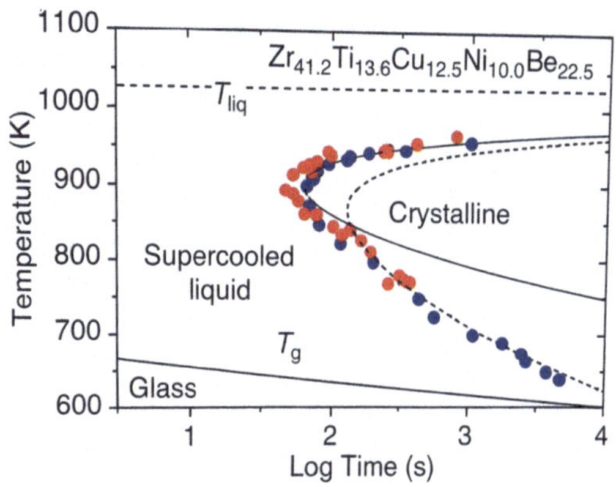

Fig. 14.43 Shows the TTT diagram for Vitreloy 1 crystallization, using data from electrostatic levitation and carbon crucibles [58]

Continuous-Cooling-Transformation (CCT) diagrams. Figure 14.43 illustrates the time–temperature–transformation (TTT) diagram for the crystallization process of the supercooled liquid $Zr_{41.2}Ti_{13.8}Cu_{12.5}Ni_{10}Be_{22.5}$ (Vitreloy 1, V1). The diagram exhibits the characteristic "nose" shape, which results from the increasing driving force for crystallization and the reduction in atomic mobility as the material is supercooled. The duration of the nose time is approximately 60 seconds, which is notably extended when compared to solid-state phase transitions such as austenite to pearlite in steels. Separating kinetic and thermodynamic factors reveals that bulk metallic glasses (BMGs) exhibit both high viscosities and low crystallization driving forces, which contribute to their enhanced glass-forming ability.

Building on this understanding, Yang et al. [59] tailored CuZr-based BMGC microstructures using multi-element micro alloying. This method decreases the driving force for B2 nucleation by reducing melting entropy and inhibits B2 growth by promoting planar crystal growth and delaying multi-component diffusion, resulting in a rightward shift of the TTT curve (Fig. 14.43). Slower B2 nucleation and growth broden the cooling rate window for BMGC formation, thereby a ensuring fine precipitate distribution. The cone-shaped casting results (Fig. 14.43) show that lower cooling rates (larger sizes) increase the second-phase size and volume fraction, making cooling rate a key factor. When cooling exceeds the critical rate (Rs), only the amorphous phase forms; below Rs, a second phase precipitates. To toughen the matrix, the second phase should have a lower shear modulus (G), specific size, and diffuse distribution. In situ phases include second glassy phases, nanocrystals, quasi-crystals, dendrites, martensite, etc. [60] (Fig. 14.44).

Phase separation is observed in Pd-, Fe-, Al-, Cu-, Mg-, and Zr-based metallic glasses (MGs). However, the ability to control the composition and glass-forming ability of the two liquid phases presents challenges, which in turn restricts the production of dual-phase bulk metallic glasses (BMGs) with dimensions of 1 mm or greater. Zr-based bulk metallic glasses (BMGs) exhibit enhanced glass-forming ability (GFA), which is often improved through the incorporation of elements that possess positive mixing enthalpy, thereby augmenting the overall GFA of the system.

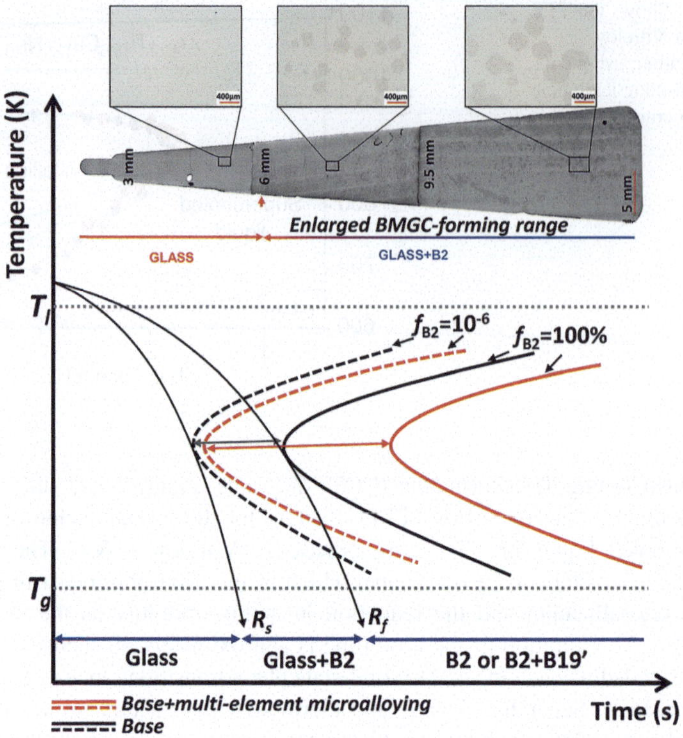

Fig. 14.44 Schematic TTT curves showing the start (fB2 = 10^{-6}, dashed) and finish (fB2 = 100%, solid) of B2–CuZr-based MGC crystallization. Inset: microstructure of cone-shaped casting of $Cu_{40.75}Zr_{42.00}Al_{4.00}Ag_{5.00}Sn_{0.75}Co_{0.50}Ti_{5.00}Hf_{1.00}Ni_{1.00}$ BMGC [60]

For example, the addition of Fe to $Zr_{63.6}Cu_{18}Ni_{10.4}Al_{8.0}$ or Mo to $Zr_{50}Cu_{44.5}Al_{5.5}$ results in the formation of dual-phase structures. Additionally, the incorporation of O facilitates Cu-rich/Copper segregation by reducing the strength of Zr–Cu bonds. Fabrication techniques encompass copper mold casting, powder metallurgy, and hot pressing. In the casting process, liquid phase separation (LPS) occurs before solidification, resulting in the formation of two amorphous phases. In powder methods, metallic glasses (MGs) amalgamate in the supercooled liquid state. LPS can occur via nucleation and growth or spinodal decomposition, particularly in Zr–Co–Al–Y systems. The cooling rate influences phase size; slower cooling results in larger separations, whereas rapid cooling yields fine nanoscale regions that enhance plasticity without compromising strength. In addition to thermal methods, mechanical deformation or ultrasonic treatment can facilitate the formation of nanoscale phases, thereby improving plasticity. Furthermore, the mixing of amorphous powders can result in the creation of multi-phase bulk metallic glass composites. It is essential to implement careful design strategies to prevent the occurrence of inhomogeneous interfaces [60].

Problems

1 In a system of liquid, the critical temperature is $T_c = 370°$ C and the critical pressure is $P_c = 22.1$ MPa. Calculate the specific volume at the critical point, given the specific volume of liquid at Tc and Pc is $v_L = 0.00113\ m^3/kg$ and that of vapor is $v_V = 0.1948\ m^3/kg$

Solution:

$v_2 = \frac{v_1 \times T_2}{T_1} = \frac{10^{-5} \times 200}{120} = 1.67 \times 10^{-5} m/s$

The specific volume at the critical point is the average of the liquid and vapor specific volumes:

$$v_{critical} = \frac{v_L + v_V}{2}$$

$$v_{critical} = \frac{0.00113 + 0.1948}{2} = 0.097965 m^3/kg$$

Thus, the specific volume at the critical point is $0.097965\ m^3/kg$

2 Given the molar gibbs free energy for component A is a binary mixture is $G_A = 10$KJ/mol and for component B it is $G_B = 20$ KJ/mol, with mole fraction $X_A = 0.6$ and $X_B = 0.4$. Calculate the total Gibbs free energy of the system

Solution:

The total Gibbs free energy of a binary system is given by:

$G_{total} = X_A G_A + X_B G_B$

Substitute the values:

$G_{total} = (0.6)(10) + (0.4)(20) = 6 + 8 = 14$ KJ/mol

Thus the total Gibbs free energy of the system is 14 KJ/mol

3 A ternary system is in equilibrium with three phases. Use Gibbs's Phase Rule to calculate the number of degrees of freedom (F)

Solution:

Gibbs Phase Rule is:

$F = C - P + 2$

Where:

C is the number of components

P is the number of phases

For a ternary system,

$F = 3 - 3 + 2 = 2$

4 In a binary alloy system, the eutectic composition is 50% of A and 50% of B. If the system is cooled from 700 °C to 650 °C, Calculate the amount of solid formed at equilibrium at 650 °C, using the lever rule. Assume the eutectic temperature is 700 °C and the solidus temperature for A is 600 °C

Solution:

The lever rule can be applied to calculate the fraction of solid phase:

Where:

$T = 650$ °C

$T_1 = 600$ °C (solidus temperature of A)

$T_2 = 700$ °C (eutectic temperature)

Substitue the values:

$\frac{S_{solid}}{S_{total}} = \frac{650 - 600}{700 - 600} = \frac{50}{100} = 0.5$

Thus the fraction of solid phase at 650 °C is 50%

References

1. White, M. A. (2005). Phases and phase equilibrium. In F. Bassani, G. L. Liedl, & P. Wyder (Eds.), *Encyclopedia of condensed matter physics* (pp. 263–272., ISBN 9780123694010). Elsevier. https://doi.org/10.1016/B0-12-369401-9/00715-4
2. Porter, D. A., Easterling, K. E., & Sherif, M. Y. (2009). *Phase transformations in metals and alloys* (3rd ed.). CRC Press.
3. Domb, C., & Green, M. S. (Eds.). (1972–2001). *Phase transitions and critical phenomena* (Vol. 1–20). Academic.
4. Austin Chang, Y., & Alan Oates, W. (2010). *Materials thermodynamics*. John Wiley & Sons.
5. Callister, W. D., Jr. (2014). *Materials science and engineering an introduction* (9th ed.). John Wiley & Sons.
6. Avner, S. H. (1997). *Introduction to physical metallurgy* (2nd ed.). McGraw-Hill Education India. ISBN 978-0074630068.
7. Gaskell, D. R. (2009). *Introduction to the thermodynamics of materials*. Taylor and Francis.
8. YuZeng, C. H. E. N., Feng, L. I. U., GenCang, Y. A. N. G., Ning, L. I. U., & YaoHe, Z. H. O. U. (2007). δ/γ transformation in non-equilibrium solidified peritectic Fe-Ni alloy. *Science in China Series G: Physics, Mechanics and Astronomy, 50*(4), 421–431.
9. Gaskell, D. . R. (**2003**). *Introduction to the thermodynamics of materials* (4th ed.). Taylor & Francis. ISBN 1591690137.
10. Pelton, A. D. (1996). Thermodynamics and phase diagrams of materials, :In physical Metallurgy:Cahn, R.W., Haasen.P.,(Eds.) Amsterdam.
11. Yeh, J. W., Chen, S. K., Lin, S. J., Gan, J. Y., Chin, T. S., Shun, T. T., Tsau, C. H., & Chang, S. Y. (2004). Nanostructured high-entropy alloys with multiple principal elements: Novel alloy design concepts and outcomes. *Advanced Engineering Materials, 6*, 299–303.
12. Cantor, B., Chang, I. T. H., Knight, P., & Vincent, A. J. B. (2004). Microstructural development in equiatomic multicomponent alloys. *Materials Science and Engineering: A, 375–377*, 213. https://doi.org/10.1016/j.msea.2003.10.257
13. Yeh, J.-W. (2016). Fundamentals of high-entropy alloys. In M. C. Gao, J. W. Yeh, P. K. Liaw, & Y. Zhang (Eds.), *High-entropy alloys: Fundamentals and applications*. Springer.
14. Tsai, K. Y., Tsai, M. H., & Yeh, J. W. (2013). Sluggish diffusion in Co–Cr–Fe–Mn–Ni high-entropy alloys. *Acta Materialia, 61*(13), 4887–4897. https://doi.org/10.1016/j.actamat.2013.04.058
15. Yeh, J. W. (2006). Recent progress in high-entropy alloys. *Annales de Chimie Science des Matériaux, 31*(6), 633–648.
16. Yeh, J. W., Chen, Y. L., Lin, S. J., & Chen, S. K. (2007). High-entropy alloys–A new era of exploitation. *Materials Science Forum, 560*, 1–9. https://doi.org/10.4028/www.scientific.net/MSF.560.1
17. Odetola, P. I., et al. (2024). Exploring high entropy alloys: A review on thermodynamic design and computational modeling strategies for advanced materials applications. *Heliyon, 10*(22), e39660.
18. Miracle, D. B., Senkov, O. N., Wilks, G. B., & Yeh, J. W. (2014). Exploration and development of high entropy alloys for structural applications. *Entropy, 16*(1), 494–525. https://doi.org/10.3390/e16010494
19. Antão, F.J.N. (2019). HEAs: High entropy alloys for advanced systems and engines. Master of Science Degree thesis, Reitoria da Universidade de Lisboa.
20. Senkov, O. . N., Wilks, G. . B., Miracle, D. . B., Chuang, C. . P., & Liaw, P. . K. (2010). Refractory high-entropy alloys. *Intermetallics, 18*(9), 1758–1765. https://doi.org/10.1016/j.intermet.2010.05.014
21. Senkov, O. . N., Wilks, G. . B., Scott, J. . M., & Miracle, D. . B. (2011). Mechanical properties of $Nb_{(25)}Mo_{(25)}Ta_{(25)}W_{(25)}$ and $V_{(20)}Nb_{(20)}Mo_{(20)}Ta_{(20)}W_{(20)}$ refractory high-entropy alloys. *Intermetallics, 19*(5), 698–706. https://doi.org/10.1016/j.intermet.2011.01.004

22. Odetola, P. I., Babalola, B. J., Afolabi, A. E., Anamu, U. S., Olorundaisi, E., Umba, M. C., et al. (2024). Exploring high entropy alloys: A review on thermodynamic design and computational modeling strategies for advanced materials applications. *Heliyon, 10*(22), e39660. https://doi.org/10.1016/j.heliyon.2024.e39660
23. Biswas, K., Gurao, N. . P., Maiti, T., & Mishra, R. . S. (2022). Functional applications of high entropy alloys. In *High entropy materials: materials horizons—from nature to nanomaterials* (pp. 259–287). Springer.
24. Biswas, K., Gurao, N. . P., & Maiti, T. (2022). Synthesis and processing of bulk high entropy materials. In *High entropy materials: materials horizons—from nature to nanomaterials* (pp. 123–176). Springer.
25. Zhang, F., Zhang, C., Chen, S. . L., Zhu, J., Cao, W. . S., & Kattner, U. . R. (2014). An understanding of high-entropy alloys from phase diagram calculations. *Calphad, 45*, 1–10.
26. Cantor, B. (2014). Multicomponent and high entropy alloys. *Entropy, 16*(1), 4749–4768. https://doi.org/10.3390/e16094749
27. Otto, F., Dlouhý, A., Somsen, C., Bei, H., Eggeler, G., & George, E. P. (2013). The influences of temperature and microstructure on the tensile properties of a CoCrFeMnNi high-entropy alloy. *Acta Materialia, 61*(15), 5743–5755. https://doi.org/10.1016/j.actamat.2013.06.018
28. Takagi, S. (1954). Electron diffraction and electron microscopy of small particles. *Journal of the Physical Society of Japan, 9*(5), 359–363.
29. Zhang, H., & Banfield, J. F. (1998). Thermodynamic analysis of phase stability of nanocrystalline titania. *Journal of Materials Chemistry, 8*(1), 1–4.
30. Ma, X., Chen, H., Bai, J., Wu, L., Xu, W., Sun, F., et al. (2018). Guiding synthesis of polymorphs of materials using nanometric phase diagrams. *Journal of the American Chemical Society, 140*(48), 16032–16038.
31. Ghasemi, M., & Johansson, J. (2015). Thermodynamic assessment of the As–Zn and As–Ga–Zn systems. *Journal of Alloys and Compounds, 638*, 95–102.
32. Ghasemi, M., Zanolli, Z., Stankovski, M., & Johansson, J. (2015). Size- and shape-dependent phase diagram of In–Sb nano-alloys. *Nanoscale, 7*(41), 17387–17396.
33. Tanaka, T., Hack, K., & Hara, S. (2000). Size- and shape-dependent phase stability in nanoscale materials: extension from bulk phase diagrams. *CALPHAD: Computers and Coupling of Phase Diagrams and Thermochemistry, 24*, 465–474.
34. Park, J., & Lee, J. (2008). Phase diagram reassessment of Ag–Au system including size effect. *Calphad: Computers and Coupling of Phase Diagrams and Thermochemistry, 32*, 135–141.
35. Lee, J., & Sim, K. J. (2014). General equations of CALPHAD-type thermodynamic description for metallic nanoparticle systems. *CALPHAD: Computer Coupling of Phase Diagrams and Thermochemistry, 44*, 129–132.
36. Kroupa, A., Káňa, T., Bursik, J., Zemanová, A., & Sob, M. (2015). Modelling of phase diagrams of nanoalloys with complex metallic phases: application to Ni–Sn. *Physical Chemistry Chemical Physics, 17*, 28200–28210. https://doi.org/10.1039/C5CP00281H
37. Bajaj, S., Haverty, M. G., Arróyave, R., & Shankar, S. (2015). Phase stability in nanoscale material systems: extension from bulk phase diagrams. *Nanoscale, 7*(21), 9868–9877.
38. Redlich, O., & Kister, A. T. (1948). Algebraic representation of thermodynamic properties and the classification of solutions. *Industrial and Engineering Chemistry, 40*(2), 345–348.
39. Pendyala, C., Vaddiraju, S., Kim, J. H., Jacinski, J., Chen, H., & Sunkara, M. K. (2010). Self-nucleation and growth of group III-antimonide nanowires. *Semiconductor Science and Technology, 25*, 024014.
40. Kim, H. G., Lee, J., & Makov, G. (2021). Phase diagram of binary alloy nanoparticles under high pressure. *Materials, 14*, 2929.
41. klement, W., Willens, R. H., & Duwez, P. O. L. (1960). Non-crystalline structure in solidified gold–silicon alloys. *Nature, 187*, 869–887.
42. Chegeni, M. R., Ma, W., Riegler, S. S., Ghavimi, A., Rohde, M., Yang, F., Seifert, H. J., Gallino, I., & Busch, R. (2025). Thermodynamic analysis and modeling of Pd-Ni-S bulk metallic glass-forming system. *Acta Materialia, 294*, 121074.

43. Lele, S., Dubey, K. S., & Ramachandrarao, P. (1985). On the temperature dependence of free energy of crystallisation. *Current Science, 54*(19), 994–995.
44. Prajapati, S. R., Kasyap, S., & Pratap, A. (2015). A thermodynamic approach towards glass-forming ability of amorphous metallic alloys. *Bulletin of Materials Science, 38*(7), 1693–1698.
45. Singh, H. . B., & Holz, A. (1983). Stability limit of supercooled liquids. *Solid State Communications, 45*(12), 985–988.
46. Ji, X. L., & Pan, Y. (2007). Relative effect of electronegativity on formation of high-entropy alloys. *Journal of Non-Crystalline Solids, 353*(24–25), 2443–2446.
47. Dubey, K. . S., & Ramachandrarao, P. (1984). On the free energy change accompanying crystallisation of undercooled melts. *Acta Metallurgica, 32*(1), 91–96.
48. Miedema, A. . R., de Châtel, P. . F., & de Boer, F. . R. (1980). Cohesion in alloys—fundamentals of a semi-empirical model. *Physica B + C, 100*(1), 1–28.
49. Davies, B. L., & Uhlmann, D. R. (**1972**). Thermodynamic and kinetic aspects of glass formation. *Journal of Non-Crystalline Solids, 7*(4), 337–352.
50. Avrami, M. (**1939**). Kinetics of phase change. I. General theory. *The Journal of Chemical Physics, 7*(12), 1103–1112.
51. Tang, C., & Zhou, H. (2011). Correlation between thermodynamics and glass forming ability in the Al–Ce–Ni system. *Journal of Alloys and Compounds, 509*(5), 2057–2063.
52. Boettinger, W. J., & Perepezko, J. H. (1993). Modeling of microstructure evolution during solidification processing. *Materials Science and Engineering: A, 178*(1–2), 1–18.
53. Bakker, W. (1988). Thermodynamic analysis of phase transformations in alloys. *Journal of Phase Equilibria, 9*(1), 12–21.
54. Bakker, H. (1988). Enthalpies of formation of liquid and solid binary alloys based on 3d metals: IV. Alloys of cobalt. *Physica B+C, 151*(3), 401–432. https://doi.org/10.1016/0378-4363(88)90296-3
55. Boer, F. R., Boom, R., Mattens, W. C. M., de Jong, G., & de Boer, D. K. G. (1988). Cohesion in metals: Transition metal alloys.
56. Kaplan, M. (2022). Designing metallic glasses. Alloying, properties, and degrees of freedom. Digital Comprehensive Summaries of Uppsala Dissertations from the Faculty of Science and Technology 2123. 84.
57. Andersson, J.-O., Helander, T., Höglund, L., Shi, P., & Sundman, B. (2002). Thermo-Calc & DICTRA, computational tools for materials science. *Calphad, 26*, 273–312. Elsevier.
58. Busch, R., Schroers, J., & Wang, W. H. (2007). Thermodynamics and kinetics of bulk metallic glass. *MRS Bulletin, 32*, 620–623.
59. Yang, M., Zhang, Y., Dong, J., Huang, Y., Lu, Z., Wang, L., et al. (2024). Developing centimeter-sized CuZr-based metallic glass composites via multi-element microalloying. *Intermetallics, 164*, 108114.
60. Dong, Q., Tan, J., Li, C., Sarac, B., & Eckert, J. (2024). Room-temperature plasticity of metallic glass composites: A review. *Composites Part B Engineering, 280*, 111453.

Chapter 15
Computational Thermodynamics and Kinetics

Contents

K. A. Padmanabhan et al., *Thermodynamics and Kinetics of Materials*, Engineering Materials, https://doi.org/10.1007/978-3-032-14527-7_15

Computational thermodynamics and kinetics form the foundation of modern materials science, providing powerful tools for understanding and predicting the behavior of materials across a wide range of conditions. These fields have evolved significantly over time, driven by the need to understand complex multi-component systems and dynamic material processes. The roots of computational thermodynamics trace back to the development of classical thermodynamics in the nineteenth century, with pioneers like Gibbs and Helmholtz laying the groundwork for understanding equilibrium states and phase behavior by describing them as thermodynamic functions that depend on temperature, pressure, and composition. The advent of the CALPHAD (CALculation of PHAse Diagrams) method in the mid-twentieth century marked a major milestone, as it allowed researchers to combine experimental data and thermodynamic principles into computational frameworks for accurate phase diagram prediction and property modeling.

In parallel, computational kinetics emerged to address the transient, time-dependent processes that drive material transformations. Early kinetic models, such as Fick's laws of diffusion and Avrami equations for phase transformations, provided foundational tools. The integration of thermodynamic driving forces into kinetic simulations gained momentum with advancements in computational power and algorithms, enabling the coupling of thermodynamic functions with kinetic models. This integration allows for the prediction of not only equilibrium states but also the pathways, rates, and mechanisms of transformations, such as diffusion, precipitation, deformation, and microstructure evolution, laying the foundation for material design, facilitating innovations in material development and manufacturing, process simulations, and developing advanced manufacturing technologies.

With these methods, now it is possible to explore complex multi-component systems with significantly improved efficiency and reduced computational time, which is transforming material design and process development. This chapter provides a detailed overview of computational thermodynamics and kinetics, exploring their principles, methodologies, and their applications in addressing scientific and industrial challenges. Through this integration, one can gain a deeper understanding of how thermodynamic and kinetic phenomena interact, paving the way for innovations in alloy design, energy-efficient materials, and advanced manufacturing technologies.

15.1 Thermodynamic Modeling Approaches

15.1.1 Gibbs Free Energy Minimization

Gibbs free energy minimization is a fundamental concept used in computational thermodynamics, serving as a cornerstone for understanding phase stability and equilibrium in material systems. At constant temperature and pressure, the equilibrium state of a system corresponds to the configuration in which its Gibbs free energy (G) is minimized. This principle is widely used in thermodynamic modeling to predict phase diagrams, phase equilibria, and thermodynamic properties of multi-component systems.

The Gibbs free energy of a system, G, is defined as,

$$G = H - TS \tag{15.1}$$

where H is the enthalpy, T is the temperature on the absolute scale, and S is the entropy. For a system compromising multiple phases, the total Gibbs free energy is a weighted sum, weighted as per the contribution in terms of mole fraction, of the free energies of individual phases. The individual phase Gibbs free energy depends on their compositions, temperatures, and pressures. To reach the equilibrium, the system adjusts its phases and compositions to minimize this total Gibbs free energy.

For a pure element, Gibbs free energy is a function of temperature and pressure, often expressed using standard thermodynamic data or equations of state and serves as a reference for understanding its phase transitions and stability. In computational thermodynamics, a polynomial function known as free energy polynomial, is used to describe the Gibbs free energy or Helmholtz free energy as a function of temperature, pressure, and composition. Typically, it is expressed as the sum of contributions, including the Gibbs free energy of pure components, Gibbs free energy of mixing for ideal solution, and excess Gibbs free energy that accounts for non-ideal interactions. For a binary system consisting of components A and B, the Gibbs free energy is expressed as,

$$G = x_A G_A^0 + x_B G_B^0 + RT(x_A \ln x_A + x_B \ln x_B) + G^E \tag{15.2}$$

where x_A and x_B are the mole fractions of the components, G_A^0 and G_B^0 are the Gibbs free energy of pure components, R is the gas constant, T is the absolute temperature, and G^E is the excess Gibbs free energy. Phase equilibrium in such a binary system is achieved when the chemical potential, which represents the change in Gibbs free energy with respect to number of atoms or molecules of a species added to the system of each component is the same in all coexisting phases. Mathematically, this can be expressed as,

$$\mu_A^\alpha = \mu_A^\beta, \mu_B^\alpha = \mu_B^\beta \tag{15.3}$$

where μ_A and μ_B are the chemical potential of component A and B, and α and β are the coexisting phases. This condition is automatically satisfied when Gibbs free energy is minimized, as any further change in composition would lead to an increase in Gibbs free energy, making the system unstable.

For a multi-component system, the Gibbs energy polynomial is extended by generalizing the binary formulation to account for interactions among all components in the system. For a system containing m components, the Gibbs free energy is given as,

$$G = \sum_{i=1}^{m} x_i G_i^0 + RT \sum_{i=1}^{m} x_i \ln x_i + G^E \tag{15.4}$$

The first part of the equation captures the contribution of the Gibbs free energy of the pure components, the second part of the equation captures the contribution of the Gibbs free energy of mixing (G_{mix}), and the last part captures the contribution of excess Gibbs free energy because of non-ideal interactions.

15.1.2 Solution Models and Activity Coefficients

In computational thermodynamics, as the number of components increases, the calculation of the Gibbs free energy presents significant challenges due to the complexity of interactions among multiple elements and the need to account for both ideal and non-ideal contributions. To overcome these complexities, solution models are developed which provide a systematic way to approximate these interactions by introducing empirical or semi-empirical parameters, which make them efficient for describing the thermodynamic behavior of multi-component systems, particularly alloys, ceramics, and liquid solutions. These models help predict phase stability, phase equilibria, and thermodynamic properties by quantifying how components interact within a solution. A key aspect of solution modeling is the use of activity coefficients, which account for non-ideal mixing behavior and deviations from Raoult's law.

A solution consists of two or more components mixed at atomic or molecular level. In thermodynamics, solutions can be classified based on their mixing behavior:

- **Ideal solutions**: Ideal solutions represent a system in which interactions between different components are assumed to be identical to those between same components. In an ideal solution, the enthalpy of mixing (ΔH_{mix}) is zero, meaning no heat is absorbed or released during the mixing process, and the volume change upon mixing is negligible. The Gibbs free energy of mixing (ΔG_{mix}), the second term in Eq. (15.4), is governed solely by the configurational entropy of mixing. The configurational entropy of mixing (ΔS_{mix}) arises solely from the random distribution of atoms or molecules in the solution, which is expressed as,

$$\Delta S_{\text{mix}} = -R \sum x_i \ln x_i \tag{15.5}$$

  where, R is the gas constant, and x_i is the mole fraction of the component i. While real solutions often deviate from ideal behavior due to interactions between dissimilar components, it serves as the foundational concept in thermodynamics, providing benchmarks for understanding more complex systems. Ideal solutions are rare, but some systems closely approximate ideal behavior under specific conditions, as mentioned in the examples below.

   - *Benzene and Toluene:* This binary mixture of these organic liquids is often cited as a near-ideal solution because their molecules have similar sizes, structures, and intermolecular forces, leading to near zero enthalpy of mixing.
   - *Hexane and Heptane:* These two alkanes form an ideal solution due to their similar molecular structure and weak van der Waals interactions, resulting in minimal deviation from ideal mixing.
   - *Liquid Metal Alloys:* Some binary metal systems, like Au-Cu at high temperatures, exhibit nearly ideal behavior due to similar atomic sizes and bonding characteristics of the components.
   - *Gas Mixtures:* Mixture of gases, such as nitrogen and oxygen at low pressure, behave ideally because the intermolecular interactions are negligible, and the entropy of mixing dominates the Gibbs free energy.

- **Non-Ideal Solutions**: Non-ideal solutions deviate from ideal behavior due to the differences in the interactions between unlike components compared to like components, leading to non-zero enthalpy of mixing. These deviations arise from factors such as differences in molecular size, polarity, or bonding characteristics, which causes preferential interactions, or ordering within a solution. The Gibbs free energy of mixing for such non-ideal solutions includes an excess energy term (G^E), third term in Eq. (15.4), accounting for these interactions. Activity coefficients (γ) are introduced to quantify the deviation of a component's behavior from ideality. The activity coefficient captures the effect of non-ideal interactions, which influences the chemical potential and thermodynamic properties of the solution. The activity coefficient can be,

   - $\gamma > 1$: Component experiences repulsive or unfavorable interactions (positive deviation from ideality), e.g., ethanol and water.
   - $\gamma < 1$: Component experiences attractive or stabilizing interactions (negative deviation from ideality), e.g., chloroform and acetone.
   - $\gamma = 1$: Ideal solution behavior.

   The Gibbs free energy of mixing for a non-ideal solution can be expressed using activity coefficients, as given below,

$$\Delta G_{\text{mix}} = RT\sum\nolimits_i x_i \ln a_i \tag{15.6}$$

Where, $a_i = \gamma_i x_i$ is the activity, x_i is the mole fraction, and γ_i is the activity coefficient. By incorporating activity coefficients, the excess Gibbs free energy of non-ideal solutions can be accurately described, enabling the prediction of phase equilibria, chemical potentials, and other critical properties. The activity coefficient deviates from 1 for non-ideal solution because of non-ideal interactions between different species, such as difference in size, charge, or intermolecular forces, which causes the actual behavior of the solution to differ from the ideal behavior.

To represent the non-ideal behavior of a multi-component systems in a systematic and mathematically consistent way, common solution models are utilized, which capture the deviations from ideal behavior by accounting for real interactions between species such as size differences, enthalpic effects, and ordering tendencies, which are critical for accurate prediction of phase equilibria, activity coefficients, and thermodynamic properties.

15.1.2.1 Common Solution Models

To model real solution behavior, various mathematical formulations exist, each incorporating excess Gibbs free energy terms to account for non-ideal mixing.

Regular Solution Model
Regular solution model assumes non-ideal mixing where interactions between different atoms contribute to the enthalpy of mixing. The model assumes that the enthalpy of mixing (ΔH_{mix}) is non-zero due to interactions between unlike atoms or molecules, while the entropy of mixing remains ideal. The excess Gibbs free energy of a binary solution is given as,

$$G^E = \Omega x_A x_B \quad (15.7)$$

where Ω is an interaction parameter, and x_A, x_B are the mole fraction of components A and B. The activity coefficients for such binary solution are derived from a relationship between the excess Gibbs free energy and mole fraction of the components, given as

$$RT \ln \gamma_A = \Omega x_B^2, RT \ln \gamma_B = \Omega x_A^2 \quad (15.8)$$

where γ_A and γ_b are the activity coefficients of the component A and B. When this model is extended to multi-component systems, it incorporates pairwise interaction parameters (Ω_{ij}) between each component pair, with the excess Gibbs free energy expressed as,

$$G^E = \sum_{i=1}^{n-1} \sum_{j>i}^{n} \Omega_{ij} x_i x_j \quad (15.9)$$

where n is the number of components. This model is useful for systems with moderate deviations from ideality but does not account for asymmetry in mixing behavior.

Redlich-Kister Subregular Models
The Redlich-Kister model is a widely used polynomial expansion to describe the excess Gibbs free energy (G^E) of binary solution, offering a flexible and systematic way to capture the non-ideal behavior. The model expresses G^E as a power series in terms of the mole fractions of the components, with coefficients that depend on temperature. The general form for a binary system is given as,

$$G^E = \sum_{i=1}^{n-1} \sum_{j>i}^{n} x_i x_j \sum_{k=0}^{m} L_k (x_i - x_j)^k \tag{15.10}$$

where L_n are the temperature-dependent interactions parameters, and m is the order of the polynomial. The first term ($n = 0$) corresponds to the regular solution model, while higher-order terms ($n \geq 1$) accounts for asymmetric and more complex interactions. Subregular model is a specific case of Redlich-Kister model, where only specific interaction parameters are considered non-zero.

The Redlich-Kister model can be extended to multi-component systems by considering pairwise interactions between each component pair. For example, in a ternary system, G^E is given as,

$$\begin{aligned} G^E = {} & x_A x_B \sum_{n=0}^{m} L_{AB,n} (x_A - x_B)^n + x_A x_C \sum_{n=0}^{m} L_{AC,n} (x_A - x_C)^n \\ & + x_B x_C \sum_{n=0}^{m} L_{BC,n} (x_B - x_C)^n \end{aligned} \tag{15.11}$$

where $L_{ij,\, n}$ represents the interaction parameters corresponding to the $i - j$ component pair. The Redlich-Kister model provides greater flexibility by including higher-order terms to fit experimental data, making it suitable for systems with complex composition-dependent behavior.

Wilson Model

The Wilson model is an advanced thermodynamic model used to describe non-ideal solutions, particularly those with strong deviation from ideality due to differences in molecular size and interactions. The Wilson model is based on a local composition concept, where arrangement of molecules in the solution is influenced by the energy difference between unlike and like interactions. The excess Gibbs free energy based on Wilson model is given as,

$$G^E = -RT \sum_i x_i \ln \left(\sum_j x_j \Lambda_{ij} \right) \tag{15.12}$$

where, x_i and x_j are mole fractions of components i and j, and Λ_{ij} are the temperature-dependent interaction parameters that describe the energy difference between unlike $(i - j)$ and like $(i - i$ or $j - j)$ interactions. This model can capture local composition effects, where the arrangements of molecules are influenced by the relative strengths of interactions between different components, such as those involving polar or associating liquids.

Associative and Quasi-chemical Models

The associative and quasi-chemical models are advanced thermodynamic approaches used to describe solutions in which specific interactions, such as hydrogen bonding, formation of complexes, or ordering, lead to significant deviations from ideal behavior. The associative model focuses on the formation of chemical complexes or clusters within solution, treating these associations as distinct species in equilibrium with free molecules. The quasi-chemical model is based on the concept of local coordination and pairwise interactions between atoms or molecules. It considers the arrangement of nearest neighbors and their energetics, making it suitable for describing ordering phenomenon in metallic alloys or strongly interacting liquids.

Solution models are indispensable tools in computational thermodynamics, providing a mathematical framework to describe the behavior of mixtures, from simple binary alloys to complex multicomponent systems. Their utility lies in their ability to quantify non-ideal interactions through activity coefficients, excess Gibbs free energy, and composition dependent parameters, enabling accurate predictions of phase equilibria, chemical potentials, and thermodynamic stability.

However, solution models also have limitations. Their accuracy heavily depends on quality of experimental data used for parameter calibration, and extrapolation the beyond fitted compositions or temperatures can lead to error. Simplified models (e.g., ideal or regular solution model) may fail to capture the complex short-range ordering or clustering effects, while more sophisticated models (e.g., quasi-chemical or associative models) require extensive computational efforts.

Despite these challenges, solution models remain foundational in materials modelling. Future advancements integrating machine learning and high-throughput data generation promise to enhance their predictive power, making them even more robust for next-generation materials engineering.

15.1.3 Compound Energy Formalism

Building upon the framework of solution models, we now turn to the Compound energy formalism (CEF), a more advanced and versatile approach for modeling phases with complex crystal structures. While traditional solution models treat phases as randomly mixed sublattices, CFE explicitly accounts for site-specific occupancies of atoms, vacancies, or ions, making it indispensable for ordered phases, interstitial solutions, and ionic compounds. Unlike simple solution models that assume a single mixing mechanism, CFE partitions the Gibbs free energy into contributions from end-member compounds and interaction parameters within sublattices, enabling precise descriptions of non-stoichiometry, defect thermodynamics, and short-range ordering. This section explores how CFE extends beyond conventional solution models to address the rich thermodynamic behavior of intermettalics, ceramics, and other structurally complex materials, bridging atomic-scale site preferences with macroscopic phase stability.

In many materials, atoms do not mix randomly but instead exhibit ordering tendencies based on atomic interactions and lattice structures. The CEF method models such systems by dividing the crystal structure into sublattices, where specific atomic species preferentially occupy certain sites. The Gibbs free energy of a phase is expressed as,

$$G = \sum\nolimits_{\alpha} y_\alpha G_\alpha + RT \sum\nolimits_{\alpha} y_\alpha \ln y_\alpha + G^E \tag{15.13}$$

Where G_α is the Gibbs free energy of a particular atomic configuration (α), y_α represents the fraction of a given atomic configuration, the second term accounts for a configurational entropy due to atomic mixing, and G^E represents interaction contributions beyond ideal mixing—see Sect. 15.1.2. By defining sublattices and interaction parameters, CEF allows for a structured and physically meaningful way to describe the complex crystal structure and site occupancies of phases such as intermetallic compounds, ceramics, and ionic solids.

Traditional solution models treat the entire phase as a single mixing entity with composition-dependent interactions, CFE revolutionizes this approach by decomposing the crystal structure into distinct sublattices, each with its own independent concentration variables. This critical innovation allows the model to capture a far broader range of phase behaviors than standard solution theories, like ordered intermetallic, non-stoichiometric compounds, and ionic solids by explicitly accounting for site-specific occupancies of atoms, ions, and even vacancies.

15.1.3.1 Structure of Sublattice in CEF

CEF models multi-component systems by introducing sublattice variables, where different atomic species can occupy distinct sites. A generic sublattice representation for a system containing elements *A*, *B*, *C*,and *D* is given by,

$$(A, B)_p (C, D)_q \tag{15.14}$$

Where $(A, B)_p$ represents a sublattice with p sites, which can be occupied by elements A or B, and $(C, D)_q$ represents a second sublattice with q sites, occupied by elements C or D. For example, in an intermetallic compound like *NiAl* (Nickel Aluminide, *B*2 structure, which is a superstructure body centered cubic lattice), *Ni* and *Al* occupy distinct sublattices,

$$(Ni, Al)_1 (Ni, Al)_1 \tag{15.15}$$

- In the ordered state, *Ni* prefers one sublattice and *Al* prefers the other, where in one sublattice the atom occupies the corner of the cube and other occupies the center of that cube.
- In the disordered state, *Ni* and *Al* mix randomly on both sublattices.
- The transition between these states is modeled using CEF.

Another example of the sublattice representation is the carbo-nitride of *Nb* and *Ti* in steels, where *Nb* and *Ti* occupy one sublattice, the corner and the center of the cube, and *C* and *N* occupy the other sublattice, which are the interstitial sites within the cubic structure.

$$(Nb, Ti)_1(C, N)_1 \tag{15.16}$$

This approach accurately captures the specific structure and compositional features of ordered phases like oxides, nitrides, carbides, etc. Ordered phases often have complex crystal structures where atoms occupy distinct sublattices in a highly organized manner. The sublattice approach allows CFE model to,

(a) **Describe Site Occupancy:** It explicitly accounts for the occupation of different species on specific sublattices.
(b) **Capture Ordering Effect:** It describes the energetic preference for certain atoms to occupy a specific sublattice.
(c) **Model Non-Stoichiometry:** The sublattice model can handle non-stoichiometric behavior by allowing vacancies or substitutions on specific sites.
(d) **Incorporate Interactions:** It includes interaction parameters within and between sublattices, enabling the description of complex mixing behavior and phase stability.

The sublattice approach also captures disordered phases by providing a flexible framework to describe the random distribution of atoms or vacancies across sublattice while still capturing the underlying crystal structure. The sublattice approach allows CFE model to,

(a) **Represent Random Mixing:** It can describe random occupation of sublattice by different species.
(b) **Incorporate Site Preference**: Even in disordered phases, certain atoms may have preferences for specific sublattice.
(c) **Handle Non-Ideal Interactions:** It includes interaction parameters to describe deviation from ideal mixing.
(d) **Seamless Transition to Ordered Phases:** Same model can describe ordered and disordered phases, enabling consistent modeling of phase transition.

Thus, the sublattice model in the CFE enables the unified description of both ordered phases by explicitly accounting for site occupancies and interactions, providing a consistent and flexible framework for modeling complex materials.

15.1.4 CALPHAD Method

The CALPHAD (CALculation of PHAse Diagrams) method is a widely used computational approach for predicting phase equilibria, thermodynamic properties, and microstructural evolution in multi-component systems. Developed in the 1970s, CALPHAD combines experimental data, theoretical models, and optimization techniques to construct self-consistent thermodynamic databases, enabling the prediction of phase evolution across various conditions [1]. It has become a cornerstone of modern materials science, facilitating alloy design, process optimization, and integration with kinetic models for non-equilibrium simulations.

The CALPHAD method is based on the principle that the Gibbs free energy (G) of all stable and metastable phases in a system can be described as functions of composition, temperature, and pressure. Using this approach, phase diagrams and thermodynamic properties can be computed by minimizing the total Gibbs free energy of the system as described in the Sect. 15.1.

The CALPHAD method involves,

- **Experimental data collection:** Experimental data on phase equilibria, thermodynamic properties, and crystal structure are the pre-requisites for CALPHAD method. Phase equilibria data includes description of phase boundaries, such as liquidus, solidus, and solvus lines, obtained from techniques such as differential thermal analysis (DTA), differential scanning calorimetry (DSC), or metallography. Thermodynamic properties data include data on enthalpy, entropy, heat capacity, and energies of formation, measured using calorimetry or adiabatic methods. Crystal structure data includes lattice parameters, and how it changes with composition and temperature, measured using X-ray diffraction (XRD) or dilatometry. Some additional data include component activities or chemical potential measured using electrochemical methods, solubility limits of components in different phases determined using XRD or electron probe microscopy (EPM), and magnetic transitions measured using magnetometry.
- **Thermodynamic models:** Mathematical functions represent Gibbs free energy for different phases, including pure elements and solutions, as described in Sects. 15.1.2 and 15.1.3, forms the core component of CALPHAD method.
- **Optimization and database development:** Parameters of the above said equations are estimated through optimization methods using experimental data, forming a self-consistent thermodynamic database.
- **Phase diagram calculation:** Algorithms are utilized to solve the set of nonlinear equations for Gibbs free energy minimization to predict equilibrium state of the system for a given composition and temperature. By systematically varying the temperature, pressure, and composition, the phase boundaries are calculated, leading into a phase diagram.

Several models are used to describe the Gibbs free energy of different phases,

- **Pure Element Model:** The Gibbs free energy of a pure substance is expressed as a function of temperature and pressure from experimental data [2].
- **Solution Model:** Describes the Gibbs free energy of multi-component solutions using terms of entropy and enthalpy of mixing, as described in Sect. 15.1.2.
- **Compound Energy Formulation:** Used for ordered phases and intermetallic compounds by describing sublattice occupation probabilities, as described in Sect. 15.1.3.
- **Gibbs Energy Polynomial Expressions:** Provide an empirical fit to solid, liquid, and gas phases (Eq. 15.4).

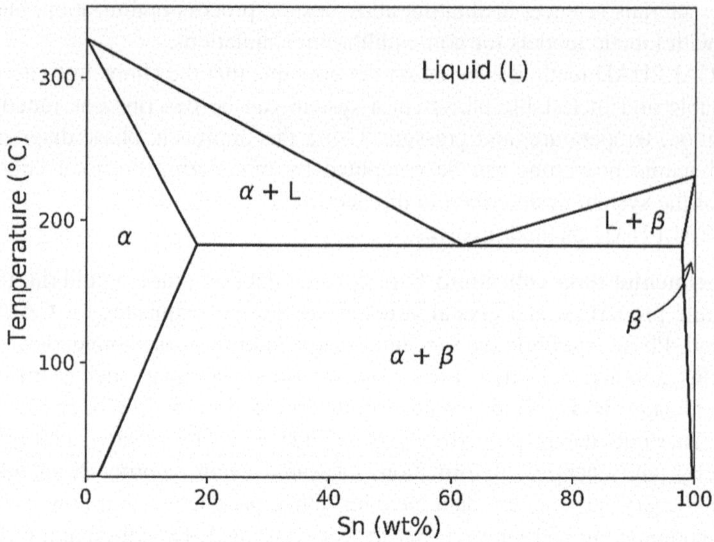

Fig. 15.1 Phase diagram of Pb-Sn binary system.[1, 2]

These models allow CALPHAD to construct thermodynamic descriptions of complex materials, extending traditional phase diagrams that are limited to binary or ternary systems, into higher-dimensional composition space. To achieve this, CALPHAD combines the following methodologies,

- **Extrapolation from Binary/Ternary Systems:** Using thermodynamic descriptions of lower-order systems to predict higher-order behavior.
- **Experimental Validations:** Comparing calculated results with experimental phase diagrams and thermodynamic measurements.
- **Coupling with First-Principles Calculations:** Using density functional theory (DFT) data to improve parameterization, particularly for systems with limited experimental data.
- **Coupling with Molecular Dynamics Simulations:** Using molecular dynamics simulations to provide additional insights into atomic-scale interactions, thermodynamic properties, and phase behavior that may be difficult to obtain experimentally.

For a binary alloy system containing *Pb* and *Sn*, a phase diagram is constructed using the CALPHAD method as shown in Fig. 15.1. The phase diagram is constructed using the thermodynamic database for solder system [3], and a thermodynamic tool PyCalphad [4].

The phase diagram consists of two solid solutions, α which is rich in *Pb* and β which is rich in *Sn*. The binary eutectic composition of this alloy is primarily used as the solder in the electronic industry, due to its low melting point and good wetting properties.

15.2 Kinetic Modeling Approaches

The study of kinetic modeling focuses on the time-dependent processes that govern the evolution of materials under various conditions. Unlike thermodynamics, which predicts equilibrium states, kinetics provides insights into how quickly a system evolves towards equilibrium and the mechanisms involved. This section briefly explores various modeling approaches that integrate thermodynamic principles to describe diffusion, phase transformations, reaction rates, and microstructure evolution to set stage for the next section on coupling with computational thermodynamics.

15.2.1 Diffusion Based Modeling

Diffusion is a fundamental kinetic process governing mass transport in material, playing a crucial role in phase transformations, microstructure evolution, and overall material behavior. In computational thermodynamics and kinetics, diffusion-based modeling provides a framework to predict how atoms or molecules move in response to concentration gradients, temperature, and external fields like, magnetic field, stress, etc. This approach is essential for simulating heat treatment, alloy design, solidification, and oxidation processes.

15.2.1.1 Fundamentals of Diffusion

Diffusion occurs due to random atomic motion driven by gradients in concentration, chemical potential, or temperature. The rate of diffusion is governed by Fick's law.

Fick's First Law (Steady-State Diffusion)
Fick's first law describes the diffusion flux in a system where the concentration gradient remains constant over time, meaning the system has reached a steady-state condition. The law is expressed as,

$$J = -D\frac{dC}{dx} \tag{15.17}$$

Where J is the diffusion flux, i.e., amount of substance transported per unit area per unit time, D is the diffusion coefficient, which is dependent on the temperature, C is the concentration of the diffusing species, and x is the position along the diffusion direction. The negative sign indicates that diffusion occurs down the concentration gradient, meaning atoms move from a region of high concentration one to low concentration.

Example Application of Fick's First Law

- In steady-state diffusion, such as gas diffusion through a thin membrane, the concentration gradient does not change with time.
- Used in coating processes, where a material is deposited at a constant rate on a surface.

Fick's Second Law (Time-Dependent Diffusion)
Fick's Second law describes how concentration changes over time when a system is in a non-steady-state condition, i.e., the concentration gradient is evolving. The law is expressed as,

$$\frac{\partial C}{\partial t} = D\frac{\partial^2 C}{\partial x^2} \tag{15.18}$$

where $\frac{\partial C}{\partial t}$ is the rate of change of concentration of species C with time, D is the diffusion coefficient, and $\frac{\partial^2 C}{\partial x^2}$ is the second spatial derivative of concentration, representing how the concentration gradient itself is changing. This law shows that diffusion is faster where the concentration gradient is steep.

Example Application of Fick's Second Law

- Used in heat treatment of alloys, such as carbon diffusion in steel during carburization.
- Important in solid-state diffusion processes, such as impurity diffusion in semiconductors.
- Describes drug release from pharmaceutical coatings, where concentration changes over time.

15.2.2 Phase Transformation Kinetics

Phase transformations are fundamental processes in material science, where a material changes from one phase to another due to change in temperature, pressure, or composition. The kinetics of phase transformations describe how fast these changes occur, which is crucial for controlling microstructure evolution and material properties. Unlike equilibrium thermodynamics, which predicts the final phase stability, phase transformation kinetics determines the time-dependent pathways and rates of these changes. Understanding phase transformation kinetics is essential for applications such as heat treatment, alloy design, and additive manufacturing.

15.2.2.1 Types of Phase Transformations

Phase transformations can be broadly classified based on the mechanism and kinetics involved as follows,

Diffusional Transformation (First-Order Transformation)
These transformations involve atomic diffusion over a finite time scale and are governed by a nucleation and growth mechanism. Diffusional transformations are broadly categorized into replacive and reconstructive types. In replacive transformations, atoms shift positions within the existing crystal lattice without altering its fundamental structure. In contract, reconstructive transformations involve the formation of a new phase with a distinctly different crystal structure, requiring significant atomic rearrangement and bond breaking.

A. Solid-State Phase Transformations

- Occur between different solid phases due to a change in temperature or composition.
- It is a reconstructive transformation.
- Examples:
  - Pearlitic transformation in steel: γ-Fe (Austenite) $\rightarrow\alpha$-Fe (Ferrite) $+Fe_3C$ (Cementite)
  - Precipitation hardening: Al-Cu alloy where $\theta^{'}$ precipitates strengthen the matrix

B. Solidification (Liquid to Solid)

- Involves nucleation of solid phase from a liquid followed by growth.
- It is a reconstructive transformation.
- Examples:
  - Dendritic growth during solidification.
  - Formation of eutectic and peritectic structures in alloy systems.

C. Recrystallization and Grain Growth

- It is a replacive transformation.
- Recrystallization is driven by stored energy from plastic deformation.
- Grain growth is driven by curvature-driven grain boundaries to reduce the total grain boundary area.
- Example:
  - Annealing in metals to reduce dislocation density and stress relieving.

Diffusionless Transformation (Second-Order Transformations)
Diffusionless transformations also known as displacive transformations occur by shear-like atomic displacements without long-range diffusion, leading to rapid transformations. The atoms do not move with independent degrees of freedom, so the change in configurational entropy is negligible. The classical example is the martensitic transformation occurring in steels. Initially referred to the microstructure

formed by quenching the steel, martensite is now referred to any diffusionless phase transformation as seen in non-ferrous and ceramic materials also.

Spinodal Decomposition

Spinodal decomposition is a mechanism by which a single thermodynamic phase spontaneously separates into two phases with distinct compositions without nucleation. Decomposition occurs when there is no thermodynamic barrier to phase separation. As a result, phase separation via decomposition does not require the nucleation events resulting from thermodynamic fluctuations, which normally trigger phase separation. Spinodal decomposition is observed in polymer blends and metallic alloys.

Magnetic and Electronic Phase Transformations

Magnetic transition refers to the change in magnetic properties that occurs in a material as a function of temperature and magnetic field. The phase transformation of a low-temperature ferroelectric phase to a high-temperature paraelectric phase upon heating a ferroelectric material is called curie transition, and the temperature where the transition occurs is called curie temperature.

Electronic phase transitions refer to changes in the electronic structure of a material due to variations in temperature, pressure, or external fields. These transitions involve a rearrangement of electron states, leading to significant changes in electrical, optical, and magnetic properties without necessarily altering the atomic structure. They are often driven by factors such as band structure modifications, electron correlation effects, or interactions with crystal lattice.

Examples:

- Metal-insulator transitions occur when a material transitions between conducting and insulating states due to electron localization or delocalization. Vanadium dioxide (VO_2) is a conductor at high temperature (68 °C), whereas at lower temperatures, it becomes an insulator due to strong electron-electron interactions.
- Superconducting transitions occur when a material transitions from a normal conductive state to a zero-resistance superconducting state below the critical temperature. Niobium (Nb) becomes superconducting below 9 K.

15.2.2.2 Nucleation and Growth Kinetics

Nucleation is the initial formation of a new phase, while growth describes its expansion over time. The kinetics of these processes determine the final phase distribution and microstructure.

Classical Nucleation Theory (CNT)

Classical nucleation theory is the most common theoretical model used to quantitatively study the kinetics of nucleation. CNT was originally derived for the formation of nuclei from supersaturated water vapor and was conceptually transferred to the nucleation of crystals from solution. The nucleation rate based on the CNT is given as,

$$N = N_0 Z \exp\left(-\frac{\Delta G^*}{k_B T}\right) \tag{15.19}$$

Where N is the rate of nucleation per unit time per unit area, N_0 is the number of available nucleation sites per unit area, Z is the Zeldovich factor which gives the probability that a nucleus at the top of the barrier will go on to form the new phase, rather than dissolve, ΔG^* is the activation energy for nucleation, k_B is the Boltzmann constant, and T is the absolute temperature.

Nucleation can be of two different types,

- Homogeneous nucleation occurs uniformly throughout the material but requires high energy due to the absence of preferred nucleation sites.
- Heterogeneous nucleation occurs at grain boundaries, dislocations nodes, or surfaces, reducing energy barriers and making nucleation easier.

Growth Kinetics

Once nucleation occurs, the newly formed nuclei grow until the system reaches equilibrium. Growth kinetics describes the rate and mechanisms by which this transformation proceeds, influenced by factors such as diffusion, interface mobility, and thermodynamic driving forces. Growth can be classified into different modes based on different limiting factors.

- **Diffusion-controlled growth:** The growth rate is governed by atomic diffusion across phase boundaries. This is common in precipitation and solidification.
- **Interface-Controlled growth:** The growth is limited by the movement of the phase boundary, often observed in martensitic transformations and rapid solidification.

The general growth law follows a parabolic trend given as,

$$r(t) = kt^n \tag{15.20}$$

where $r(t)$ is the phase radius at time t, k is the growth rate constant, and n is the growth rate exponent depending on the mechanism.

Coarsening and Ostwald Ripening

As the phase growth progresses, smaller particles dissolve while larger particles grow, a process known as coarsening. This phenomenon occurs due to the reduction of total system energy, driven by differences in chemical potential between particles of different sizes.

Ostwald ripening is a type of coarsening where larger particles grow at the expense of smaller ones due to the differences in surface curvature and solubility.

This is governed by the Lifshitz-Slyozov-Wagner (LSW) theory, where the average particle size is given as,

$$r^3 - r_0^3 = kt \tag{15.21}$$

where r is the mean particle radius, r_0 is the initial particle radius, k is the material-dependent constant, and t is time.

15.2.2.3 Johnson-Mehl-Avrami-Kolmogorov (JMAK) Model

The JMAK model is widely used as a framework for describing the kinetics of phase transformations combining both nucleation and growth. It quantifies how the fraction of the transformed product evolves over time, making it essential for modeling solid-state transformations, such as recrystallization, precipitation, and phase decomposition. The JMAK model expresses the transformed fraction of a material, $X(t)$, as,

$$X(t) = 1 - \exp(-kt^n) \tag{15.22}$$

where $X(t)$ is the volume fraction of the transformed phase at time t, k is the rate constant dependent on the temperature and material properties and n is the Avrami exponent which is dependent on the nucleation and growth mechanism. The model assumes random nucleation of the new phases and isotropic growth of nuclei in an infinite medium.

The Avrami exponent, n, provides insight into the transformation mechanism.

- $n = 1 \rightarrow$ Interface-controlled transformation with constant nucleation rate
- $n = 2 - 3 \rightarrow$ Diffusion-controlled growth in two or three dimensions
- $n = 4 \rightarrow$ Site-saturated nucleation with three-dimensional growth.

The JMAK model is widely used for solid-state phase transformations, such as recrystallization and precipitation.

15.2.3 Reaction Rate Models

Reaction rate models play a crucial role in computational thermodynamics and kinetics by describing how the concentration of reactants and products evolves over time in a chemical or phase transformation process. These models help predict the dynamics of diffusion, precipitation, phase transitions, and solid-state reactions, providing insights into microstructural evolution and material processing.

15.2.3.1 Rate Law

The most fundamental approach to reaction rate modelling is the rate law, which expresses the rate of a reaction as a function of reactant concentrations.

$$r = k[A]^m[B]^n \tag{15.23}$$

where r is the reaction rate, k is the rate constant, $[A]$ and $[B]$ are the molar concentrations of the reactants, and exponent m and n are the reaction orders. The rate constant, k, and the reaction orders are determined experimentally by observing how the rate of reaction changes as the concentrations of the reactants are changed. The rate constant, k, is independent of the concentrations of the reactants, and is dependent on the temperature at which the reaction takes place.

The order of a reaction is expressed as the sum of the exponents of the concentration term in the rate law equation. A reaction can be zero-order, where the rate of reaction is independent of reactant concentration, so changing the reactant concentrations has no effect on the rate of reaction. Decomposition of ammonia on a hot platinum surface is an example of a zero-order reaction.

$$NH_3 \rightarrow N_2 + H_2 \tag{15.24}$$

A first-order reaction is one where the rate of reaction is directly proportional to one reactant's concentration, and the other reactants can be present, but their concentrations will not have any effect on the rate of the reaction. Radioactive decay of isotopes is an example of first-order reaction.

$$^{14}N + \gamma_{\text{rays}} \rightarrow {}^{13}N + n \tag{15.25}$$

A second-order reaction is one where the rate of reaction is dependent on the square of one reactant's concentration or the product of two different reactant concentrations. Some examples of second-order reaction are as follows.

$$2NO_2 \rightarrow 2NO + O_2 \tag{15.26}$$

$$2HI \rightarrow H_2 + I_2 \tag{15.27}$$

Higher order reactions, such as third-order reactions, are less common but do exist. For example, the oxidation of nitrogen monoxide,

$$2NO + O_2 \rightarrow 2NO_2 \tag{15.28}$$

Understanding the reaction order is crucial for predicting the reaction behavior. The rate law approach is widely used in modeling simple reactions such as oxidation, reduction, and catalytic processes in material science.

15.2.3.2 Transition State Theory

Transition state theory (TST) describes the reaction rate based on the formation of a high-energy intermediate state, known as an activated complex or transition state, between reactants and products. According to TST, reactant molecules must acquire sufficient energy to reach this unstable transition state, which then rapidly converts into product (Fig. 15.2). This theory introduces the concept of the activation energy, which is the minimum energy required to reach the transition state. The rate of reaction depends on how many atoms/molecules can overcome this energy barrier, which is influenced by the temperature and the catalyst. The reaction rate is given as,

$$r = \frac{k_B T}{h} \exp\left(-\frac{\Delta G^*}{RT}\right) \tag{15.29}$$

where r is the reaction rate, k_B is Boltzmann's constant, h is Planck's constant, T is the absolute temperature, ΔG^* is the activation energy, and R is the gas constant. The presence of catalysts lowers the activation energy by providing an alternative reaction pathway, stabilizing the reaction state, and increasing the reaction rate. TST is particularly useful in modeling solid-state diffusion, precipitation reactions, rate of deformation in solid state, phase transformations and so on.

15.2.3.3 Arrhenius Equation

The Arrhenius equation describes how the reaction rate constant k varies with the temperature.

$$k = A \exp\left(-\frac{E_a}{RT}\right) \tag{15.30}$$

Where A is the pre-exponential factor, E_a is the activation energy, R is the gas constant, and T is the absolute temperature. The equation highlights that as the temperature increases, the exponential term becomes larger, leading to a higher rate

Fig. 15.2 Evolution of the Gibbs Free energy of a Transition state reaction

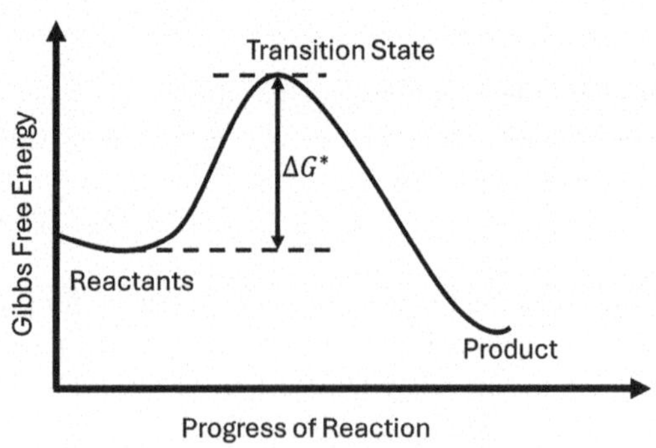

constant and consequently a faster reaction. In collision theory, Arrhenius equation helps explain how increasing temperature enhances the fraction of atoms/molecules with sufficient energy to overcome the activation barrier. Arrhenius equation is widely used to predict temperature-dependent behavior in processes such as combustion reactions, enzyme kinetics, and material degradation. This model is also widely applied in predicting diffusion-controlled reactions and thermally activated processes such as recrystallization, grain growth, and high temperature deformation.

15.2.3.4 Parabolic Rate Law

Parabolic rate law is the basis of an empirical kinetic model that describes diffusion-controlled reactions, particularly in high-temperature oxidation, corrosion, and phase transformation processes. It is commonly observed in systems where growing reaction product layer acts as a diffusion barrier, slowing down the rate of the reaction over time. Unlike linear kinetics, where the reaction rate remains constant, the parabolic rate law predicts that the rate decreases as the reaction progresses due to the increasing thickness of the product layer. The parabolic rate law is expressed as,

$$x^2 = kt \tag{15.31}$$

Where x is the thickness of the reaction product layer, k is the parabolic rate constant, and t is time.

The parabolic rate law arises in systems where the reaction is diffusion-controlled, meaning that the rate is governed by the movement of the reactant species through a solid or liquid reaction product layer rather than the availability of reactants at the surface. This phenomenon is commonly seen in high-temperature oxidation of metals, solid-state diffusion reactions like growth of intermetallic compounds at the metal interface, silicon oxidation in semiconductor processing, and corrosion of metals where a passive oxide layer forms among others.

Reaction rate models provide fundamental understanding of how materials transform over time under different conditions. Whether through empirical rate law, statistical mechanisms, or diffusion-controlled models, these approaches enable accurate predictions of microstructure evolution, aiding in the design and optimization of materials for industrial applications. The choice of the approach depends on the nature of transformation, the governing mechanism, and the available experimental data.

15.2.4 *Interface Controlled Processes*

Interface-controlled processes play a crucial role in phase transformation, grain growth, and microstructure evolution in materials. Unlike diffusion-controlled processes, where the rate limiting step is mass transport, interface-controlled processes are governed by the movement of phase boundaries, grain boundaries, or interphase interfaces. The kinetics of these transformations depend on factors such as interfacial energy, mobility, crystallographic orientations, and external driving forces such as temperature, pressure, and stress.

15.2.4.1 Mechanisms of Interface-Controlled Processes

(a) Nucleation and Growth

In many solid-state transformations, a new phase first nucleates and then grows by interfacial motion. If atomic attachment at the interface is slow due to structural or energetic constraints, the process is interface controlled. The velocity of the moving interface is determined by the rate of atomic rearrangement and interfacial mobility rather than long-range diffusion.

(b) Interface Migration and Grain Growth

Grain growth is a classic example of an interface-controlled process. Grain boundaries move to reduce the total energy of the system by minimizing the overall grain boundary area. The rate of growth is dictated by the boundary mobility and the curvature driven capillary force, given as,

$$v = M\gamma\kappa \tag{15.32}$$

where v is the velocity of the grain boundary, M is the grain boundary mobility, γ is the grain boundary energy, and κ is the local curvature of the grain boundary.

(c) Coherent and Incoherent Interfaces

The nature of the interface significantly affects the kinetics of the phase transformations. Coherent interfaces have a well-matched atomic arrangement, leading to slower interface migration. Incoherent interfaces have significant lattice mismatch, which promotes faster migration due to higher interfacial energy and interface mobility.

(d) Discontinuous Precipitation

In some alloy systems, precipitation occurs by the motion of a migrating interface rather than by bulk diffusion. The process is interface-controlled when the nucleation and growth of the new phase are governed by the interfacial attachment kinetics rather than solute diffusion in the bulk. An example of discontinuous precipitation is the precipitation of Ag in Cu-Ag alloy.

(e) Martensitic Transformation

Martensitic transformations are diffusionless and occur via coordinated atomic displacements across the interface. Since no long-range diffusion is involved, the kinetics of martensitic transformations are governed entirely by interface mobility and applied stress.

15.2.4.2 Mathematical Description of Interface-Controlled Growth

For an interface-controlled transformation, the growth rate of the interface is expressed as,

$$v = M\left(\Delta G_{\text{driving}} - \Delta G_{\text{resistance}}\right) \tag{15.33}$$

where v is the interface growth rate, M is the interface mobility, $\Delta G_{\text{driving}}$ is the driving force for the transformation, and $\Delta G_{\text{resistance}}$ accounts for factors opposing the interface motion, such as elastic strain or solute drag.

In grain growth, the rate follows a parabolic law, given by,

$$r^2 - r_0^2 = kt \tag{15.34}$$

where r is the grain radius at time t, r_0 is the initial grain radius, and k is the rate constant that depends on the temperature and material properties. The kinetics of the interface-controlled transformation are influenced by temperature, interfacial energy, external stress, and solute drag effect.

15.3 Coupled Computational Thermodynamic and Kinetic Modeling Approaches

Coupled computational thermodynamic and kinetic modeling approaches provide a comprehensive framework for predicting material behavior by integrating equilibrium thermodynamics with time-dependent kinetic processes. While thermodynamics defines the stable phases and equilibrium states of a system, kinetics governs the rate and pathways through which these states are reached. By combining these two aspects, computational models can accurately describe phase transformations, diffusion, nucleation, and microstructure evolution in a complex multi-component system. These coupled models are essential for designing advanced materials, optimizing industrial processes, and understanding the dynamic nature of material behavior under real-world conditions. Different types of these coupled approaches are described in this section.

15.3.1 *Mean-Field Theories (MFTs)*

Mean-Field theories (MFTs) provide a powerful framework for modelling phase transformations, diffusion, and microstructural evolution in materials by approximating the effect of interactions between atoms, defects, or phases. These models replace the complex local interactions with an averaged or mean-field representation, making it computationally feasible to study large-scale material behavior. MFTs are widely applied in computational thermodynamics and kinetics to bridge atomic-scale interactions with continuum-scale phenomena, enabling prediction of phase equilibria, order-disorder transitions, and reaction kinetics.

MFTs assume that each particle or phase in a system experiences an average interaction field generated by all other particles rather than considering explicit local interactions. This approximation simplifies the governing equations and allows for analytical and numerical solutions to otherwise complex thermodynamic and kinetic problems.

15.3.1.1 Types of Mean-Field Approaches

(a) Bragg-Williams Approximation

The Bragg-Williams model is a simple mean-field approach used to describe order-disorder transformations in alloys. It simplifies the complex interactions between atoms by assuming that each atom experiences an average field due to its neighbors, neglecting local fluctuations and correlations. In this approximation, the configurational entropy is calculated based on random distribution of atoms, while the enthalpy of mixing is derived from average pairwise interactions. This model provides insights into phase stability and is often coupled with CALPHAD-based thermodynamic descriptions.

(b) Ginzburg-Landau Theory

Ginzburg-Landau theory extends the mean-field approach by expressing the free energy of a system as a function of an order parameter (η), which characterizes the degree of order in the system. For a second-order phase transition, the free energy (F) is typically expanded as a polynomial in η, including terms that reflect symmetry and stability of the system. The simplest form of the Ginzburg-Landau free energy equation is given as,

$$F = F_0 + \int \left[\frac{a}{2}\eta^2 + \frac{b}{4}\eta^4 + \frac{\kappa}{2}(\nabla\eta)^2 \right] dV \qquad (15.35)$$

where, F_0 is the free energy of a disordered phase, a and b are temperature-dependent coefficients, and κ is a gradient term coefficient that accounts for spatial variations in the order parameter. The term $\frac{a}{2}\eta^2$ represents the energy cost of ordering, while $\frac{b}{4}\eta^4$ stabilizes the ordered phase. The gradient term $\frac{\kappa}{2}(\nabla\eta)^2$ penalizes rapid spatial variations in η, ensuring a smooth transition between ordered and disordered regions. This theory provides a powerful and versatile framework for understanding continuous phase transitions like in cases of superconducting, ferroelectric, and pattern forming materials.

(c) Cahn-Hilliard Equation

The Cahn-Hilliard equation is a fundamental mean-field approach used to model diffusion-driven phase separation and microstructure evolution, such as spinodal decomposition. It couples thermodynamic driving forces, derived from the Gibbs free energy, with a diffusional equation to describe the time evolution of a composition field, given as,

$$\frac{\partial c}{\partial t} = \nabla . \left(M \nabla \left(\frac{\partial F(c)}{\partial c} \right) \right) \tag{15.36}$$

where, c is the composition field, M is the mobility coefficient, and $F(c)$ is the composition dependent free energy functional, which typically includes a bulk free energy term and a gradient energy term. The bulk free energy, often expressed using Ginzburg-Landau-type expression, describes the thermodynamic driving force for phase separation, while gradient term accounts for the energy cost of composition gradients at interfaces. The term $\frac{\partial F(c)}{\partial c}$ represents the chemical potential, which drives the diffusion process.

(d) Allen-Cahn Equation

The Allen-Cahn equation is another fundamental mean-field approach used to model interface-driven processes, such as phase boundary motion, grain growth, and order-disorder transitions. It describes the time evolution of an order parameter (ϕ), which represents the phase or structural state of the system, by coupling thermodynamic driving forces with kinetic principles, given as

$$\frac{\partial \phi}{\partial t} = -M \frac{\partial F}{\partial \phi} \tag{15.37}$$

where, M is the mobility coefficient, and F is the free energy functional which is a function of order parameter. The term $\frac{\partial F}{\partial \phi}$ represents the variational derivative of the free energy functional with respect to order parameter, which drives the system towards equilibrium. The free energy functional typically includes a bulk free energy term, often expressed using Ginzburg-Landau-type expression, and a gradient term that penalizes sharp interfaces, ensuring a smooth transition between phases.

(e) Cluster Vibration Method

The Cluster Vibration Method (CVM) is an advanced mean-field approach which improves upon the Bragg-Williams approximation by considering short-range order and correlations within cluster of atoms, while still maintaining a mean-field framework. The CVM accounts for local arrangement of atoms by dividing the crystal lattice into small clusters (e.g. pairs, triangles, or tetrahedra) and calculating the configurational entropy and energy based on these clusters.

The CVM provides a more accurate description of ordered phases, phase transitions, and short-range order than simpler mean field models, making it particularly useful for studying complex systems such as intermetallic compounds and solid solutions.

15.3.2 Coupled Thermodynamic-Kinetic Models

Coupled thermodynamic-kinetic models are essential for understanding and predicting the time-dependent evolution of materials, as they integrate thermodynamic driving forces with kinetic mechanisms governing atomic or molecular processes. These models bridge the gap between equilibrium thermodynamics, which describes the stable state of the system, and kinetics, which capture the rates and pathways for transformations. The thermodynamic component, often derived from Gibbs free energy minimization, provides a driving force for phase transformations, diffusion, or chemical reactions, and the kinetic component, described by equations such as Fick's laws of diffusion, or reaction rate theories, governs how quickly these processes occur. For example, DICTRA (Diffusion-Controlled Transformation), a software tool based on CALPHAD method, couples thermodynamic databases with diffusion and phase transformation kinetics to predict phase transformation, time-temperature-transformation (TTT), continuous cooling transformation (CCT) diagrams among others. By combining thermodynamic and kinetic principles, these models enable the prediction of complex phenomena, such as nucleation, growth, and coarsening, providing comprehensive framework for designing and optimizing material processing and performance.

15.3.3 Phase-Field Modeling

Phase-Field modeling is a powerful computational approach for simulating microstructure evolution in materials by coupling thermodynamic driving forces with kinetic equations. It describes the system using one or more continuous order parameters (ϕ) that represent phases, grain orientations, or other structural features, eliminating the need to explicitly track interfaces. The evolution of these order parameters is governed by the Cahn-Hilliard (15.36) and Allen-Cahn (15.37) equations, which link thermodynamics and kinetics.

Phase-field modeling distinguishes itself from traditional sharp-interface models by employing a diffuse interface approach, where the transition between phases or grains is represented by a smooth variation of order parameters across a finite interface region. The diffuse interface approach eliminates the need to explicitly track the interface, making it computationally efficient for complex microstructures and multi-phase systems. For example, in a two-phase system, ϕ might vary from 0 in one phase to 1 in the other phase, with intermediate values representing the interface. The width of the interface is controlled by the gradient energy term in the free energy functional, ensuring a physically realistic representation.

The free energy functional used here typically includes two main contributions: a bulk energy term, and a gradient energy term. For example, in a binary system undergoing phase separation, the bulk free energy can be represented by a double-well potential.

$$F = \int \left[f(\phi) + \frac{\kappa}{2} (\nabla \phi)^2 \right] dV \tag{15.38}$$

Where, $f(\phi)$ is the bulk free energy density, often represented by a double-well potential (e.g. $f(\phi) = A\phi^2(1 - \phi)^2$), which ensures that ϕ tends to 0 or 1 in bulk phases. The gradient $\frac{\kappa}{2}(\nabla\phi)^2$ penalizes sharp variations in ϕ, controlling the width of the interface. The parameter κ determines the interfacial energy and thickness, allowing the model to replicate realistic interface properties.

15.3.3.1 Applications of Phase-Field Modeling

(a) **Solidification:** Phase-Field models are used to simulate dendritic growth during solidification, where ϕ represents the solid and liquid phases. Coupled with temperature and concentration fields, these models predict complex patterns like dendrite morphology and micro-segregation.
(b) **Spinodal Decomposition:** In alloys, phase-field modeling captures spontaneous separation of a homogeneous solution into two distinct phases, driven by composition fluctuations and diffusion.
(c) **Grain Growth:** The Allen-Cahn equation is used to model the evolution of grain boundaries in a polycrystalline material, predicting how grain grows and coarsen over time. Figure 15.3 showcases the simulation of grain growth of strain-free nuclei during static recrystallization. A multi-phase field approach is utilized, where the strained matrix, recrystallized grain, and grain boundaries are treated as three different order parameters and evolved over time.
(d) **Precipitation and Coarsening:** Phase-field models simulate the nucleation, growth, and coarsening of precipitates in alloys, providing insights into the precipitation kinetics.

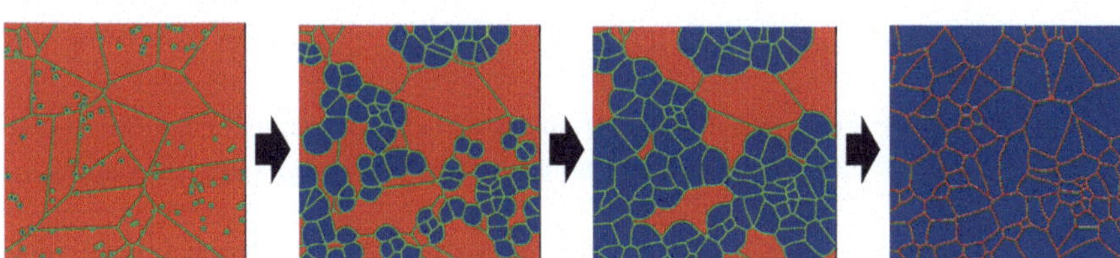

Fig. 15.3 Grain growth during recrystallization captured using phase-field modeling

(e) **Electrochemistry:** In battery materials, phase-field models describe the evolution of electrode microstructure during charging and discharging.
(f) **Fracture Mechanics:** Phase-field models simulate the crack nucleation and propagation in metals and alloys, providing insights into fracture mechanics.

15.3.4 Cellular Automata Models

Cellular automata (CA) models are discrete computational approaches used to simulate microstructure evolution in materials by coupling thermodynamic driving forces with kinetic rules. In CA models, the material is represented as a grid of cells, each of which can exist in a finite number of states (e.g. phases, grain orientations, or compositions). The evolution of the system is governed by local rules that determine how the state of a cell changes based on its current state and the state of its neighboring cells. These rules often incorporate thermodynamic principles, such as minimizing free energy, and kinetic mechanisms, such as nucleation, growth, and diffusion. CA models are particularly useful for simulating complex processes like grain growth, recrystallization, and solidification, where the interplay between thermodynamics and kinetics determines the final microstructure.

15.3.4.1 Key Components of Cellular Automata Model

(a) **Grid Representation:** The material is discretized into a grid of cells, typically in 2D or 3D. Each cell has a state (S_i) that represents its phase, orientation, or composition.
(b) **Neighborhood Definition**: The behavior of a cell is influenced by its neighbors, which can be defined using different neighborhood configurations, such as von Neumann's or Moore's.
(c) **State Update Rules:** The state of a cell at the next timestep is determined by a transition function (f) that depends on its current state and the states of its neighbors ($S_{i,\ j}$),

$$S_i(t+1) = f\left(S_i(t), S_{i,j}(t)\right) \tag{15.39}$$

This function encodes the thermodynamic and kinetic rules governing the systems.
(d) **Growth Rules:** For processes like grain growth or phase transformations, a cell may change its state to match a neighboring state if it reduces the free energy of the system.
(e) **Nucleation Rules:** New phase or grains can nucleate based on probabilistic rules that depend on local conditions, such as undercooling, supersaturation, or stored energy.

(f) **Diffusion Rules:** In a system where diffusion plays a role, such as solute diffusion during solidification, CA models can incorporate discrete diffusion rules to update the composition of cells based on the concentration gradients with their neighbors.

15.3.4.2 Advantages of CA Models

(a) **Computational Efficiency:** CA models are computationally efficient due to their discrete nature and local update rules, making them suitable for large-scale simulations.
(b) **Flexibility:** The rules can be tailored to incorporate complex thermodynamic and kinetic phenomena, such as anisotropic growth, heterogeneous nucleation, or multi-phase interactions.
(c) **Visualization:** The grid-based structure of CA models allows for intuitive visualization of microstructure evolution, providing insights into mechanisms governing the processes.
(d) **Scalability:** CA models can be easily scaled with parallel computing to simulate systems of different sizes.

15.3.5 Monte Carlo Method

The Monte Carlo (MC) method is a powerful stochastic approach for modeling thermodynamic and kinetic processes in materials at atomic and mesoscopic scales. Unlike deterministic methods, MC simulations rely on random sampling to explore the configurational space of a system, making them particularly effective for studying equilibrium and non-equilibrium phenomena where probabilistic transitions dominate. The method is fundamentally based on the Metropolis algorithm, where system configurations are generated through random perturbations and accepted or rejected based on a probability criterion derived from thermodynamic principles. This makes them particularly suitable for studying statistical mechanics problems, rare events, and systems where thermal fluctuations dominate. The MC method is widely applied in thermodynamic equilibrium simulations like phase transitions, adsorption, or magnetic ordering, kinetic processes like diffusion, nucleation, grain growth, and microstructure evolution like spinodal decomposition, or recrystallization.

15.3.5.1 Types of Monte Carlo Simulations

Canonical (NVT) Ensemble
The canonical (NVT) ensemble Monte Carlo method is a fundamental computational approach for studying thermodynamic equilibrium in systems with a fixed number of particles (N), volume (V), and temperature (T). In this ensemble,

configurations are sampled according to the Boltzmann distribution, where the probability of a microstate with energy E is given by

$$P = \min\left(1, e^{-\frac{\Delta E}{k_B T}}\right) \tag{15.40}$$

where, ΔE is the energy change, k_B is the Boltzmann constant, and T is the absolute temperature. The NVT ensemble is particularly useful for simulating phase transitions, such as melting or magnetic ordering, where thermal fluctuations drive the system between states. This approach provides direct access to thermodynamic averages (e.g., heat capacity) while naturally incorporating entropic effects, making it indispensable for equilibrium property prediction in materials science.

Grand Canonical (μ VT) Ensemble

The grand canonical (μ VT) ensemble Monte Carlo method is a powerful computational tool for studying open systems that exchange particles with a reservoir at fixed chemical potential (μ), volume (V), and temperature (T). In this ensemble, the simulation cell can undergo particle insertion and deletion moves in addition to traditional displacements, allowing the particle number to fluctuate while maintaining equilibrium with the external reservoir. The acceptance criteria for these moves are derived from the grand canonical partition function, where insertion and deletion attempts are accepted with probability,

$$P_{\text{insert}} = \min\left(1, \frac{V}{\Lambda^3 (N+1)} e^{\beta(\mu - \Delta E^+)}\right) \tag{15.41}$$

$$P_{\text{delete}} = \min\left(1, \frac{\Lambda^3 N}{V} e^{-\beta(\mu + \Delta E^-)}\right) \tag{15.42}$$

where, Λ is the thermal de Broglie wavelength, and $\Delta E^{\pm}$ are the energy changes. This approach is particularly valuable for modeling adsorption phenomenon in porous materials, studying phase coexistence, such as liquid-vapor equilibrium, and electrochemical system, such as ion insertion in battery electrode.

Kinetic Monte Carlo (kMC)

The kinetic Monte Carlo method is a stochastic computational technique designed to simulate the time evolution of dynamic processes in materials by explicitly accounting for transition rates between different states. Unlike traditional MC methods that sample equilibrium configurations, kMC focuses on the kinetics of events such as atomic diffusion, chemical reactions, or surface growth, where each event is assigned a rate, given as,

$$r_i = \nu_i \exp\left(-\frac{E_i}{k_B T}\right) \tag{15.43}$$

where, v_i is the attempt frequency, and E_i is the activation energy. The algorithm selects events probabilistically based on their rates and advances the system clock by $\Delta t = -\frac{\ln(\xi)}{\sum_i r_i}$, where ξ is the uniform random number, ensuring proper temporal evolution. This method is particularly effective for modeling vacancy-mediated diffusion in alloys, thin film deposition, and catalytical reactions on surfaces.

Monte Carlo methods provide a versatile and powerful framework by combining statistical sampling with physically meaningful transition rules that bridge the gap between atomic-scale mechanisms and macroscopic material behavior.

15.4 Data Required for Thermodynamic and Kinetic Modelling

Computational thermodynamics and kinetics rely on high-quality experimental and theoretical data to build accurate models. The choice of data depends on the modeling approach (e.g. CALPHAD, phase-field, kinetic Monte Carlo) and the material system under study. Below, we categorize the essential data types and provide examples of their applications.

15.4.1 Thermodynamic Data

Thermodynamic modeling requires input parameters that describe the free energy of phases as a function of composition, temperature, and pressure. Table 15.1 lists some examples of thermodynamic data required by thermodynamic and kinetic modeling, along with the methods to obtain such data.

15.4.2 Kinetic Data

Kinetic modeling requires parameters that govern the rates of processes such as diffusion, phase transformation, and reactions. Table 15.2 list some examples of the kinetic data required by thermodynamic and kinetic modeling, along with the methods to obtain such data.

15.4.3 Data for Multi-component and Multi-phase Systems

For complex systems (e.g. high-entropy alloys, ceramics, etc.), additional data are needed apart from the ones mentioned above. Some examples of such data are as follows,

Table 15.1 Examples of thermodynamic data required along with experimental and theoretical methods to obtain such data

Examples	Experimental or theoretical methods
Phase equilibria data	
Liquidus/Solidus temperature	Differential Thermal Analysis (DTA)
Solubility Limit	Differential Scanning Calorimetry (DSC)
Invariant reactions (eutectic, peritectic, monotectic)	Metallography combined with Electron Probe Micro-analysis (EPMA)
Thermochemical properties	
Enthalpy of formation	Calorimetry
Heat capacity	Phonon calculations in Density Function Theory (DFT)
Entropy	
Activity and chemical potential data	
Activity coefficients	Electrochemical cell measurements
Partial molar quantities	Vapor pressure techniques like Knudsen effusion
Magnetic, electronic, and vibrational contributions	
Curie Temperature	Magnetometry
Debye Temperature	Inelastic neutron scattering

Table 15.2 Examples of kinetic data required along with experimental and theoretical methods to obtain such data

Examples	Experimental or theoretical methods
Diffusion coefficients	
Tracer diffusion coefficients	Radiotracer experiments
Interdiffusion coefficients	Diffusion couples analyzed via EPMA or Secondary Ion Mass Spectroscopy (SIMS)
Impurity diffusion	
Interface mobility and transformation kinetics	
Grain boundary mobility	In-situ synchrotron X-ray diffraction,Molecular Dynamics (MD) simulations
Phase transformation rates	Orientation Imaging Microscopy (OIM)
Reaction rate constants	
Oxidation rates	Thermogravimetric Analysis (TGA)
Precipitation kinetics	Small-angle X-ray scattering (SAXS)
Activation energies	
Vacancy migration energy	Arrhenius fitting of experimental data
Dislocation climb activation energy	Atomistic simulations (MD, DFT)

- Ternary and higher-order interaction parameters (e.g. $L_{Fe-Cr-Ni}$ in stainless steel)
- Sublattice site occupancies
- Cross-diffusion coefficients (e.g. D_{ij} in Al-Co-Cr-Ni system)

15.4.4 Validation Data

To ensure the reliability of the developed models, independent validation datasets are critical, such as

- Time-Temperature-Transformation (TTT) diagrams
- Microstructure evolution (e.g., grain size distribution, precipitate size distribution)
- Mechanical Properties

15.5 Thermodynamic and Kinetic Databases

Computational thermodynamics and kinetics rely on comprehensive databases that store critically assessed thermodynamic and kinetic parameters for material systems, enabling the prediction of phase equilibria, transformation kinetics, and microstructure evolution. These databases compile experimental and computational data such as Gibbs free energy functions, diffusion coefficients, and interfacial properties into self-consistent frameworks that can be integrated with modeling tools like CALPHAD, phase-field, and kinetic Monte Carlo simulations. For example, thermodynamic databases describe the stability of phases in multicomponent alloys, while kinetic databases provide diffusion mobilities and reaction rate constants essential for simulating time-dependent processes. These databases are built through rigorous assessments that reconcile first-principles calculations, experimental measurements, and empirical correlations. This section explores the structure, development, and application of these databases, highlighting their role in bridging fundamental material properties with predictive modeling for alloy design, process optimization, and materials discovery.

Thermodynamic databases, such as those following the CALPHAD framework, contain Gibbs free energy expressions for each phase, modeled using polynomial expansions (Eq. 15.11) or sublattice formalisms (Eq. 15.13). These databases include coefficients for temperature and composition dependent terms, along with interaction parameters for multi-component systems. Kinetic databases complement these by storing diffusion mobilities, pre-exponential factors, and activation energies, often in Arrhenius type equations, to describe atomic transport and phase transformation rates. Both types of databases are structured in machine-readable formats (e.g., for thermodynamic **TDB** format which stands for Thermodynamic DataBase, and for mobility data **MOB** format which stands for MOBility database) and are designed for interoperability with simulation software (e.g. Thermo-Calc, DICTRA). Metadata such as experimental sources, uncertainty estimates, and assessment references, are often embedded to ensure traceability and facilitate updates.

The creation of thermodynamic and kinetic databases involves a rigorous, multi-step process that integrates experimental data, computational methods, and critical assessments to ensure accuracy and predictive capability. For thermodynamics databases, the process begins with collecting high-quality experimental phase equilibria, thermochemical properties and activity measurements from literature or direct experiments, as described in Sect. 15.4.1. These data are then modeled using formalisms such as the CALPHAD approach, where Gibbs free energy functions for each phase are parameterized using polynomial expansions or sublattice models. First-principles calculations (e.g. DFT) may supplement missing data, particularly for metastable phases or high-temperature regimes. For kinetic databases, key inputs include diffusion coefficients, interface mobilities, and reaction rate constants, often derived from tracer experiments, diffusion couples, or atomistic simulations, as described in Sect. 15.4.2. The data are fitted to Arrhenius-type equations to describe temperature and composition dependence. Both types of databases undergo iterative optimization using software like Thermo-Calc or PARROT, where parameters are adjusted to minimize discrepancies between calculated and experimental values. Finally, databases are validated against independent datasets (e.g. TTT diagrams, microstructure observations) and formatted for compatibility with simulation tools. This structured approach enables seamless integration with computational tools, supporting predictive modeling of phase stability, reaction kinetics, and microstructure evolution in complex material systems.

Thermodynamic and kinetic databases can be classified into proprietary and open-source categories. Proprietary databases, such as Thermo-Calc's TDB (e.g. TCFE for steel, TCHEA for high entropy alloys) [5] and FactSage's FACT (e.g. Ftoxid for oxides, FScopp for copper alloys) [6] databases, offer rigorously assessed, industry-standard parameters for Gibbs free energy and phase equilibria, often with commercial software integration. Similarly, kinetic databases like DICTRA's MOB2 for diffusion mobilities are proprietary and optimized for use with specific simulation tools. In contrast, open-source databases like OpenCalphad's database [7], SGTE's pure element database [2], and MTDATA's public datasets [8] provide transparent, community-driven parameterizations for thermodynamics, while platforms such as Materials Project [9] and NOMAD [10] offers DFT-calculated diffusion barriers and reaction kinetics for border accessibility.

15.6 Tools Available

Computational thermodynamics and kinetics rely on specialized software tools to predict material behavior, phase transformations, and microstructure evolution. These tools vary in their theoretical foundations, numerical approaches, and application domains. This section provides an overview of widely used computational tools, comparing their capabilities in handling thermodynamic and kinetic problems.

15.6.1 Thermodynamic Tools

Table 15.3 lists various available thermodynamic tools and compares their features.

15.6.2 Phase-Field Tools

Table 15.4 lists various available phase-field tools and compares their features.

15.6.3 Monte-Carlo Tools

Table 15.5 lists various available Monte-Carlo tools and compares their features.

15.7 Optimization and Machine Learning

The integration of optimization techniques and machine learning (ML) has revolutionized computational thermodynamics and kinetics, enabling faster, more accurate prediction of material properties and behavior. These methods address key challenges in traditional approaches, such as high computational costs, incomplete datasets, and complex multi-scale phenomenon. This section explores how optimization and ML are applied to enhance thermodynamic modeling, kinetic simulations, and material discovery.

15.7.1 Optimization in Thermodynamics and Kinetics

Optimization techniques play a pivotal role in computational thermodynamics and kinetics by refining model parameters to achieve consistency with experimental data and theoretical constraints. In thermodynamic modeling, optimization is central to the CALPHAD methodology, where Gibbs free energy parameters for individual phases are calibrated to reproduce phase equilibria, thermochemical properties, and chemical potentials. Advanced algorithms such as the Levenberg-Marquardt method and genetic algorithm are employed to minimize the discrepancy between calculated and measured phase diagrams, with error functions often weighted by experimental uncertainties. For instance, in optimizing the Fe-Cr-Ni system, interaction parameters (L_{Fe-Cr}, L_{Cr-Ni} etc.) are iteratively adjusted until the predicted miscibility gaps and ordering transitions align with neutron scattering and calorimetry data. In kinetic modeling, optimization focuses on determining rate constants, diffusion

Table 15.3 Comparison of thermodynamic simulation software

Tool	License	Key features	Database support	Strengths	Limitations	Typical applications
Thermo-Calc [5]	Proprietary	• Multicomponent systems • GUI and scripting (Python) available • DICTRA for diffusion calculation	Thermo-calc developed TDB	• Industry standard accuracy • Extensive databases • Robust optimization tools	• Expensive licenses • Steep learning curve	Alloy design, heat treatment, oxidation, precipita-tion
FactSage [6]	Proprietary	• Integrated thermo-chemical databases • Reaction equilibrium, phase equilibria • Slag/oxide chemistry	FactSage developed databases for compounds and oxides	• Specialized in ceramics and slags • User friendly UGI • Process simulations	• Limited kinetic modeling • Database gaps for newer alloys	Extractive Metallurgy, battery materials, slags chemistry
Pandat [11]	Proprietary	• PanEngine CALPHAD kernel • PanPrecipita-tion for kinetics • Custon database tools	Pandat developed databases	• Good for precipitation modeling • Batch calculation	• Smaller database ecosystem • Less community support	Aging simula-tions
MatCalc [12]	Proprietary	• CALPHAD and kinetic precipitation • Dislocation density modeling • GUI with visualization	MatCalc developed databases	• Strong in precipitation kinetics • Microstructure-property linkages • Scripting in Python	• Limited support for oxides and ceramics • Requires calibration	Aging studies, nucleation studies
Open Calphad [7]	Open-source	• Python integrated CALPHAD • Free energy minimization • Custom database creation	Supports TDB format	• No license cost • Transparent algorithm • Active GitHub community	• Limited availability of quality databases • Less polished GUI	Research, education, code development
PyCalphad [4]	Open-source	• Python library for CALPHAD • Symbolic	Supports TDB format	• Flexible • Works with materials project data	• Requires coding skills • Limited documentation	

		Gibbs free energy praising • Machine learning integration				Data-driven alloy discovery, academic, education
MTDATA [8]	Open-source	• UK National Physical Lab's tool • Multicomponent phase equilibria	Supports TDB format	• Free for academic use • Command line and GUI options • Reliable assessments	• Interface is old • Limited updates	Legacy alloys systems

Table 15.4 Comparison of phase-field simulation software

Tool	License	Multi-physics coupling	Numerical method	Strengths	Limitations	Typical applications
MOOSE [13]	Open-source	• Thermodynamics • Mechanics • Electromagnetics • Fluid flow	Finite Element Method (FEM)	• Extremely flexible for multi-physics • Large community • Modular C++ code • Adaptive mesh refinement • MPI based parallelize-tion	• Steep learning curve • Requires programing knowledge • Minimal GUI	Nuclear materials, fuel cells, fracture
MICRESS [14]	Proprietary	• CALPHAD Thermodynamics • Fluid flow • Stress	Finite Volume Method (FVM)	• Optimized for material science • Direct TDB database integration • Specialized for solidification • GUI available • OpenMPI based parallelization	• Expensive license • Limited to materials problem	Solidification, additive manufacturing, grain growth
PRISMS-PF [15]	Open-source	• Electrochemistry • Elasticity • Thermal	Finite Difference Method (FDM)	• Optimized for HPC • Strong materials focus • Python integration • MPI and CUDA based parallelization	• Less mature • Fewer physics modules • No GUI	Batteries, spinodal decomposition, precipitation
MicroSim [16]	Open-source	• Thermodynamics • Multi-physics	Finite Volume Method (FVM) and Finite Difference Method (FDM)	• Specialized for solidification • GUI available • MPI, CUDA, and OpenCL based parallelization	• Less mature • Limited to few applications	Solidification, grain growth, precipitation

Table 15.5 Comparison of Monte Carlo simulation software

Tool	License	Key features	MC type	Strengths	Limitations	Typical applications
KMCLib [17]	Open-source	• Event based KMC • Custom rate calculation • Python interface	Kinetic Monte Carlo	• Optimized for sparse systems • Exact KMC algorithm • MPI parallelization	• No built-in potential • Requires coding	Surface diffusion, catalysis
LAMMPS [18]	Open-source	• Atomistic MC moves • Replica exchange • Embedded atom method (EAM) potentials	Hybrid Monte Carlo/Molecular Dynamics	• Combines MD and MC • Extensive force fields • MPI/GPU parallelization	• Not a pure KMC • Steep learning curve	Alloy ordering, defect evolution
SPPARKS [19]	Open-source	• Sparce KMC solver • Potts/Ising models • Phase-field KMC coupling	Lattice/off-lattice KMC	• Multi-algorithm support • MPI/GPU parallelization	• Limited documentation • Steep learning curve	Grain growth, thin film deposition, solidification
MCCube [20]	Proprietary	• Graphical workflow builder • Pre-built material templates	Kinetic Monte Carlo	• User-friendly GUI • Experimental data import • OpenMP parallelization	• Closed source code • Limited customization	Semiconductor processing
CASM [21]	Open-source	• Cluster expansion • Grand canonical MC	Statistical Monte Carlo	• VASP integration • Automated workflows	• Specialized for alloys	Cluster expansion, phase stability

coefficients, and activation energies. Markov Chain Monte Carlo (MCMC) methods are used to infer kinetic parameters from time-resolved data, such as concentration profile in diffusion couples or precipitate growth kinetics observed via in-situ TEM. Adjoint-based optimization, which effectively computes gradients of objective functions, is particularly powerful for fitting mobility databases in tools like DICTRA. Challenges include handling multi-objective optimization (e.g., simultaneously matching phase fractions and solute concentrations) and addressing parameter correlations that lead to non-unique solutions. Recent advances in Bayesian optimization to quantify uncertainties in CALPHAD assessments and kinetic Monte Carlo rate catalogs, ensuring robust predictions. By systematically optimizing thermodynamic and kinetic models, predictive accuracy of these tools is significantly enhanced.

15.7.2 Machine Learning Approaches in Thermodynamics and Kinetics

Machine learning (ML) has emerged as a transformative tool in computational thermodynamics and kinetics, enabling data-driven predictions of material properties, accelerated simulations, and the discovery of novel materials. In thermodynamics, ML models such as neural networks and Gaussian processes (GPs) are trained on datasets from experiments, density function theory (DFT), or CALPHAD databases to predict Gibbs free energies, phase stabilities, and chemical potentials with high efficiency. For example, surrogate ML models can approximate the free energy surfaces of multicomponent alloys, bypassing computational expensive ab-initio calculations while maintaining desirable accuracy. ML also enhances CALPHAD assessments by identifying optimal interaction parameters through gradient-based optimization or genetic algorithm, reducing the need for manual tuning. In kinetics, ML accelerates MD and kMC simulations by predicting atomic migration barriers, diffusion coefficients, and reaction rates from structural descriptors. Graph neural networks (GNN's) are particularly effective for modeling atomic-scale processes, such as vacancy-mediated diffusion in metals or ion transport in solid electrolytes. Additionally, reinforcement learning (RL) has been applied to optimize reaction pathways in catalysis, while convolutional neural networks (CNN's) analyze microstructure images to predict phase transformation kinetics. A major advancement is the development of physics-informed neural network (PINN's), which embed thermodynamic constraints (e.g., the Cahn-Hilliard equation) directly into ML architectures, ensuring physically consistent predictions. Challenges remain, including the need for high-quality training data and the interpretability of ML models, but ongoing innovations in active learning and hybrid ML-physics approaches are bridging these gaps.

15.8 Engineering Applications

15.8.1 Precipitate Dissolution Kinetics During Reheating

In High Strength Low Alloy (HSLA) steel carbonitride precipitates significantly influence microstructure and mechanical properties, primarily through strengthening mechanisms. These precipitates influence the yield and ultimate tensile strength of the material when present as finely dispersed nanosized precipitates in the matrix. During the casting of these steels into slabs, these precipitates often grow bigger in size, thereby reducing their efficiency in inhibiting the grain growth and thereby strength of the material. To achieve their desired strengthening effect, the cast slab is reheated to dissolve back the precipitates sufficiently to enrich the matrix with precipitate forming elements and subsequently precipitating them during the hot working process as fine nanosized precipitates to control the grain growth during recrystallization. The study of their dissolution behavior during reheating becomes important to optimize the reheating schedule by identifying the critical temperature-time combinations.

An HSLA steel having a composition as mentioned in Table 15.6, contains $(NbTi)C$ precipitates in the cast slab having the precipitate volume fraction of $1.4 * 10^{-3}/m^3$ and the mean radius of the precipitate 56 nm. The precipitates present are quite bulky in size and the matrix does not contain sufficient Nb, Ti, and C for subsequent finer precipitation to occur. By controlling the reheating schedule, this bulky precipitate can be dissolved back into the matrix. Precipitation dissolution kinetics at a reheating temperature of 1200 °C, is captured using the computational thermodynamics and kinetics approach and is shown in Fig. 15.4. This precipitation kinetics model is simulated using a code developed in python language and the thermodynamic data required to simulate the precipitate kinetics such as equilibrium solubility of different solutes in the matrix, and diffusivity of the solutes at reheating temperature are calculated using Thermo-Calc software.

Figure 15.4 shows the precipitation dissolution kinetics as the evolution of the precipitate volume fraction, precipitate mean radius, and solute concentrations in the matrix with time. It can be observed that the precipitate volume fraction decreases with time as the dissolution takes place, and correspondingly the precipitate mean radius also decreases, and the solute concentration increases. But beyond a certain holding time the precipitate mean radius increases, as coarsening takes place with the increase in the holding time. The coarsening kinetics is enhanced by the high diffusivities at the holding temperature. The black dashed line marks the point at which the major dissolution of the precipitation has taken place based on the equilibrium solubilities of the solutes, and holding beyond this will not influence further dissolution. This point is at 1500 s or 45 min, which is estimated to be the ideal reheating time for the reheating temperature of 1200 °C.

Table 15.6 Chemical composition of the HSLA stee in wt%

C	Mn	Nb	Ti	Fe
0.0595	1.91	0.039	0.055	Bal.

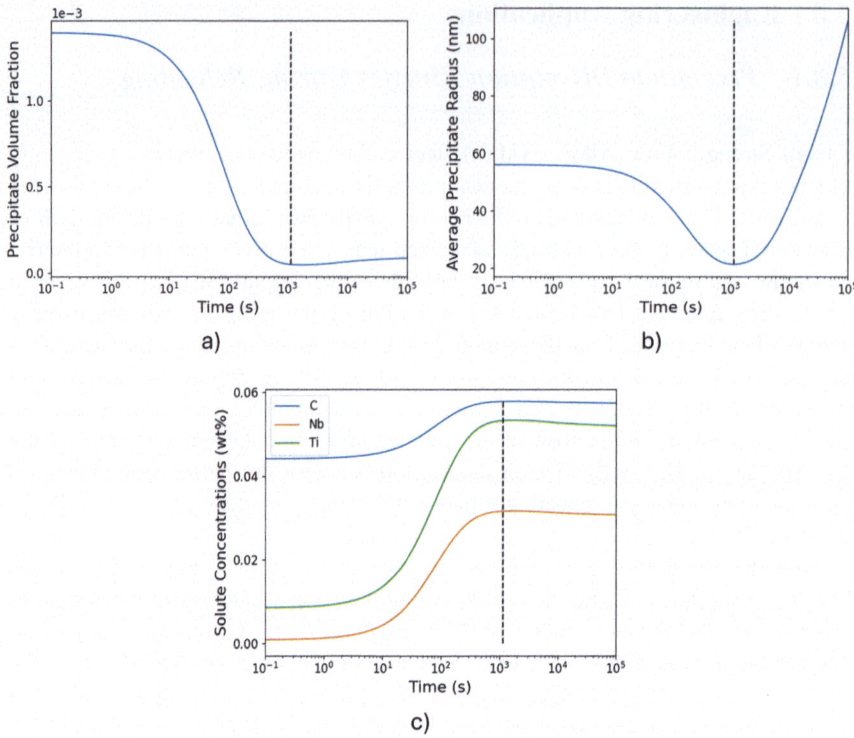

Fig. 15.4 Precipitate dissolution kinetics (**a**) precipitate volume fraction, (**b**) precipitate mean radius, and (**c**) solute concentration evolution

15.8.2 Static Recrystallization

During hot rolling of steel, between the adjacent roll passes, the deformed material undergoes static recrystallization. The deformation during hot rolling introduces high density of dislocations, hardening of the material, and reduction of the ductility for the subsequent roll passes. Static recrystallization leads to formation of strain-free grains, which are equiaxed and smaller, and thereby restoring the workability of the material before it enters the next pass. This directly affects grain refinement, texture evolution, and strength and toughness of the hot-rolled strip. Industrial mills require precise models to predict whether complete, partial, or no recrystallization will occur between roll passes, a critical factor in designing rolling schedules that achieve target properties while minimizing energy consumption.

The CA framework discussed earlier provides an ideal platform for simulating static recrystallization in hot rolling conditions by capturing both thermodynamic driving force and kinetic limitations characteristic of high-temperature deformation processes. The chemical composition of the steel used to simulate the recrystallization is shown in Table 15.7.

A steel is subjected to a deformation of strain 0.3 and strain rate 1 s^{-1} at a temperature of 1000 °C. The static recrystallization in the steel that is held at the same temperature is captured using CA, and the evolution of the recrystallization is shown in Figs. 15.5 and 15.6. Figure 15.5 showcases the recrystallization fraction evolution for static recrystallization, with complete recrystallization taking place within 4.7 s. Figure 15.6 showcases microstructure evolution during static recrystallization. The blue region represents the strained grains, red region represents the grain boundaries, and grey region represents the recrystallized grains. The initial microstructure has average grain size of 83 μm, and the final recrystallized microstructure has an average grain size of 41.8 μm. The initial representative microstructure covering an area of 500 × 500 μm is generated using Voronoi tessellations for the initial average grain size and is then deformed based upon the applied strain. The microstructure evolution is simulated using CA code written on the Fortran language.

15.8.3 Phase Transformation During Cooling

The austenite to ferrite phase transformation during continuous cooling is a critical metallurgical process that determines the final microstructure and mechanical properties of the steel. In industrial heat treatment or thermomechanical processing, controlled cooling rates are applied to achieve target phase fractions, grain sizes,

Table 15.7 Chemical composition of the steel in wt%

C	Mn	Si	Nb	Ti	Al	N	Fe
0.06	1.25	0.325	0.05	0.024	0.035	0.004	Bal.

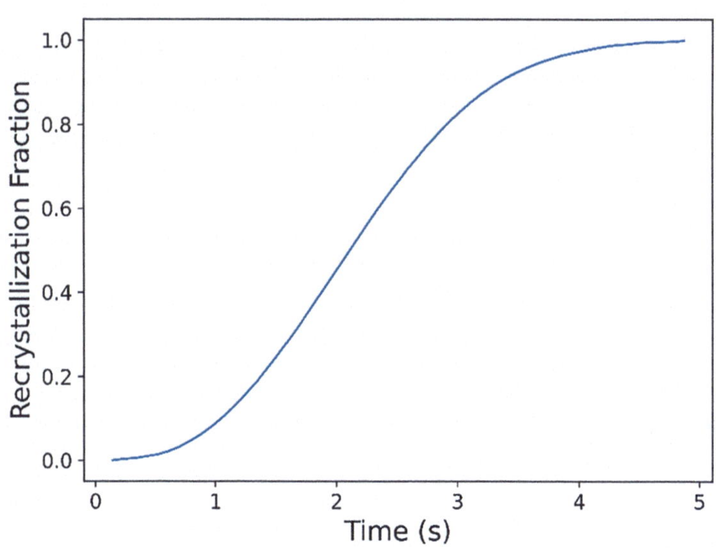

Fig. 15.5 Evolution of the recrystallization fraction with time

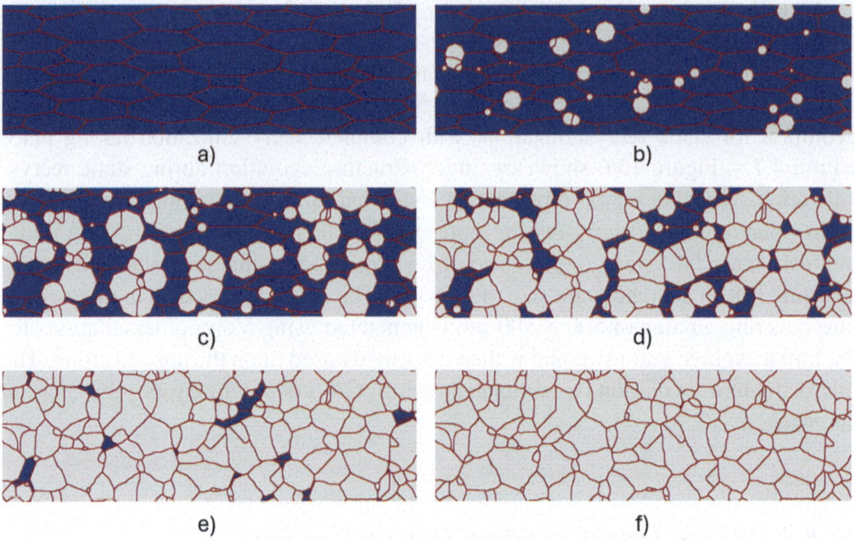

Fig. 15.6 Evolution of the microstructure during static recrystallization

and hardness. However, predicting the transformation kinetics and resulting microstructure is complex due to competing factors such as nucleation barriers, solute partitioning, and interfacial effects. Phase-field modeling has emerged as a powerful tool to simulate this transformation by coupling thermodynamic driving forces with kinetic mechanisms, enabling virtual design of cooling strategies for applications like automotive body panels or pipeline steels.

A steel having the chemical composition as shown in Table 15.8, is cooled after austenization from 900 °C to the room temperature at a rate of 2 °C/s. The average grain size after austenization is 40 μm. The initial representative microstructure of 75 × 75 μm is generated using Voronoi tessellations for the initial average grain size. Thermodynamic data such as carbon solubility in austenite and ferrite phases, free energy of austenite and ferrite phases as a function of temperature, composition is obtained from Thermo-Calc. The phase transformation is simulated using phase field code written on C++ language.

The evolution of the microstructure during phase transformation and distribution of carbon in the matrix is shown in Fig. 15.7. The Fig. 15.7a–c shows the evolution of ferrite phase in austenite matrix, with the austenite matrix represented by blue, grain boundaries represented by green and ferrite grains represented by red color. The Fig. 15.7d–f shows the evolution of carbon composition distribution with austenite phase getting enriched in carbon as low carbon ferrite phase is growing. The extent of phase transformation depends on the thermodynamic driving force and at the end of transformation the remaining austenite phase will transform to pearlite.

Table 15.8 Chemical composition of the steel in wt%

C	Mn	Si	Al	Nb	S	P	N	Fe
0.06	1.49	0.2	0.038	0.047	0.002	0.009	0.0094	Bal.

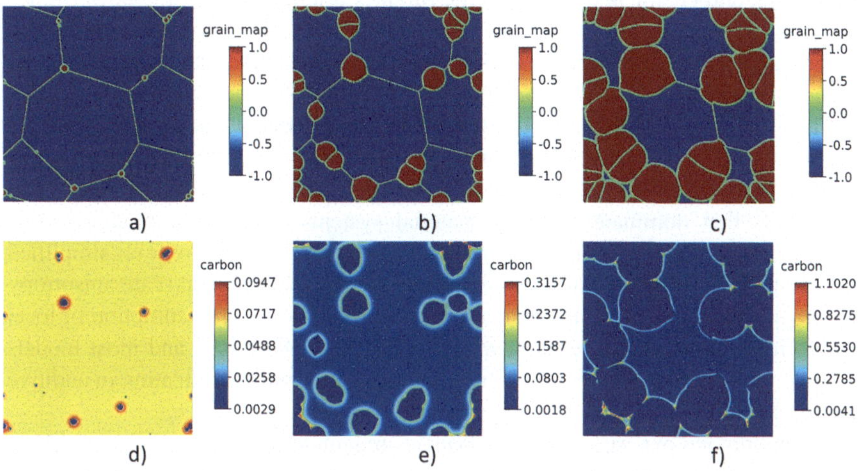

Fig. 15.7 Evolution of the ferrite phase (**a–c**) and carbon concentration distribution (**d–f**) during transformation

15.9 Challenges and Limitations

While computational thermodynamics and kinetics have become indispensable tools in modern materials science, several fundamental challenges and limitations persist that need to be acknowledged. These constraints arise from theoretical approximations, computational bottlenecks, and gaps in experimental validation that collectively impact the predictive power and practical utility of these models.

Data limitations and parameter sensitivity pose significant challenges across all modeling approaches. The accuracy of CALPHAD-based thermodynamic models is fundamentally constrained by the availability and quality of experimental data. For multicomponent systems common in industrial alloys, the exponential growth of required interaction parameters often outstrips reliable measurements, forcing modelers to make simplifying assumptions. Kinetic models face even greater uncertainties, particularly for interface properties like grain boundary mobilities or nucleation barriers that are notoriously difficult to measure experimentally. The problem is compounded by the strong parameter correlations in many models, where different combinations of values can produce equally good fits to limited experimental data, leading to non-unique solutions. This issue is particularly acute when extrapolating models beyond their calibrated ranges, such as predicting phase stability at extreme temperatures or pressure when experimental validation is scarce.

Computational costs and scaling limitations present another major barrier to broader adoption. While atomic-scale methods like density function theory (DFT) provide fundamental insights, their computational expense restricts simulations to small system (typically less than 1000 atoms) and short timescales (typically nanoseconds). Mesoscale methods such as phase-field modeling can access larger length scales but still become prohibitively expensive when modeling realistic polycrystalline microstructure or when incorporating multiple coupled physics, such as thermal, mechanical and chemical effects. Kinetic Monte Carlo methods, while powerful for studying rare events, require complete catalogs of possible transitions that may be unknown for complex materials. These limitations create persistent gaps between the scales accessible to simulation and those relevant for industrial applications.

Theoretical approximations inherent in current models introduce systematic errors that are often difficult to quantify. Mean-filed approaches like the Braggs-Williams approximation neglect important local fluctuations and short-range ordering effects that dominate in many material systems. Phase-field models, while elegant in their treatment of microstructural evolution, typically rely on simplified representations of interface physics that may not capture atomistic-scale anisotropy or complex drag effects from solute atmospheres. The common assumption of local equilibrium at interfaces breaks down for rapid transformations, and most models struggle with strongly non-equilibrium processes like rapid solidification in additive manufacturing.

Validation and experimental discrepancies remain an enduring challenge. Many key kinetic parameters, particularly those describing nucleation phenomena or interface properties, cannot be measured directly and must be inferred from indirect observations. Even when experimental data exists, the gap between idealized simulation conditions and real material behavior, containing defects, impurities, and processing artifacts can lead to significant discrepancies. The problem is particularly acute for industrial materials where proprietary compositions and processing histories limit the availability of validation data.

Looking forward, the field is addressing these challenges through several promising directions. Machine learning approaches are helping bridge scales by developing surrogate models trained on both simulated and experimental data. High-throughput computational workflows are enabling systematic exploration of parameter spaces to identify critical gaps in understanding. Emerging multiscale frameworks aim to seamlessly connect quantum, atomistic, and continuum descriptions. However, these advances require renewed emphasis on uncertainty quantification and development of standardized validation protocols.

For students entering the field, developing a critical perspective on model limitations, understanding where and why simulations might diverge from reality is just as important as mastering the computational tools themselves. The most impactful computational materials scientists and engineers will be those who can judiciously combine simulations with targeted experiments, while maintaining clear communication about the boundaries of predictive capability.

Problems

1	The temperature difference across a material is 40 K. If the thermal conductivity of the material is 200 W/m K and the length is 0.2 m, calculate the heat flux. **Solution:** $q = -k\frac{\Delta T}{L} = -200\frac{40}{0.2} = -40{,}000\ \mathrm{W/m^2}$
2	A reaction occurs with a rate constant $k = 3$ mol/L and initial concentrations $C_A = 0.2$ mol/L and $C_B = 0.4$ mol/L. Calculate the rate of reaction. **Solution:** $r = k \times C_A \times C_B = 3 \times 0.2 \times 0.4 = 0.24$ mol/L
3	The rate law for a reaction is $\frac{dC_A}{dt} = -k \cdot C_A$. If $C_A(0) = 1.0$ mol/L and $k = 0.11$/s, Calculate the concentration of A after 5 s. **Solution:** $C_A(t) = C_A(0) \cdot e^{-kt} = 1.0 \cdot e^{-0.1 \times 5} = 1.0 \cdot e^{-0.5} \approx 0.6065$ mol/L
4	Write the expanded form of the Gibbs free energy for a ternary system of Fe-Cr-Ni assuming regular solution model. Answer The Gibbs free energy for a multi-component system is given as, refer Eq. (15.4) in the Sect. 15.1.1. $G = \sum_{i=1}^{m} x_i G_i^0 + RT\sum_{i=1}^{m} x_i \ln a_i + G^E$ For a ternary system of Fe-Cr-Ni, the Gibbs free energy becomes, $G = x_{Fe}G_{Fe}^0 + x_{Cr}G_{Cr}^0 + x_{Ni}G_{Ni}^0 + RT(x_{Fe}\ln a_{Fe} + x_{Cr}\ln a_{Cr} + x_{Ni}\ln a_{Ni}) + G^E$ For a regular solution model, the excess Gibbs free energy is given as, refer Eq. (15.9) in the Sect. 15.1.2.1. $G^E = \sum_{i=1}^{n-1}\sum_{j>i}^{n} \Omega_{ij}x_i x_j = \Omega_{Fe-Cr}x_{Fe}x_{Cr} + \Omega_{Fe-Ni}x_{Fe}x_{Ni} + \Omega_{Cr-Ni}x_{Cr}x_{Ni}$
5	Draw a phase diagram of Al-Zn binary system using any thermodynamic software. Here is a code snippet to draw the phase diagram using pycalphad. Code Snippet:

(continued)

```
import matplotlib.pyplot as plt
from pycalphad import Database, binplot
import pycalphad.variables as v

# Load database and choose the phases that will be considered
db_alzn = Database('alzn_mey.tdb')
my_phases_alzn = ['LIQUID', 'FCC_A1', 'HCP_A3']

# Create a matplotlib Figure object and get the active Axes
fig = plt.figure(figsize=(9,6))
axes = fig.gca()

# Compute the phase diagram and plot it on the existing axes using the `plot_kwargs={'ax': axes}` keyword argument
binplot(db_alzn, ['AL', 'ZN', 'VA'] , my_phases_alzn, {v.X('ZN'):(0,1,0.02), v.T: (300, 1000, 10), v.P:101325, v.N: 1}, plot_kwargs={'ax': axes})

plt.show()
```

Pre-requisites: python 3.9+, pycalphad
Download the thermodynamic database ("*alzn_mey.tdb*") from:https://github.com/pycalphad/pycalphad.
The binary phase diagram of Al-Zn is as follows,

	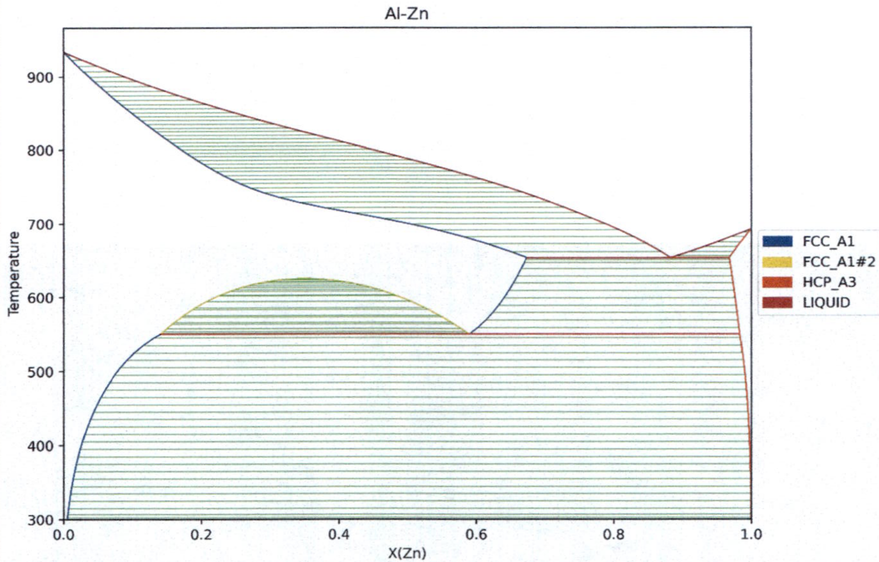
6.	For an Al-Zn binary system, plot the Gibbs free energies of all the phases at 800 K using any thermodynamic software. **A3:** Here is a code snippet to plot the Gibbs fee energies using pycalphad. Code Snippet:

(continued)

```python
from pycalphad import Workspace, variables as v
from pycalphad.property_framework.metaproperties import IsolatedPhase
import matplotlib.pyplot as plt

wks2 = Workspace('alzn_mey.tdb', ['AL', 'ZN'], ['FCC_A1', 'HCP_A3', 'LIQUID'], {v.X('ZN'):
(0,1,0.02), v.T: 800, v.P:101325, v.N: 1})

fig = plt.figure()
ax = fig.add_subplot()

x = wks2.get(v.X('ZN'))
ax.set_xlabel("X(Zn)")
ax.set_ylabel("Gibbs Free Energy (J/mol)")

for phase_name in wks2.phases:
    # Workaround for poor starting point selection in IsolatedPhase
    metastable_wks = wks2.copy()
    metastable_wks.phases = [phase_name]
    prop = IsolatedPhase(phase_name, metastable_wks)(f'GM({phase_name})')
    prop.display_name = f'GM({phase_name})'
    ax.plot(x, wks2.get(prop), label=prop.display_name)
ax.legend()
plt.tight_layout()
plt.show()
```

The Gibbs free energy plot at 800 K is as follows,

	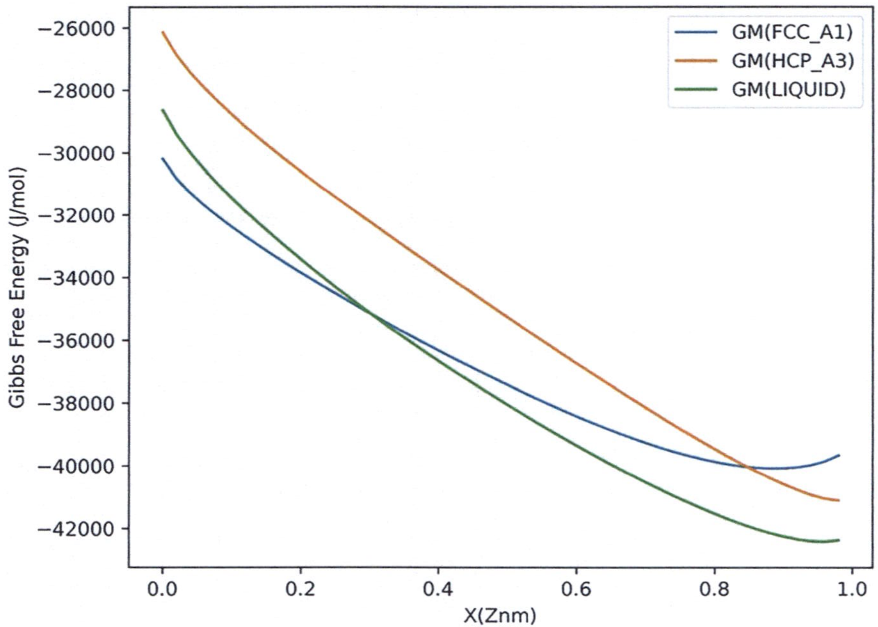
7.	For a Fe-Mn system having 0.25 mole fraction Mn, calculate the activity coefficient of Mn if its activity is 0.8. **A4:** The activity of an element is given as, refer to Eq. (15.6) in the Sect. 15.1.2. $a_{Mn} = x_{Mn}\gamma_{Mn}$ $\gamma_{Mn} = \frac{a_{Mn}}{x_{Mn}} = \frac{0.8}{0.25} = 3.2$

(continued)

8. In a microstructure having an average grain size of 25 μm is subjected to a temperature of 1000 °C where grain coarsening follows Ostwald ripening. Calculate the average grain size of the microstructure if it is held for 1 h at that temperature. The material dependent rate constant for this temperature is 10 $\mu m^3/s$.

A5: According to Ostwald ripening, the average grain size is, referring to Eq. (15.21) in the Sect. 15.2.2.2.

$r = \sqrt[3]{r_0^3 + kt} = \sqrt[3]{25^3 + 10 * 3600} = 37.2\ \mu m$

9. The initial data for the rate of equation, $A + B \rightarrow C$, is given below, calculate the rate constant for the reaction.

Trial	$[A]$ mol/L	$[B]$ mol/L	Rate of $[C]$ mol/L s
1	0.15	0.2	0.00108
2	0.15	0.5	0.00675
3	0.45	0.5	0.18225

A6: The reaction rate for the above reaction is given as, refer Eq. (15.23) in the Sect. 15.2.3.1.

$r = k[A]^m[B]^n$

To calculate the rate constant, first the order of the reaction is to be determined. The value of m and n are calculated from the available data as,

$\frac{\frac{r}{[A]^m[B]^n}}{\frac{r}{[A]^m[B]^n}} = \frac{\frac{0.00108}{0.15^m * 0.2^n}}{\frac{0.00675}{0.15^m * 0.5^n}} = \frac{0.00108}{0.00675} * \left(\frac{0.5}{0.2}\right)^n = 1 \rightarrow n = 2$

$\frac{\frac{r}{[A]^m[B]^n}}{\frac{r}{[A]^m[B]^n}} = \frac{\frac{0.00675}{0.15^m * 0.5^n}}{\frac{0.18225}{0.45^m * 0.5^n}} = \frac{0.00675}{0.18225} * \left(\frac{0.45}{0.15}\right)^m = 1 \rightarrow m = 3$

The rate constant for the equation is,

$k = \frac{r}{[A]^m[B]^n} = \frac{0.00108}{0.15^3 * 0.2^2} = 8\ L/mol\ s$

10 A plain carbon steel is held at 950 °C, calculate the grain boundary velocity if the curvature of the grain is 0.035/m. The grain boundary mobility prefactor is $2.5 * 10^{-3}$ m^4/Js, and the activation energy for grain boundary mobility is 150 kJ/mol. The grain boundary energy is 0.5 J/m^2.

A7: The grain boundary velocity is given as, refer Eq. (15.32) in the Sect. 15.2.4.1.

$v = M\gamma\kappa$

The temperature dependent grain boundary mobility is, refer Eq. (15.30) in the Sect. 15.2.3.3.

$M = M_0 \exp\left(-\frac{Q}{RT}\right) = 2.5 * 10^{-3} * \exp\left(-\frac{150000}{8.314 * 1223}\right) = 9.8 * 10^{-10}$

The grain boundary velocity is,

$v = M\gamma\kappa = 9.8 * 10^{-10} * 0.5 * 0.035 = 1.7 * 10^{-11}$ m/s = 0.06 μm/h

11 For an interface-controlled transformation with a constant nucleation rate, calculate the amount of volume fraction of a phase transformed in 13 min using JMAK method. The rate constant for this phase transformation is 0.001 s^{-1}.

A8: The volume fraction of a transformed phase represented by JMAK model is given as, refer to Eq. (15.22) in the Sect. 15.2.2.3.

$X(t) = 1 - \exp(-kt^n)$

And for an interface-controlled transformation with constant nucleation rate, $n = 1$.

The volume fraction of transformed phase is

$X(t) = 1 - \exp(-kt) = 1 - \exp(0.001 * 13 * 60) = 0.54$

12	Calculate the entropy of mixing for an ideal solution containing 3 components having the composition of $x_A = 0.2$, $x_B = 0.35$, and $x_C = 0.45$. **A9:** The entropy of mixing for an ideal solution is given as, refer to Eq. (15.5) in Sect. 15.1.2. $\Delta S_{\text{mix}} = -R\sum x_i \ln x_i = -R(x_A \ln x_A + x_B \ln x_B + x_C \ln x_C) = -8.314 * (0.2 * \ln 0.2 + 0.35 * \ln 0.35 + 0.45 * \ln 0.45) = 8.72$ J/mol K
13	For Allen-Cahn equation, calculate the rate of change of order parameter if the free energy function is assumed to be a double-well potential, $F = A\phi^2(1-\phi)^2$, at $\phi = 0.25$. The mobility coefficient is 0.01, and $A = 0.2$. **A10:** As per the Allen-Cahn equation the evolution of the order parameter is given as, refer to Eq. (15.37) in Sect.15.3.1. $\frac{\partial\phi}{\partial t} = -M\left(\frac{dF}{d\phi}\right) = -M\left(\frac{d\left(A\phi^2(1-\phi)^2\right)}{d\phi}\right) = -2MA\phi\left(1-3\phi+2\phi^2\right) = -2*0.01*0.2*0.25*\left(1-3*0.25+2*0.25^2\right) = -0.000375$

References

1. Kaufman, L., & Bernstein, H. (1970). *Computer calculation of phase diagrams. With special reference to refractory metals*. Academic.
2. Dinsdale, A. (1991). SGTE data for pure elements. *Calphad, 15*(4), 317–425.
3. NIST. *Phase diagrams and computational thermodynamics of solder systems*. Retrieved from https://www.metallurgy.nist.gov/phase/solder/solder.html
4. Otis, R., & Liu, Z.-K. (2017). pycalphad: CALPHAD-based computational thermodynamics in python. *Journal of Open Research Software, 5*(1), 1.
5. Andersson, J.-O., Helander, T., Höglund, L., Shi, P., & Sundman, B. (2002). Thermo-Calc & DICTRA, computational tools for materials science. *Calphad, 26*(2), 273–312.
6. Bale, C. W., Bélisle, E., Chartrand, P., Decterov, S. A., Eriksson, G., Gheribi, A., Hack, K., Jung, I. H., Kang, Y. B., Melançon, J., Pelton, A. D., Petersen, S., Robelin, C., Sangster, J., Spencer, P., & Van Ende, M.-A. (2016). FactSage thermochemical software and databases—2010–2016. *Calphad, 54*, 35–53.
7. Sundman, B., Kattner, U. R., Palumbo, M., & Fries, S. G. (2015). OpenCalphad—A free thermodynamic software. *Integrating Materials and Manufacturing Innovation, 4*, 1–15.
8. Davies, R., Dinsdale, A., Gisby, J., Robinson, J., & Martin, S. (2002). MTDATA—Thermodynamic and phase equilibrium software from the national physical laboratory. *Calphad, 26*(2), 229–271.
9. Jain, A., Ong, S., Hautier, G., Chen, W., Richards, W., Dacek, S., Cholia, S., Gunter, D., Skinner, D., Ceder, G., & Persson, K. (2013). Commentary: The materials project: A materials genome approach to accelerating materials innovation. *APL Materials, 1*(1), 011002.
10. Scheidgen, M., Himanen, L., Ladines, A., Sikter, D., Nakhaee, M., Fekete, Á., Chang, T., Golparvar, A., Márquez, J., Brockhauser, S., Brückner, S., Ghiringhelli, L., Dietrich, F., Lehmberg, D., Denell, T., Albino, A., Näsström, H., Shabih, S., Dobener, F., Kühbach, M., Mozumder, R., Rudzinski, J., Daelman, N., Pizarro, J., Kuban, M., Salazar, C., Ondračka, P., Bungartz, H. J., & Draxl, C. (2023). NOMAD: A distributed web-based platform for managing materials science research data. *Journal of Open Source Software, 8*(90), 53888.
11. Chen, S. L., Zhang, F., Xie, F. Y., Daniel, S., Yan, X. Y., Chang, Y. A., Schmid-Fetzer, R., & Oates, W. A. (2003). Calculating phase diagrams using PANDAT and panengine. *JOM, 55*, 48–51.
12. Kozeschnik, E. (2022). Mean-field microstructure kinetics modeling. *Encyclopedia of Materials: Metals and Alloys, 4*, 521–526.
13. Giudicelli, G., Lindsay, A., Harbour, L., Icenhour, C., Li, M., Hansel, J., German, P., Behne, P., Marin, O., Stogner, R., Miller, J., Schwen, D., Wang, Y., Munday, L., Schunert, S., Spencer, B., Yushu, D., Recuero, A., Prince, Z., Nezdyur, M., & Permann, C. (2024). 3.0—MOOSE: Enabling massively parallel multiphysics simulations. *SoftwareX, 26*, 101690.
14. MICRESS. Retrieved from https://www.micress.de/
15. DeWitt, S., Rudraraju, S., Montiel, D., Andrews, W., & Thornton, K. (2020). PRISMS-PF: A general framework for phase-field modeling with a matrix-free finite element method. *npj Computational Materials, 6*, 29.
16. Dutta, T., Mohan, D., Shenoy, S., Attar, N., Kalokhe, A., Sagar, A., Bhure, S., Pradhan, S., Praharaj, J., Mridha, S., Kushwaha, A., Shah, V., Gururajan, M., Shenoi, V., Phanikumar, G., Bhattacharyya, S., & Choudhury, A. (2025). MicroSim: A high-performance phase-field solver based on CPU and GPU implementations. *Computational Materials Science, 246*, 113438.
17. Leetmaa, M., & Skorodumova, N. (2014). KMCLib: A general framework for lattice kinetic Monte Carlo (KMC) simulations. *Computer Physics Communications, 185*(9), 2340–2349.
18. Thompson, A., Aktulga, H., Berger, R., Bolintineanu, D., Brown, W., Crozier, P., in 't Veld, P., Kohlmeyer, A., Moore, S., Nguyen, T., Shan, R., Stevens, M., Tranchida, J., Trott, C., & Plimpton, S. (2022). LAMMPS—A flexible simulation tool for particle-based materials modeling at the atomic, meso, and continuum scales. *Computer Physics Communications, 271*, 108171.

19. Mitchell, J., Abdeljawad, F., Garcia-Cardona, C., Holm, E., Homer, E., Madison, J., Rodgers, T., Thompson, A., Tikare, V., Webb, E., & Plimpton, S. (2023). Parallel simulation via SPPARKS of on-lattice kinetic and Metropolis Monte Carlo models for materials processing. *Modelling and Simulation in Materials Science and Engineering, 31*, 055001.
20. {MCC}ube: Markov chain cubature via {JAX}. (2023). Retrieved from https://mccube.readthedocs.io/en/latest/
21. CASM. (2021). Retrieved from https://prisms-center.github.io/CASMcode_docs/

Chapter 16
Non-equilibrium Thermodynamics and Kinetics

Contents

Non-equilibrium thermodynamics and kinetics applies to systems that are driven away from equilibrium by external forces, gradients, or chemical reactions, leading to continuous flows of energy, mass, or charge. Unlike equilibrium thermodynamics, where state variables like temperature, pressure, and chemical potential are uniform and time-independent, non-equilibrium systems exhibit spatial and temporal variations, resulting in irreversible processes such as heat conduction, diffusion, and chemical reactions, where they involve energy transformations that cannot be reversed. A fundamental distinction lies in entropy; while equilibrium thermodynamics considers entropy as static, maximized quantity, non-equilibrium systems continuously produce entropy due to dissipative processes, quantified by entropy production rate (σ) as mentioned below,

K. A. Padmanabhan et al., *Thermodynamics and Kinetics of Materials*, Engineering Materials, https://doi.org/10.1007/978-3-032-14527-7_16

$$\sigma = \sum J_i X_i \tag{16.1}$$

where J_i are fluxes (e.g., heat flux, diffusion flux, etc.), and X_i are the conjugate driving forces (e.g., temperature gradient, chemical potential gradient, etc.).

From a thermodynamic perspective, equilibrium systems are governed by the minimization of free energy and the maximization of entropy at a fixed state, whereas non-equilibrium systems operate under dynamic constraints where energy and mass flows sustain the system away from equilibrium. The local equilibrium hypothesis allows the application of classical thermodynamic relations to small subsystems, but the global behavior emerges from interactions between these subsystems, often resulting in complex phenomena such as pattern formation or oscillating reactions.

From a kinetic perspective, equilibrium systems have forward and reverse reaction rates that balance each other, resulting in no net change in composition. In contrast, non-equilibrium kinetics deal with irreversible reactions where the net flux is nonzero, and the reaction rates depend on how far the system is displaced from equilibrium. While near-equilibrium kinetics can often be described by linear laws (e.g. Arrhenius-type rate equations), far-from equilibrium processes exhibit nonlinear behavior, leading to phenomena such as autocatalysis, bi-stability, and chaos, which cannot be predicted from the equilibrium principles alone.

Thus, non-equilibrium thermodynamics and kinetics provide a broader framework for understanding real-world processes, such as phase transformation under rapid cooling, heat transfer, mass diffusion, electrochemical energy conversion, and biological metabolism, where equilibrium assumptions fail, and dynamic, irreversible behavior dominates. This distinction is crucial for designing advanced materials and technologies that operate under persistent energy and mass flows.

This chapter provides a broad overview of the subject, aiming to introduce the core concepts and selected applications of non-equilibrium thermodynamics and kinetics, without delving into detailed derivations or the full theoretical rigor of the field.

16.1 Departure from Equilibrium Thermodynamics and Kinetics

16.1.1 Local Thermodynamic Equilibrium

The fundamental bridge between equilibrium thermodynamics and non-equilibrium systems is the Local Thermodynamic Equilibrium (LTE) hypothesis. This powerful assumption states that while a system may be globally out of equilibrium, it can be conceptually divided into infinitesimal volume elements that are themselves in a state of near-equilibrium. Each of these microscopic subsystems can be described using conventional equilibrium thermodynamics, but with intensive parameters like temperature, chemical potential, pressure, etc., that vary smoothly across the macroscopic system. These variations are smooth enough that any infinitesimal volume element can be in thermodynamic equilibrium for timescales much shorter than those characterizing the timescale of evolution of the global system. This assumption allows us to retain the powerful machinery of equilibrium thermodynamics while extending its applicability to systems undergoing irreversible processes.

The validity of LTE rests on several physical criteria. First, the timescale for local molecular equilibration (e.g., via collisions or vibrational relaxation) must be much shorter than the timescale over which macroscopic gradients evolve. Second, the mean free path of particles must be small compared to the characteristic length scales of the system, ensuring that local interactions dominate over non-local effects. Third, thermodynamic forces (gradients of temperature, chemical potential, etc.) must remain sufficiently small so that the linear response theory can be applied at the local level. When these conditions are met, intensive variables retain their physical meaning locally, even as they vary across the system. This enables the definition of local entropy production rates and the formulation of constitutive relations between thermodynamic fluxes and forces, which are central to non-equilibrium thermodynamics.

However, the LTE framework also highlights key departure from classical equilibrium thermodynamics. Most notably, global extremum principles (e.g., entropy minimization at equilibrium) no longer apply, as the system continuously generates entropy through irreversible processes. The breakdown of detailed balance means that the microscopic reversibility is lost, and the reciprocal relations between coupled processes become dynamic rather than static. Furthermore, while equilibrium systems are fully characterized by a small set of state variables, non-equilibrium systems under LTE require additional information about spatial and temporal evolution of these variables. This necessitates the introduction of flux-force relationships, where thermodynamic fluxes (e.g., heat flow, diffusion currents, reaction rates) are driven by conjugate forces (e.g., temperature gradient, chemical potential gradients, reaction affinities).

The LTE framework finds widespread application in materials science, where it enables the analysis of processes such as heat treatment, solid-state diffusion, and electrochemical reactions. For example, during the quenching of an alloy, LTE allows us to describe the local phase equilibria and diffusion fluxes that govern microstructure evolution, even as the system globally departs from equilibrium. Similarly, in the study of battery electrodes, LTE provides the foundation for modeling charge transfer kinetics and ion transport, where local electroneutrality and chemical potential gradients determine the rates of interfacial reactions. However, as systems are driven further from equilibrium, whether through large external fields, rapid phase transformations, or nanoscale confinement, the assumptions underlying LTE begin to break down. In such cases, more advanced theoretical approaches become necessary to capture the emergent behavior of far-from-equilibrium systems, a broad overview of which will be presented later in this chapter.

16.1.2 Flux-Force Relationship

The fundamental departure from equilibrium thermodynamics manifests through emergence of thermodynamic forces that drive irreversible processes. In non-equilibrium thermodynamics, thermodynamic forces (X_i) represent the driving mechanism that induces fluxes of extensive quantities, acting as generalized gradients of intensive parameters. These forces emerge from spatial or temporal inhomogeneities in the system and are derived from an entropy production rate, which must remain non-negative according to the Second law. The fluxes (J_i) represent the flow of extensive quantities (e.g., energy, mass, charge) driven by the thermodynamic forces.

The mathematical foundation for the relationship between fluxes and forces is established through entropy production rate which serves as the quantitative measure of irreversibility in the system, as represented in Eq. (16.1). This bilinear form of entropy production, arising from the fact that the thermodynamic fluxes are proportional to the thermodynamic forces, maintains consistency with the Second law of Thermodynamics, requiring $\sigma \geq 0$ for all physically realizable processes, where σ is the entropy production rate. Under the local equilibrium hypothesis, these fluxes are expressed as linear combinations of thermodynamic forces for small departure from equilibrium, given as,

$$J_k = \sum_l L_{kl} X_l \tag{16.2}$$

where L_{kl} are the phenomenological coefficients that characterize the system's response to applied thermodynamic force.

The thermodynamic forces and fluxes are demonstrated below using an example of a multicomponent system with heat, mass and charge transport, such as electrochemical cells. For example, in an electrochemical reaction occurring within a battery or fuel cell, ionic species migrate under electric fields (charge transport), react chemically (mass transport), and generate or absorb heat (heat transport), all driven by gradients in electrochemical potential, concentrations, and temperature. Throughout this chapter, this example will be utilized to describe the equations in different sections. Along with the coupled heat, mass, and charge transport, force and flux associated with the chemical reactions are also utilized in the section below in this chapter. The thermodynamic forces and fluxes for the heat, mass, and charge transport in the example described above are described below.

The thermal force, X_q, is given by,

$$X_q = \nabla\left(\frac{1}{T}\right) \tag{16.3}$$

where T is the local temperature. This force drives the heat flow (J_q) and reflects the tendency of the system to equilibrate thermal energy. The heat flux driven by the temperature gradient is,

$$J_q = -\kappa \nabla T \tag{16.4}$$

where κ is the thermal conductivity.

The chemical force, $X_{m,\,i}$, is given by,

$$X_{m,i} = -\nabla\left(\frac{\mu_i}{T}\right) \tag{16.5}$$

where $X_{m,\,i}$ represents the chemical force of species i, μ_i is the chemical potential of species i. This force governs diffusion (J_i) and phase separation. The mass flux of species i driven by the chemical potential gradient is,

$$J_i = -\sum_j D_{ij} \nabla c_j \tag{16.6}$$

where D_{ij} is the diffusivity tensor, and c_j is the concentration of species j.

The electrical force, X_e, is given by,

$$X_e = -\nabla\left(\frac{\phi}{T}\right) \tag{16.7}$$

where ϕ is the electric potential. This force drives charge transport (J_e) in electrolytes or electronic conductors. The charge flux driven by electrochemical potential gradient is,

$$J_e = -\sigma \nabla \tilde{\mu}_e \tag{16.8}$$

where σ is electrical conductivity, and $\tilde{\mu}_e$ is the electrochemical potential.

For a chemical reaction, the affinity (A) acts as the thermodynamic force for the reaction flux. The corresponding force is given as,

$$X_{r,j} = \frac{A_j}{T} \tag{16.9}$$

where $X_{r,\,j}$ represents the thermodynamic force for the reaction for the reactant j, A_j is the affinity of the reactant j, and T is the temperature at which reaction takes place.

The flux associated with the chemical reaction is the reaction rate (v), given as,

$$v = \frac{1}{\nu_i}\frac{d\xi}{dt} \tag{16.10}$$

where ν_i is the stoichiometric coefficient, and ξ is the extent of the reaction.

The coupling between different processes (e.g., thermoelectric effects) is described by the cross coefficients, L_{kl} ($k \neq l$). These coupled equations are described later in this chapter in Sect. 16.2.2, Eqs. (16.16)–(16.18). These relationships form the foundation for analyzing transport phenomena in materials under non-equilibrium conditions.

In systems close to equilibrium, thermodynamic forces and fluxes are related in a simple, linear way, meaning that as the force increases, the flux increases proportionally. However, when the system is far from equilibrium, this linear relationship no longer holds, and the response becomes more complex, and higher-order (nonlinear) effects must be considered. This means that we need more advanced, nonlinear models to accurately describe how the system behaves under strong driving forces. The magnitude of the force determines the validity of local equilibrium assumptions, with the linear regime typically requiring $\frac{|X_k| l}{k_B T} \ll 1$, where l is the characteristic microscopic length.

16.1.3 Entropy Production

The concept of entropy production serves as a fundamental bridge between equilibrium thermodynamics and the analysis of real-world, non-equilibrium processes in materials. In contrast to equilibrium states where entropy reaches a maximum and remains constant, non-equilibrium systems continuously generate entropy through irreversible processes. This entropy production rate (σ) provides a quantitative measure of how far a system departs from equilibrium and represents the thermodynamic cost of maintaining non-equilibrium conditions.

The foundation for understanding entropy production in non-equilibrium systems begins with the Gibbs relaxation extended to local thermodynamic equilibrium.

$$Tds = du - \sum_i \mu_i dn_i + PdV \tag{16.11}$$

Where T is the local temperature, s is the local entropy density, u is the internal energy density, μ_i are the chemical potentials, n_i are the molar densities, P is the local pressure, and V is the specific volume. Combining this with conservation laws for mass and energy leads to the entropy balance equation,

$$\sigma = \frac{\partial s}{\partial t} + \nabla \cdot J_s \tag{16.12}$$

where J_s represents the entropy flux. The entropy production rate takes the canonical bilinear form as shown in Eq. (16.1), which reveals that entropy production arises from the combined effect of various thermodynamic fluxes driven by their conjugate forces.

The entropy balance equation is then transformed into extropy production using the first and second law of thermodynamics. For example, a general multicomponent system undergoing heat and mass transfer along with chemical reactions, as in the case of a solid oxide fuel cell, the entropy production decomposes into distinct contributions as given below.

$$\sigma = J_q \nabla\left(\frac{1}{T}\right) - \sum_i J_i \nabla\left(\frac{\mu_i}{T}\right) + \sum_j \frac{v_j A_j}{T} \tag{16.13}$$

The first term on the right-hand side represents the entropy production from heat conduction, where J_q is the heat flux and $\nabla\left(\frac{1}{T}\right)$ is its conjugate thermal force. The second term accounts for the diffusion processes, with J_i being the molar flux of the component i and $-\nabla\left(\frac{\mu_i}{T}\right)$ the corresponding chemical force. The final term describes the contribution of chemical reaction, where v_j is the reaction rate of component j and $\frac{A_j}{T}$ is the affinity-based driving force.

The principle of minimum entropy production plays a crucial role in determining the steady-state configurations of non-equilibrium systems. For fixed boundary conditions, systems evolve towards states that minimize the total entropy production,

$$P = \int_V \sigma dV \tag{16.14}$$

with the time derivative satisfying $\frac{dP}{dV} \leq 0$. This variational principle has profound implications for pattern formation in materials systems, explaining why certain dissipative structures (e.g., Lissegang rings, dendritic growth patterns, etc.) emerge preferentially under non-equilibrium conditions. Minimization occurs within the constraints imposed by the boundary conditions and conservation laws, leading to the development of characteristic length scales in non-equilibrium systems.

In practical applications, entropy production analysis enables the calculation of energy dissipation rates in transport processes, determination of efficiency limits in energy conversion devices, prediction of stability criteria for non-equilibrium states, and analysis of pattern formation during phase transformation among others.

The framework remains valid when local relaxation times are much shorter than the observation times, system dimensions greatly exceed molecular mean free path, and thermodynamic forces are sufficiently small. When these conditions break down (e.g., nanoscale systems or under extreme driving forces), extended theories become necessary, but the fundamental concepts of entropy production continue to provide valuable insights into non-equilibrium material behavior.

16.2 Linear Non-equilibrium Thermodynamics and Kinetics

16.2.1 Foundation of Linear Response Theory

Linear non-equilibrium thermodynamics provides the essential framework for analyzing systems that are slightly perturbed from equilibrium. The theory is built on two fundamental postulates. (1) The local equilibrium hypothesis, which assumes that thermodynamic quantities can be defined locally even when the global system is out of equilibrium, as described in Sect. 16.1.1, and (2) linear flux-force relationship, which holds when the system's departure from equilibrium is sufficiently small. The mathematical foundation stems from a Taylor expansion of the fluxes about their equilibrium values, retaining only the first-order terms, as shown in Eq. (16.2),

$$J_k = \sum\nolimits_l \left(\frac{\partial J_k}{\partial X_l} \right)_0 X_l = \sum\nolimits_l L_{kl} X_l \tag{16.15}$$

where L_{kl} are the phenomenological coefficients that characterize the system's response to applied thermodynamic force. The linear approximation remains valid when the dimensionless force satisfies $\frac{|X_l|\lambda}{k_B T} \ll 1$, where λ is the characteristic length scale of the process for transport processes, and the affinity (A) satisfies $|A| \ll RT$ for a chemical reaction. For typical materials systems, this corresponds to temperature gradients below ~10^6 K/m, concentration gradients below ~10^{28} particles/m^3, and electric fields below ~10^7 V/m.

16.2.2 Coupled Transport Phenomenon

Coupled transport phenomena become particularly important in materials systems where multiple irreversible processes occur simultaneously. Linear non-equilibrium thermodynamics excels at describing these coupled transport phenomena. The general constitutive equations for a system with heat, mass, and charge transport, as described in Sect. 16.1.2, take the form,

$$J_q = L_{qq} \nabla \left(\frac{1}{T} \right) + L_{qm} \left(- \nabla \frac{\mu}{T} \right) + L_{qe} \left(- \nabla \frac{\phi}{T} \right) \tag{16.16}$$

$$J_m = L_{mq} \nabla \left(\frac{1}{T} \right) + L_{mm} \left(\nabla \frac{\mu}{T} \right) + L_{me} \left(- \nabla \frac{\phi}{T} \right) \tag{16.17}$$

$$J_e = L_{eq} \nabla \left(\frac{1}{T} \right) + L_{em} \left(\nabla \frac{\mu}{T} \right) + L_{ee} \left(- \nabla \frac{\phi}{T} \right) \tag{16.18}$$

These coupled equations describe phenomena such as thermoelectric effects like Seeback (generation of electric voltage across the material when there is a temperature difference between its ends, illustrating the coupling between heat and charge transport), and Peltier (absorption or release of heat at the junction of two different conductors when an electric current passes through, demonstrating the interplay between electric current and heat flow) effect, electrokinetic phenomena like electroosmosis (movement of liquid through a porous material or capillary under the influence of an applied electric field, illustrating the coupling between electric force and fluid flow), and streaming potential (electric potential generated when a liquid flows through a charged porous medium or capillary, illustrating the coupling between fluid flow and charge separation), and thermo-diffusion effect like Soret effect (temperature gradient causes the redistribution of concentration gradients, illustrating the coupling between heat and mass transport).

16.2.3 Onsager Reciprocal Relations and Their Implications

A cornerstone of linear non-equilibrium thermodynamics is the Onsager reciprocal relations, which states that the matrix or phenomenological coefficients are symmetric,

$$L_{kl} = L_{lk} \tag{16.19}$$

This symmetry originates from microscopic reversibility and time-reversal symmetry of equilibrium fluctuations. This relation profoundly reduces the number of independent transport coefficients needed to characterize coupled processes. For example, in a system with coupled heat and mass transport, Eqs. (16.16) and (16.17), the reciprocal relation of $L_{qm} = L_{mq}$ means that the same coefficient governs both mass flux due to a temperature gradient (Soret effect) and heat flux due to a concentration gradient (Dufour effect).

The Onsager reciprocal relations reduce the number of independent coefficients needed to describe these coupled phenomena. These coefficients can be related to measurable transport properties, mentioned in Table 16.1.

Here, κ is the thermal conductivity, D is the diffusion coefficient, c is the concentration, ρ is the electrical resistivity, s is the Soret coefficient characterizing thermo-diffusion, S is the Seeback coefficient, and μ_{E_O} is the Electroosmotic mobility. These coupling coefficients provide a unified framework for analyzing coupled transport in materials. They link thermodynamic forces to measurable transport properties (e.g., κ, D, σ, S) and are crucial for designing energy materials, microfluidic devices, and electrochemical systems. The Onsager relation ensures consistency with microscopic reversibility, while the Second law bounds their

Table 16.1 Coupling coefficients for coupled transport phenomena of heat, mass and charge transport

Coefficient	Physical meaning	Driving force	Relationship
L_{qq}	Thermal conductivity	Heat flow due to ∇T	$L_{qq} = \kappa T^2$
L_{mm}	Diffusivity	Mass flow due to $\nabla \mu$	$L_{mm} = \frac{DcT}{\frac{\partial \mu}{\partial c}}$
L_{ee}	Electrical conductivity	Charge flow due to $\nabla \phi$	$L_{ee} = \sigma T$
$L_{qm} = L_{mq}$	Thermo-diffusion	Mass flux (J_m) due to ∇t	$L_{qm} = sDcT^2$
$L_{qe} = L_{eq}$	Thermoelectric effect	Charge flux (J_e) due to ∇T	$L_{qe} = S\sigma T^2$
$L_{me} = L_{em}$	Electrokinetic effect	Mass flux (J_m) due to $\nabla \phi$	$L_{me} = \mu_{E_O} \rho T$

magnitudes. Beyond the linear regime, these couplings become nonlinear, requiring more advanced theories.

16.3 Far-From-Equilibrium Systems

Rayleigh-Benard convection exemplifies the pattern formation in fluid mechanics, where a temperature gradient across a fluid layer induces buoyancy-driven convection rolls once a critical threshold (Rayleigh number $Ra > 1708$) is surpassed. The resulting patterns of hexagonal cells or spiral defects are dissipative structures that export entropy while maintaining order. Similarly, the Belousov-Zhabotinsky reaction is a chemical oscillator where nonlinear kinetics produces traveling waves or spatial patterns in concentrations of reactants like cerium and malonic acid. These oscillations arise from autocatalysis and feedback loops, violating linear kinetics and demonstrating how far-from equilibrium systems can sustain temporal and spatial order. These pattern-forming phenomena are critical in materials processing, energy conversion, and biological systems, illustrating how far-from equilibrium thermodynamics bridges microscopic kinetics and macroscopic order.

Far-from-equilibrium systems occur when thermodynamic forces become sufficiently large that linear flux-force relationships break down, leading to complex, nonlinear behavior not observed near equilibrium. The departure from linear response is characterized by the failure of Onsager's reciprocal relations ($L_{kl} \neq L_{lk}$) and the emergence of new phenomena such as dissipative structures, chemical oscillations, and chaotic dynamics. The entropy production rate (σ) remains positive, remains same as Eq. (16.13), but adopts a nonlinear form, where the fluxes (J_k) now depend nonlinearly on the forces (X_k). The heat and mass transport fluxes become dependent on temperature gradient and the rate at which temperature changes over time, as in non-Fourier heat conduction.

$$J_q = -\kappa \nabla T + \tau \frac{\partial J_q}{\partial t} \tag{16.20}$$

where τ is relaxation time that accounts for thermal inertia.

Far-from equilibrium systems often exhibit instabilities and pattern formation, described by reaction–diffusion equations,

$$\frac{\partial c_i}{\partial t} = D_i \nabla^2 c_i + f_i \tag{16.21}$$

where f_i are nonlinear kinetic terms. Under certain conditions (e.g., Turing instability), small perturbations grow exponentially over time leading to formation of spatial patterns. The stability of such systems is governed by the eigenvalues (λ) of the Jacobian matrix of the kinetic equation (16.21). Specifically, if at least one eigenvalue has a positive real part (>0), the system becomes unstable and self-organizing.

Examples of such systems include combustion fronts, dendritic growth in solidification, and active matter systems, where energy input sustains ordered states.

16.4 Reaction Kinetics in Non-equilibrium Systems

In non-equilibrium systems, reaction kinetics deviate fundamentally from near-equilibrium behavior due to large thermodynamic driving forces, nonlinear rate laws, and coupling between multiple processes. Unlike equilibrium systems where the principle of detailed balance, where the forward and reverse reaction rates are equal holds true, non-equilibrium reactions exhibit complex dynamics governed by the interplay between affinity and reaction flux. For elementary reactions, the net reaction rate, $v = v^+ - v^-$ is no longer a linear function of affinity but follows nonlinear form such as the mass-action law,

$$v = k^+ \prod_i [C_i]^{v_i^+} - k^- \prod_i [C_i]^{v_i^-} \tag{16.22}$$

where $k^\pm$ are the rate constants, $\prod_i [C_i]^{v_i^+}$ represents the total forward reaction rate where the operator Π represents the product of a group, $\prod_i [C_i]^{v_i^-}$ represents the total backward reaction rate, and C_i is the concentration of element i. In electrochemical systems, Butler-Volmer equation replaces linear kinetics with exponential dependence on the driving force (overpotential),

$$v = j_0 \left[\exp\left(\frac{(1-\alpha)F\eta}{RT} \right) - \exp\left(-\frac{\alpha F\eta}{RT} \right) \right] \tag{16.23}$$

where j_0 is the standard rate constant in terms of exchange current density, F is the Faraday constant, η is the overpotential, and α is the charge transfer coefficient. The equation (16.23) is connected with equation (16.22) in such a way that $\exp\left(\frac{(1-\alpha)F\eta}{RT}\right)$ represents the forward reaction rate and $\exp\left(-\frac{\alpha F\eta}{RT}\right)$ represents the backward reaction rate, with j_0 being the rate constant for both the forward and backward reaction.

The Belousov-Zhabotinsky reaction is a classic example of a non-equilibrium chemical oscillator that exhibits temporal and spatial pattern formation in far-from-equilibrium systems. Far-from-equilibrium conditions enable autocatalytic feedback where certain reaction intermediates accelerate their own production, leading to oscillations and complex spatial behavior. The entropy production rate ($\sigma = vA/T$) remains positive but becomes a non-quadratic function of affinity, reflecting the breakdown of Onsager reciprocity. Such systems often exhibit multiple steady states, bi-stability, or spatiotemporal patterns, as described by the reaction diffusion equation in Eq.(16.21). These phenomena are critical in material synthesis (e.g., combustion, electrodeposition) and biological networks, where non-equilibrium kinetics sustains life processes. Understanding these non-equilibrium kinetic principles is crucial for controlling complex materials processes and designing functional non-equilibrium materials systems.

16.5 Application of Non-equilibrium Thermodynamics and Kinetics

Non-equilibrium thermodynamics and kinetics provide a powerful framework for understanding and engineering complex systems maintained far from equilibrium, where nonlinear dynamics, coupled with transport processes, and self-organization play critical roles. Non-equilibrium thermodynamics and kinetics have broad applications across diverse fields, where it explains the transport of energy, matter, and momentum, as well as the rates of chemical reactions in systems driven by gradients. Below, we provide a glimpse of its key applications, highlighting how these approaches enable the design, analysis, and optimization of real-world processes.

16.5.1 Applications in Materials Science

Non-equilibrium thermodynamics plays a central role in materials synthesis and processing, where rapid quenching or external fields can trap systems in metastable states. Metallic glass formation, for example, occurs when a molten alloy is cooled so quickly that crystallization is bypassed, resulting in an amorphous solid. The "glass transition" is inherently a non-equilibrium process, with the system evolving toward equilibrium through slow structural relaxation (aging).

In thin-film deposition techniques such as chemical vapor deposition (CVD) or sputtering, the growth of nanostructures is governed by the competition between reaction kinetics and surface diffusion. Non-equilibrium conditions enable the creation of unique microstructures, such as strained heterostructures or defect-engineered materials, which are inaccessible under equilibrium conditions. Similarly, battery materials like lithium-ion phosphate ($LiFePO_4$) exhibit phase separation during charging/discharging, a process described by reaction–diffusion models that incorporate nonlinear thermodynamics.

16.5.2 Applications in Chemical Engineering

In chemical engineering, non-equilibrium thermodynamics is essential for modeling reaction and transport processes in industrial systems. For example, catalytical reactors often operate under large thermodynamic driving forces, where the linear flux-force relationship for non-equilibrium thermodynamics breaks down. The oxidation of sulfur dioxide (SO_2) to sulfur trioxide (SO_3) in the contact process is governed by nonlinear kinetics, with reaction rate depending exponentially on temperature and reactant concentrations. The entropy production rate in such reactors must be carefully controlled to minimize energy dissipation while maximizing yield. Similarly, distillation columns, which separate mixtures based on volatility gradients, rely on coupled heat and mass transfer phenomena described by non-equilibrium thermodynamics. The efficiency of these systems depends on managing entropy generation due to viscous flow, thermal conduction, and diffusion mixing.

Another key application is electrochemical engineering, where the Butler-Volmer equation describes charge transfer kinetics in fuel cells and electrolyzers. For instance, in proton-exchange membrane (PEM) fuel cells, the oxygen reduction reaction (ORR) at the cathode exhibits nonlinear current–voltage behavior due to high overpotentials. Optimizing these devices requires balancing reaction kinetics, ionic transport, and electron conduction, all of which are governed by far-from-equilibrium thermodynamics.

16.5.3 Applications in Biological Systems

Living organisms are quintessential non-equilibrium systems, relying on continuous energy input to sustain order and function. Cellular metabolism is a prime example, where ATP hydrolysis drives biochemical reactions against their equilibrium directions. The glycolysis pathway, for instance, exhibits oscillatory behavior under certain conditions, a phenomenon explained by nonlinear kinetics and feedback inhibition. Entropy production in such systems is tightly regulated to maintain homeostasis while allowing adaptive response to environmental changes.

Biological membranes also operate far from equilibrium, with ion pumps (e.g., the Na^+/K^+-ATPase) maintaining electrochemical gradients essential for nerve impulse transmission. These active transport processes are described by non-equilibrium models that account for the coupling between chemical reactions (ATP hydrolysis) and ion fluxes. Similarly, protein folding under kinetic control often leads to metastable states, with the folding pathway influenced by competing thermodynamic and kinetic factors.

16.5.4 Applications in Environmental Science

Environmental systems are often driven by natural gradients (e.g., temperature, concentration) and anthropogenic perturbations, making them inherently non-equilibrium. Atmospheric chemistry provides a compelling example, where formation and decomposition of ozone (O_3) in the troposphere and stratosphere involve complex reaction networks far from equilibrium. The photochemical smog mechanism, for instance, includes autocatalytic steps that lead to nonlinear concentration dynamics under sunlight exposure.

In water treatment, non-equilibrium kinetics dictates the removal of pollutants via advanced oxidation processes (AOPs). The degradation of organic contaminants by hydroxyl radicals follows rate laws that depend on local radical concentrations and reaction intermediates, requiring models that couple reaction and diffusion. Similarly, soil remediation strategies, such as electrokinetic extraction of heavy metals, rely on applying electric fields to drive ion migration in porous media, a process described by coupled Nernst-Plank and Poisson equations.

16.5.5 Applications in the Steel Industry

16.5.5.1 Microstructure Evolution in the Heat-Affected Zone during Welding

A quintessential example of non-equilibrium thermodynamics is the microstructure evolution in the heat-affected zone (HAZ) during welding, where rapid heating and cooling drive the system far from equilibrium. The HAZ experiences steep thermal gradients and high cooling rates, leading to complex phase transformations that cannot be predicted by equilibrium phase diagrams alone. Instead, Continuous Cooling Transformation (CCT) diagrams are constructed under non-isothermal conditions and are used to corelate cooling rates with resulting microstructures. To demonstrate microstructure evolution, an example of welding of Cr-Mo low alloy steel, having the chemical composition as mentioned in Table 16.2, is described below.

JMatPro software is utilized to calculate the CCT diagram of this steel [1] and different cooling rates are overlapped on this CCT diagram to obtain the final microstructure, as shown in Fig. 16.1.

The CCT diagram shows the phase transformation start and finish for different phases by solid and broken lines respectively. Ac_3 temperature represents the critical temperature during heating above which the ferrite to austenite phase transformation is completed. M_s, B_s, F_s, P_s represents the start of phase transformation from austenite phase to martensite, bainite, ferrite and pearlite respectively, and corresponding M_f, B_f, P_f represents the end of transformation. During welding, the

Table 16.2 Chemical composition of Cr-Mo low alloy steel [1]

C	Mn	Si	P	S	Cr	Mo	Ni	Nb + V + Ti	Al	Fe
0.295	0.892	0.225	0.015	0.002	0.82	0.186	0.017	0.011	0.024	Bal.

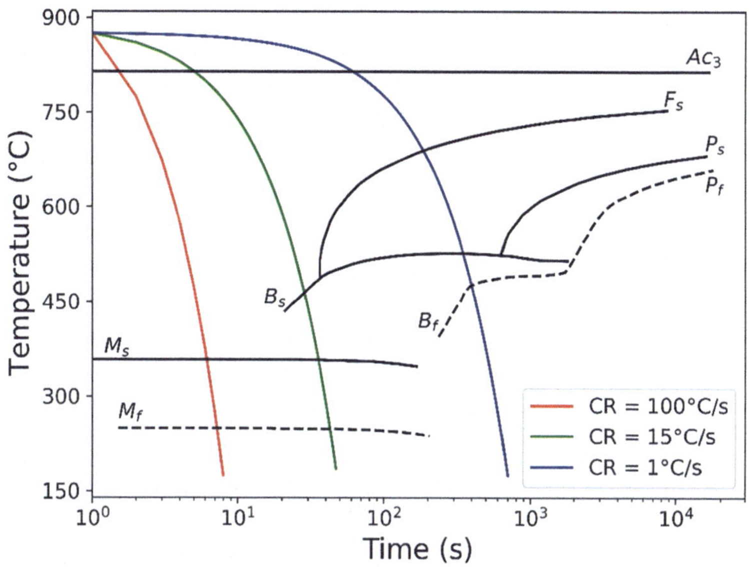

Fig. 16.1 CCT diagram of Cr-Mo low ally steel [1]

area of the metal surrounding the weld is referred to as the HAZ, and here the metal is heated to a temperature above the Ac_3, and is subsequently cooled to room temperature. For Cr-Mo low alloy steel, the equilibrium microstructure at room temperature consists of Ferrite and Pearlite phases, but as the HAZ experiences high cooling rate, non-equilibrium microstructures form in the HAZ. As represented by the red line in Fig. 16.1, if the cooling rate experienced in HAZ is 100 °C/s, the final microstructure obtained is completely martensitic, but if the cooling rate is controlled to 15 °C/s as represented by the green line, the final microstructure obtained is a mixture of martensite and bainite, and controlling the cooling to even lower cooling rate of 1 °C/s as represented by the blue line, the final microstructure obtained is a mixture of ferrite and bainite.

Non-equilibrium phase transformation during cooling in HAZ induces volumetric changes resulting in generation of residual stresses, which can cause cracking in HAZ. By tailoring the welding parameters, and preheating the metal parts before welding, the cooling rates in the HAZ can be controlled and thereby engineering the microstructure to optimize the hardness, toughness, and weldability of the HAZ zone. An ideal microstructure in the HAZ is one that ensures a balance of strength, toughness, and ductility without forming brittle or hard phases. Typically, this includes a fine-grained ferrite-pearlite or bainitic microstructure, which provides good mechanical properties and minimizes the risk of cracking.

16.5.5.2 Design of Dual Phase Steels

Dual phase (DP) steels represent an excellent example of the use of non-equilibrium thermodynamics in materials design, where controlled thermal processing creates a microstructure with superior mechanical properties. DP steels are low alloy steels, having a duplex ferrite-martensite microstructure. The cold rolled steel sheets are annealed via an inter-critical annealing treatment and then quenched to obtain the duplex microstructure. The inter-critical annealing process used to produce DP steels demonstrates how far-from-equilibrium conditions can be harnessed to achieve optimal phase distribution.

The mechanical properties of the DP steels are determined by the phase fractions of ferrite and martensite phases present. Martensite phase fraction controls the ultimate tensile strength of the DP steel, and the ferrite phase fraction governs the elongation, thereby achieving a strength–ductility balance. To demonstrate the design of the DP steels to achieve a desired balance between strength and ductility, an example of a low alloy steel having the chemical composition as mentioned in Table 16.3, is described below.

Designing the DP steel is based on the design of the microstructure, i.e. the phase fractions of ferrite and martensite to achieve certain mechanical properties. The control of the phase fractions is dependent on the inter-critical annealing schedule [3]. To showcase an example, Fig. 16.2 shows two cases where the holding temperature for the inter-critical annealing is selected as 740 °C, and two different inter-critical annealing hold times are selected, $t_1 = 30$ s, and $t_2 = 100$ s. Here, Ac_1 and Ac_3 represents the temperature at which the ferrite/pearlite to austenite transformation starts and finishes respectively during heating.

For the case of hold time t_1, the martensite phase fraction obtained is 10%, and for the case of hold time t_2, the martensite phase fraction obtained is 20%. The yield strength, ultimate tensile strength and % uniform elongation obtained for case 1 are 298.4 MPa, 690.5 MPa, and 21.34% respectively, and for case 2 are 319.9 MPa, 794.7 MPa, and 18.37% respectively. As the martensite phase fraction obtained in case 2 is more, case 2 showcases higher strength and lower ductility compared to case 1.

Similarly, Fig. 16.3 shows two cases where the hold time for inter-critical annealing for both the cases is 100 s, and two different inter-critical annealing holding temperatures are selected, the orange line represents a holding temperature of 740 °C and the blue line represents a holding temperature of 780 °C.

For the case of the orange line, the martensite phase fraction obtained is 20%, and for the case of blue line, the martensite phase fraction obtained is 35%. The yield strength, ultimate tensile strength and % uniform elongation obtained for the case of orange line are 319.9 MPa, 794.7 MPa, and 18.37% respectively, and for the case of

Table 16.3 Chemical composition of low alloy DP steel [2]

C	Mn	Si	S	P	Al	Fe
0.12	1.47	0.24	0.01	0.009	0.008	Bal.

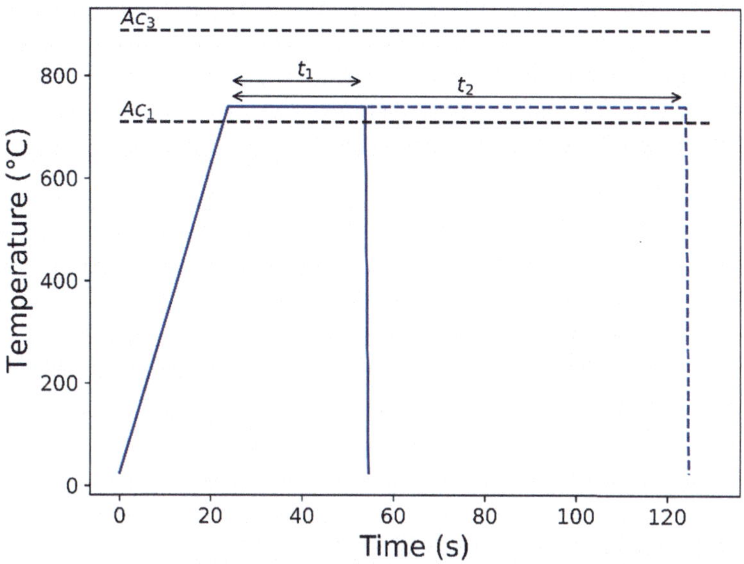

Fig. 16.2 Designing a DP steel by changing the hold time during inter-critical annealing

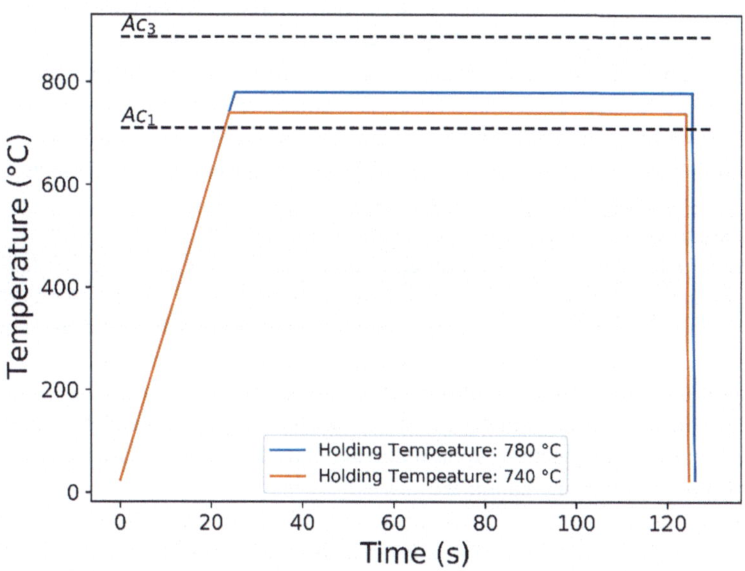

Fig. 16.3 Designing a DP steel by changing the holding temperature during inter-critical annealing

blue line are 360.8 MPa, 956.2 MPa, and 15.78% respectively. As the martensite phase fraction obtained in the case of the orange line is more, this case showcases higher strength and lower ductility compared to the case of blue line.

By precisely controlling the inter-critical annealing time, and temperature, manufacturers manipulate the phase fractions of ferrite and martensite. The selected inter-critical annealing temperature determines the thermodynamic driving force for phase partitioning, where higher temperature increases the austenite phase fraction at the expense of ferrite, while annealing time governs the chemical homogenization of the alloying elements in austenite. Upon rapid cooling, this non-equilibrium process transforms the austenite into martensite while preserving the ferrite matrix, resulting in a duplex microstructure. This deliberate control of phase fractions through inter-critical annealing enables steels with superior strength–ductility combinations, which demonstrates how far-from-equilibrium processing can unlock advanced material properties for automotive and structural applications.

16.6 Emerging Trends in Non-equilibrium Thermodynamics and Kinetics

Recent advances in non-equilibrium thermodynamics and kinetics are revolutionizing our understanding of complex systems and enabling groundbreaking applications across multiple disciplines. These developments are enabling new applications in nanotechnology, active matter, quantum systems, and energy conversion, while also providing deeper insights into biological processes and environmental dynamics. These emerging trends also reflect the growing recognition that many natural and engineered systems operate far from equilibrium, requiring new theoretical frameworks and computational approaches to describe their behavior.

One significant trend is the development of stochastic thermodynamics for small systems, which extends classical non-equilibrium thermodynamics to the nanoscale where fluctuations dominate. This approach has been particularly transformative in single-molecule biophysics, where thermodynamic quantities like work and entropy production can now be measured for individual molecular motors such as kinesin walking along microtubes. The discovery of thermodynamic uncertainty relations, which establish fundamental bounds on the precision of biomolecular processes in terms of entropy production, represents a major theoretical advancement with implications for designing nanomachines.

Non-equilibrium statistical mechanics is rapidly evolving as the cornerstone of modern thermodynamics, offering powerful tools to describe systems driven far from equilibrium. Techniques such as large deviation theory now enable the prediction of rare high-entropy-production trajectories in nanoscale devices, while non-equilibrium Green's functions and Keldysh field theory provide rigorous descriptions of quantum systems under continuous driving. By bridging microscopic dynamics with macroscopic observables, non-equilibrium statistical mechanics unlocks new strategies for controlling energy dissipation in nanotechnologies, optimizing catalytic reactions, and even understanding the thermodynamic constraints on emergent phenomena in living systems.

In materials science, there's growing interest in active matter systems that continuously dissipate energy to maintain non-equilibrium states. Examples include self-propelled colloidal particles that exhibit collective behaviors like motility-induced phase separation, and chemically active droplets that serve as prototypes for primitive cellular life. These systems challenge traditional equilibrium statistical mechanics and have led to new concepts such as "time crystals" and periodically driven systems that break time-translation symmetry. Researchers are now using these principles to develop adaptive materials with life-like properties, such as self-healing polymers and shape-morphing metamaterials.

Machine learning is emerging as a powerful tool for tackling complex non-equilibrium problems. Neural networks are being used to learn effective thermodynamic descriptions of many-body systems directly from simulation data, bypassing the need for explicit modeling of all microscopic degrees of freedom. This approach has shown promise in predicting the behavior of glass-forming liquids and other systems with slow relaxation dynamics. Neural networks are being trained to predict reaction pathways in complex chemical networks, identify order parameters for non-equilibrium phase transitions, and optimize control protocols for minimizing entropy production in nanoscale devices. Meanwhile, reinforcement learning algorithms are being applied to optimize non-equilibrium processes in chemical engineering, such as controlling reaction networks in a flow reactor.

The field of quantum thermodynamics is extending non-equilibrium concepts to quantum systems, with potential applications in quantum computing and energy harvesting. Recent experiments have demonstrated quantum heat engines and refrigerators that exploit coherence and entanglement to achieve performance beyond classical limits. These developments are accompanied by new theoretical tools such as fluctuation theorems for quantum trajectories and resource theories for quantum thermodynamics.

In environmental science, non-equilibrium approaches are being applied to understand and mitigate climate change impacts. For instance, researchers are developing thermodynamic models of urban heat islands that account for the complex interplay between energy flows, material cycles, and human activities. Similarly, new kinetic models of atmospheric chemistry are improving predictions of aerosol formation and its climate effects.

Energy applications are benefiting from advances in non-equilibrium electrocatalysis, where operando techniques combined with theoretical modeling reveal how dynamic surface reconstructions and local field effects enhance reaction rates. This knowledge is guiding the design of next-generation catalysts for sustainable fuel synthesis and carbon dioxide utilization. The emerging field of "electrified" chemical engineering seeks to precisely control reaction pathways by manipulating non-equilibrium conditions through pulsed electric fields and other external stimuli. Non-equilibrium models are also used to study battery degradation, where phase separation dynamics under fast charging are governed by nonlinear diffusion.

These trends share a common theme, moving beyond linear response theory to embrace the complexity and opportunity inherent in far-from-equilibrium systems. As measurement techniques improve and computational power grows, we can expect even deeper insights into how energy flows and entropy production shape the behavior of systems across all scales, from quantum devices to living organisms to planetary ecosystems. This knowledge will be crucial for addressing the grand challenges in energy, sustainability, and advanced manufacturing in the coming decades.

This chapter has explored the fundamental principles of non-equilibrium thermodynamics and kinetics, providing a framework to understand systems driven far from equilibrium by external forces, gradients, or chemical reactions. Beginning with the theoretical foundations, it is established how irreversible processes govern energy, mass and charge transport. Key examples are provided which illustrate how non-equilibrium thermodynamics connects theory to real-world applications. Emerging trends highlight the dynamic and rapid evolution of this field. These advances are critical for designing next generation materials, from high-strength alloys to energy storage systems, where controlled deviations from equilibrium enable unprecedented functionality. As experimental and computational tools continue to grow, non-equilibrium thermodynamics and kinetics will remain indispensable for tackling challenges in materials science, biotechnology, and sustainable energy.

Problems

1	For a first-order reaction with a rate constant $k = 0.051/\mathrm{s}$, Calculate the concentration C_A after 10 s, given that $C_A(0) = 1$ mol/L **Solution:** $C_A(t) = C_A(0) \cdot e^{-kt} = 1.0 \cdot e^{-0.05 \times 1.0} = 1.0 \cdot e^{-0.5} \approx 0.6065$ mol/L
2	In a coupled system, the reaction rate r is influenced by the Gibbs free energy ΔG and temperature T. The rate constant follows, where $A = 10^6\ \mathrm{S}^{-1}$ and $\Delta G = -100$ KJ/mol. Calculate k at $T = 350$ K. **Solution:** $k = 10^8 \cdot e - \frac{100 \times 10^3}{8.314 \times 350} = 10^6 \cdot e^{3.87} \approx 10^6 \cdot 48.4 \approx 4.84 \times 10^7 \mathrm{s}^{-1}$
3	If a thermodynamic database provides the following values for two components, A and B: $\Delta H_f^A = -50\ \mathrm{KJ/mol}$ $\Delta H_f^B = -30\ \mathrm{KJ/mol}$ Calculate the enthalpy change for the reaction $A + B \rightarrow C$, where $\Delta H_f^C = -60\ \mathrm{KJ/mol}$ **Solution:** $\Delta H_f^C - \left(\Delta H_f^A + \Delta H_f^B\right) = -60 - (-50 + (-30)) = -60 + 80 = 20\ \mathrm{KJ/mol}$
4	Given a binary system with components A and B, the CALPHAD model predicts a liquidus temperature of 1600 °C and solidus temperature of 1400 °C. If the system is at 1500 °C, calculate the fraction of the liquid phase. **Solution:** $fl_{\text{iquid}} = \frac{T_{\text{solidus}} - T}{T_{\text{solidus}} - T_{\text{liquidus}}} = \frac{1400 - 1500}{1400 - 1600} = \frac{-100}{-200} = 0.5$

(continued)

5	In a specific application, a material has an initial concentration of 2.0 mol/L and the rate of change is governed by the equation: $\frac{dC}{dt} = -k \cdot C$ If $k = 0.1$1/s and the initial concentration is 2.0 mol/L. Calculate the concentration after 5 s. **Solution:** $C(t) = C(0) \cdot e^{-kt} = 2.0 \cdot e^{-0.1 \times 5} = 2.0 \cdot e^{-0.5} \approx 1.213$ mol/L

References

1. Dornelas, P. H. G., Filho, J. d. C. P., Farias, F. W. C., Moraes e Oliveira, V. H. P., Moraes, D. d. O., & Júnior, P. Z. (2020). FEM-thermodynamic simulation methodology to predict the influence of t8/5 on the coarse grain heat-affected zone of a Cr-Mo low-alloy steel pipe. *Journal of Manufacturing Processes, 60*, 520–529.
2. Speich, G. R., Demarest, V. A., & Miller, R. L. (1981). Formation of austenite during intercritical annealing of dual-phase steels. *Metallurgical Transactions A, 12*, 1419–1428.
3. Khan, D., Suhane, A., Srimannarayana, P., Bhattacharjee, A., Tennyson, G., Zagade, P., & Gautham, B. P. (2017). Integrated microstructure based modelling of process-chain for cold rolled dual phase steels. In *Proceedings of the 4th world congress on integrated computational materials engineering (ICME 2017)* (The minerals, metals & materials series). Springer.

Further Reading

Cahn, J. W., & Balluffi, R. W. (1979). Diffusional mass-transport in polycrystals containing stationary r migrating grain boundaries. *Scripta Metallurgica et Materialia, 13*(6), 499–502.

Dinsdale, A. (1991). SGTE data for pure elements. *Calphad, 15*(4), 317–425. https://doi.org/10.1016/0364-5916(91)90030-N

Factsage. (n.d.). *Factsage software and databases*. Retrieved from https://factsage.com/database-documentation/

Harrison, L. G. (1961). Influence of dislocations on diffusion kinetics in solids with particular refernec to the alkali halides. *Transactions of the Faraday Society, 57*(7), 1191–1199.

Kaufman, L., & Bernstein, H. (1970). *Computer calculation of phase diagrams. With special reference to refractory metals*. Academic Press Inc.

Khan, D., Suhane, A., Srimannarayana, P., Bhattacharjee, A., Tennyson, G., Zagade, P., & Gautham, B. (2017). Integrated microstructure based modelling of process-chain for cold rolled dual phase steels. In *Proceedings of the 4th World Congress on Integrated Computational Materials Engineering (ICME 2017)* (The Minerals, Metals & Materials Series) (pp. 15–22). Springer. https://doi.org/10.1007/978-3-319-57864-4_2

MTDATA. (n.d.). *MTDATA databases*. Retrieved from https://www.phase-trans.msm.cam.ac.uk/mtdata/intro_databases.html

NIST. (n.d.). *Phase diagrams and computational thermodynamics of solder systems*. Retrieved from Material Science and Engineering Division of NIST https://www.metallurgy.nist.gov/phase/solder/solder.html

NOMAD. (n.d.). *NOMAD*. Retrieved from https://nomad-lab.eu/prod/v1/gui/search/entries

OpenCalphad. (n.d.). *Available open databases*. Retrieved from https://www.opencalphad.com/databases.php

Otis, R., & Liu, Z.-K. (2017). Pycalphad: CALPHAD-based computational thermodynamics in python. *Journal of Open Research Software, 5*(1), 1. https://doi.org/10.5334/jors.140

Speich, G., Demarest, V., & Miller, R. (1981). Formation of austenite during intercritical annealing of dual-phase steels. *Metallurgical Transactions A, 12*, 1419–1428. https://doi.org/10.1007/BF02643686

Turnbull, D. (1951). Grain boundary and surface diffusion. In J. H. Holloman (Ed.), *Atom movements* (pp. 129–151). American Society for Metals Special Volume of ASM.

K. A. Padmanabhan et al., *Thermodynamics and Kinetics of Materials*, Engineering Materials, https://doi.org/10.1007/978-3-032-14527-7

Index

K. A. Padmanabhan et al., *Thermodynamics and Kinetics of Materials*, Engineering Materials, https://doi.org/10.1007/978-3-032-14527-7

The manufacturer's authorised representative in the EU is Springer Nature Customer Service Centre GmbH, Europaplatz 3, 69115 Heidelberg, Germany. If you have any concerns regarding our products, please contact ProductSafety@springernature.com

Printed and bound by CPI Group (UK) Ltd, Croydon, CR0 4YY
27/08/2026
02204938-0004